Soft Systems Methodology: a 30-year retrospective

Peter Checkland

D0074477

JOHN WILEY & SONS, LTD

Chichester · New York · Weinheim · Brisbane · Singapore · Toronto

Other Wiley Editorial Offices

John Wiley & Sons Inc., 111 River Street, Hoboken, NJ 07030, USA

Jossey-Bass, 989 Market Street, San Francisco, CA 94103-1741, USA

Wiley-VCH Verlag GmbH, Boschstr. 12, D-69469 Weinheim, Germany

John Wiley & Sons Australia Ltd, 33 Park Road, Milton, Queensland 4064, Australia

John Wiley & Sons (Asia) Pte Ltd, 2 Clementi Loop #02-01, Jin Xing Distripark, Singapore
129809

John Wiley & Sons Canada Ltd, 22 Worcester Road, Etobicoke, Ontario, Canada M9W 1L1

British Library Cataloguing in Publication Data

A catalogue record for this book is available from the British Library

ISBN-13 978-0-471-98606-5 (PB)

Typeset in 10/12 pt Times by C/K.M. Typesetting, Salisbury, Wiltshire
Printed and bound in Great Britain by TJ International Ltd, Padstow, Cornwall

To

Louise and Conor Taaffe

We are all in their debt

Contents

Preface

It is appropriate, on the republication of the first two books about soft systems methodology (SSM), to add new material written from the perspective of 1999. It would not be appropriate to change the text of the books themselves; they stand, telling the first part of the story. It seems appropriate to bring the story up to date by reflecting upon the 30 years of research on, and use of, SSM, and to offer a summing up of what is now a mature and much-tested process of inquiry into problem situations in human affairs. That is what is done in this chapter, which extends the earlier books.

The chapter examines: the emergence of soft systems thinking as systems engineering failed in problems which were not technically defined; the evolving expression of the methodology as a whole from 1972 to the 1990s; and then its parts, covering all the significant developments in the intellectual 'apparatus' of SSM as these have been honed in action. Next, in the light of these developments the chapter returns to look again at the methodology as a whole, first examining the implications of 'methodology' not being the same as 'method' (something almost completely ignored in the secondary literature), then re-examining the question of what must be done if a claim to be 'using SSM' is to be sustainable. Finally, what happens as the methodology is internalized and use moves from naïvety to sophistication is discussed. The last section then relates SSM to its larger context of social inquiry and research.

This structure is itself a small reflection of the cyclic learning which has always been involved in the development of SSM, and will no doubt continue.

Very many people have been involved in the development of SSM, and the acknowledgements made in *Systems Thinking, Systems Practice* (1981), *Soft Systems Methodology in Action* (1990) and *Information, Systems and Information Systems* (1998) are reiterated, with renewed thanks. For discussions relevant to the new material in this chapter I am greatly indebted to a particular group of people. *Primus inter pares* is Sue Holwell, research collaborator *par excellence*, with whom I have had a rich ongoing conversation over 10 years. We quickly learned always to have a notebook handy, and I have a collection of notes on our deliberations with headings like: 'M6, Feb 92' or 'Bullet train to Osaka, July 94'. Others in this special group include academic colleagues Nimal Jayaratna and Mark Winter, together with an exceptional group of reflective practitioners: Steve Clarke, Mike Haynes, Kees van der Heiden, Luc Hobeke, Jaap Leemhuis, John Poulter, Peter Wood.

Finally, for their ever-willing professional help in the production of this chapter, I am very grateful to Jenny Seddon and Martin Lister.

Peter Checkland
Lancaster, April 1999

INTRODUCTION

Although the history of thought reveals a number of holistic thinkers—Aristotle, Marx, Husserl among them—it was only in the 1950s that any version of holistic thinking became institutionalized. The kind of holistic thinking which then came to the fore, and was the concern of a newly created organization, was that which makes explicit use of the concept of 'system', and today it is 'systems thinking' in its various forms which would be taken to be the very paradigm of thinking holistically. In 1954, as recounted in Chapter 3 of *Systems Thinking, Systems Practice*, only one kind of systems thinking was on the table: the development of a mathematically expressed general theory of systems. It was supposed that this would provide a meta-level language and theory in which the problems of many different disciplines could be expressed and solved; and it was hoped that doing this would help to promote the unity of science.

These were the aspirations of the pioneers, but looking back from 1999 we can see that the project has not succeeded. The literature contains very little of the kind of outcomes anticipated by the founders of the Society for General Systems Research; and scholars in the many subject areas to which a holistic approach is relevant have been understandably reluctant to see their pet subject as simply one more example of some broader 'general system'!

But the fact that general systems theory (GST) has failed in its application does not mean that systems thinking itself has failed. It has in fact flourished in several different ways which were not anticipated in 1954. There has been development of systems ideas as such, development of the use of systems ideas in particular subject areas, and combinations of the two. The development in the 1970s by Maturana and Varela (1980) of the concept of a system whose elements generate the system itself provided a way of capturing the essence of an autonomous living system without resorting to use of an observer's notions of 'purpose', 'goal', 'information processing' or 'function'. (This contrasts with the theory in Miller's *Living Systems* (1978), which provides a general model of a living entity expressed in the language of an observer, so that what makes the entity autonomous is not central to the theory.) This provides a good example of the further development of systems ideas as such. The rethinking, by Chorley and Kennedy (1971), of physical geography as the study of the dynamics of systems of four kinds, is an example of the use of systems thinking to illuminate a particular subject area.

The two books to which this chapter is an adjunct provide an example of the third kind of development: a combination of the two illustrated above. We set out to see if systems ideas could help us to tackle the messy problems of 'management', broadly defined.

In trying to do this we found ourselves having to develop some new systems concepts as a response to the complexity of the everyday problem situations we encountered, the kind of situations which we all have to deal with in both our professional and our private lives. The aim in the research process we adopted was to make neither the ideas nor the practical experience dominant. Rather the intention was to allow the tentative ideas to inform the practice which then became the source of enriched ideas—and so on, round a learning cycle. This is the action research cycle whose emergence is described in *Systems Thinking, Systems Practice* and whose use and further development is the subject of *SSM in Action.*

The action research programme at Lancaster University was initiated by the late Gwilym Jenkins, first Professor of Systems at a British university, and Philip Youle, the perspicacious manager in ICI who saw the need for the kind of collaboration between universities and outside organizations which the action research programme required. Thirty years later that programme still continues, and with the same aim: to find ways of understanding and coping with the perplexing difficulties of taking action, both individually and in groups, to 'improve' the situations which day-to-day life continuously creates and continually changes. Specifically, the programme explores the value of the powerful bundle of ideas captured in the notion 'system', and they have not been found wanting, though both the ideas themselves and the ways of using them have been extended as a result of the practical experiences.

The progress of the 30 years of research has been chronicled and reflected upon since 1972 in about 100 papers and four books—which will be referred to in the remainder of this chapter by the initials of their titles. The nature of the books is summarized briefly below.

Systems Thinking, Systems Practice (STSP) (Checkland 1981) makes sense of systems thinking by seeing it as an attempt to avoid the reductionism of natural science, highly successful though that is when investigating natural phenomena; it describes early experiences of trying to apply 'systems engineering' outside the technical area for which it was developed, the rethinking of 'systems thinking' which early experience made necessary, and sets out the first developed form of SSM as a seven-stage process of inquiry.

Systems: Concepts, Methodologies and Applications (SCMA) (Wilson 1984, 2nd Edn 1990) describes the response of a professional control engineer to experiences in the Lancaster programme of action research; less concerned with the human and social aspects of problem situations, it cleaves to the functional logic of engineering and presents an approach which Holwell (1997) argues is best viewed as classic systems engineering with the transforming addition of human activity system modelling.

Soft Systems Methodology in Action (SSMA) (Checkland and Scholes 1990) describes the use of a mature SSM in both limited and wide-ranging situations in both public and private sectors; it moves beyond the 'seven-stage' model of the methodology (still useful for teaching purposes and—occasionally—in some real situations) to see it as a sense-making approach, which, once internalized, allows exploration of how people in a specific situation create for themselves the meaning of their world and so act intentionally; the book also initiates a wider discussion of the concept of 'methodology', a discussion which will be extended below.

Information, Systems and Information Systems (ISIS) (Checkland and Holwell 1998) stems from the fact that in very many of the Lancaster action research projects the creation of 'information systems' was usually a relevant, and often a core, concern; it attempts some conceptual cleansing of the confused field of IS and IT, treating IS as being centrally concerned with the human act of creating meaning, and relates experiences based on a mature use of SSM to a fundamental conceptualization of the field of IS/IT; it carries forward the discussion of SSM as methodology but less explicitly than will be attempted here.

It is important to understand the nature of these books if the aim of this chapter is itself to be properly understood. The less than impressive but nevertheless sprawling literature of 'management' caters in different ways for several different audiences. There is an apparent insatiable appetite for glib journalistic productions, offering claimed insights for little or no reader effort—*Distribution Management in an Afternoon*: that kind of thing. Such books are more often purchased than actually read. There is also a need for textbooks which systematically display the conventional wisdom of a subject for aspiring students. These need to be updated periodically in new editions. And also, more austerely, there are books which carry the discussion which is the real essence of any developing subject, and try to extend the boundaries of our knowledge. The books described above are of this kind. It is not usually appropriate—as it is with textbooks—to update them in new editions. They are 'of their time'. But it is useful on republication, as here, to offer reflections on the further development of the ideas as new experiences have accumulated since the books were written. That is what is done here for *STSP* and *SSMA*.

A particular structure is adopted. First, the emergence of soft systems thinking is briefly revisited. Then the methodology as a whole is considered, since the way in which it is thought about now is very different from the view of it in the 1970s, when it was a redefined version of systems engineering. This consideration of the methodology as a whole frames reflection on the separate parts which make up the whole (Analyses One, Two, Three; CATWOE; rich pictures; the three Es, etc.). This in turn yields a richer understanding of both the whole and its context. Such a structure, in which an initial consideration of the whole leads to an understanding of the parts, which in turn enables a richer understanding of the whole to be gained, is itself an example of Dilthey's 'hermeneutic circle' (Mueller-Vollmer 1986; Morse 1994, Chs 7 and 8). Here, it is a modest reflection of the same process through which SSM was itself developed, a process which tried to ensure that both whole and parts were continually honed and refined in cycles of action.

THE EMERGENCE OF SOFT SYSTEMS THINKING

The Starting Position

In the culture of the UK the word 'academic' is more often than not used in a pejorative sense. To describe something as 'academic' is usually to condemn it as unrelated to the rough and tumble of practical affairs. This was certainly the outlook of Gwilym Jenkins when he moved to Lancaster University in the mid-1960s to found the first systems department in a UK university. He did not want a department which could be dismissed as 'academic'. He rejected the idea that the name of the department should be Systems Analysis, in favour of a Department of Systems Engineering. 'Analysis is not enough', he used to say heretically. 'Beyond analysis it is important to put something together, to create, to "engineer" something.' Given this attitude it was not surprising that he initiated the programme of action research in real-world organizations outside the university. The intellectual starting point was Optner's concept (1965) that an organization could be taken to be a system with functional sub-systems—concerned with production, marketing, finance, human resources, etc. Jenkins' idea was that the real-world experiences would enable us gradually to build up knowledge of systems of various kinds: production systems, distribution systems, purchasing systems, etc. and that this knowledge would support the better design and operation of such systems in real situations. History did not, however, unfold in this way. Instead, the practical experiences led us to reject the taken-as-given assumption underlying the initial expectation, so taking the thinking in a very different direction. In doing this we had to distinguish between two fundamentally different stances within systems thinking: the two outlooks now known as 'hard' and 'soft' systems thinking.

At the outset, by formulating a research aim to uncover the fundamental characteristics of systems of various kinds, we were making the unquestioned assumption that the world contained such systems. Along with this went a second assumption that such systems could be characterized by naming their objectives. It seems obvious, for example, that 'a production system' will have objectives which can be expressed as: to make product X with a certain quality, at a certain rate, with a certain use of resources, under various constraints (budgetary, legal, environmental, etc.). Given such an explicit definition of an objective, then a system can in principle be 'engineered' to achieve that end. This is the stance of classic systems engineering (as described in Chapter 5 of *STSP*). This was what constituted 'systems thinking' at the time our research started, and its origins, as far as application to organizations goes, lie in the great contribution to management science made by Herbert Simon in the 1950s and 1960s (Simon 1960, 1977), which propounded the clarifying (but ultimately limited) concept that managing is to be thought of as decision-taking in pursuit of goals or objectives.

The Learning Experience

We found that although we were armed with the methodology of systems engineering and were eager to use its techniques to help engineer real-world systems to achieve their objectives, the management situations we worked in were always too complex for straightforward application of the systems engineering approach. The difficulty of answering such apparently simple questions as: What is the system we are concerned with? and What are its objectives? was usually a reason why the situation in question had come to be regarded as problematical. We had to accept that in the complexity of human affairs the unequivocal pursuit of objectives which can be taken as given is very much the occasional special case; it is certainly not the norm. A current long-running example of the surprising difficulty in using the language of 'objectives' in human affairs is provided by the arguments which wax and wane over the Common Agricultural Policy (CAP) of the EEC. The Treaty of Rome boldly declares that the CAP has three equally important objectives: to increase productivity in the agricultural industry; to safeguard jobs in the industry; and to provide the best possible service to the consumer. No wonder the CAP is a constant source of never resolved issues: progress towards any one of its (equally important) objectives will be at the expense of the other two! This is typical of the complexity we meet in human affairs as soon as we move out of the more straight-forward area in which problems can be technically defined: e.g. 'increase as much as possible the productivity of this phthalic anhydride plant', or 'make a device to produce radio waves with a 10 cm wavelength'. (If you insisted on using the language of 'objectives', you would have to conclude that the objective of the CAP is con-stantly to maintain and adjust a balance between the three incompatible objectives which is politically acceptable—which is not a very useful definition for 'engineer-ing' purposes.)

It was having to abandon the classic systems engineering methodology which caused us to undertake the fundamental thinking described in Chapters 2–4 of *STSP*. And it was this rethink which led ultimately to the distinction between 'hard' and 'soft' systems thinking.

Four Key Thoughts

The process of learning by relating experience to ideas is always both rich and confusing. But as long as the interaction between the rhetoric and the experi-enced 'reality' is the subject of conscious and continual reflection, there is a good chance of recognizing and pinning down the learning which has occurred. Looking back at the development of SSM with this kind of reflective hindsight, it is possible to find four key thoughts which dictated the overall shape of the development of SSM and the direction it took (Checkland 1995).

Firstly, in getting away from thinking in terms of some real-world systems in need of repair or improvement, we began to focus on the fact that, at a higher level, every situation in which we undertook action research was a human situation in which people were attempting to take purposeful action which was meaningful for them. Occasionally, that purposeful action might be the pursuit of a well-defined objective, so that this broader concept *included* goal seeking but was not restricted to it. This led to the idea of modelling purposeful 'human activity systems' as sets of linked activities which together could exhibit the emergent property of purposefulness. Ways of building such models were developed.

Secondly, as you begin to work with the idea of modelling purposeful activity—in order to explore real-world action—it quickly becomes obvious that many inter-pretations of any declared 'purpose' are possible. Before modelling can begin choices have to be made and declared. Thus, given the complexity of any situation in human affairs, there will be a huge number of human activity system models which *could* be built; so the first choice to be made is of which ones are likely to be most relevant (or insightful) in exploring the situation. That choice made, it is then necessary to decide for each selected purposeful activity the perspective or viewpoint from which the model will be built, the *Weltanschauung* upon which it is based. Thus when David Farrah, a director of the then British Aircraft Corporation asked us to use our systems engineering approach to see how the Concorde project might be improved, possible relevant systems might have included 'a system to manage relations with the British Government' (since they were funding it) or 'a system to sustain a European precision engineering industry' (since Concorde would help to stimulate such activity). Thinking like systems engineers at the time (What is the system? What are its objectives?) Dave Thomas and I in fact proceeded only with the most basic and obvious of possible choices: 'a system to carry out the project'. Neither did the second choice give us pause: how would we conceptualize that project? Again, with our systems engineering blinkers firmly in place, it did not occur to us to think of it as anything other than an engineering project. But given its origins, at a time when President de Gaulle of France was vetoing British entry into the European Common Market, a defensible alternative world-view would be to treat it as a *political* project. On the day the Concorde project agreement was signed the British Government let it be known that it expected Britain to join the European Community within a year, while de Gaulle a few weeks later told a press conference that it was probable that negotiations for British entry might not succeed; in fact he made the supersonic aircraft project a touchstone of Britain's sincerity in applying for membership (Wilson 1973, pp. 31–32). So a model of the project based on a political world-view might be as useful as—or perhaps more useful than—the more obvious one based on a technical world-view.

The learning here was that in making the idea of modelling purposeful activity a usable concept, we had to accept that it was necessary to declare both a world-view which made a chosen model relevant, and a world-view which would then determine the model content. Equally, because interpretations of purpose will always be many and various, there would always be a number of models in play, never simply one model purporting to describe 'what is the case'.

This moved us a good way away from classic systems engineering, and the next key thought in understanding our experience recognized this. It was the thought which can now be seen to have established the shape of SSM as an inquiring process. And that in turn established the 'hard/soft' distinction in systems thinking, though that too was not immediately recognized at the time.

We had moved away from working with the idea of an 'obvious' problem which required solution, to that of working with the idea of a *situation* which some people, for various reasons, may regard as problematical. We had developed the idea of building models of concepts of purposeful activity which seemed relevant to making progress in tackling the problem situation. Next, since there would always be many possible models it seemed obvious that the best way to proceed would be to make an initial handful of models and—conscious of them as embodying only pure ideas of purposeful activity rather than being descriptions of parts of the real world—to use them as a source of questions to ask of the real situation. SSM was thus inevitably emerging as an organized *learning system*. And since the initial choice of the first handful of models, when used to question the real situation, led to new knowledge and insights concerning the problem situation, this leading to further ideas for relevant models, it was clear that the learning process was in principle ongoing. What would bring it to an end, and lead to action being taken, was the development of an accommodation among people in the situation that a certain course of action was both desirable in terms of this analysis and feasible for these people with their particular history, relationships, culture and aspirations.

SSM thus gradually took the form shown in Figure 1.3 of *SSMA* (p. 7), repeated with some embellishment here as Figure A1. This was the form of representation of SSM which eventually took hold, and is the one now normally used. The initial version of it was the 'seven-stage model' which is shown in Figure 6 in *STSP*, p. 163 and Figure 2.5 in *SSMA*, p. 27. This version, though still often used for initial

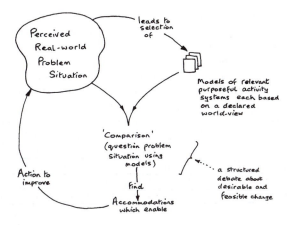

Figure A1 The inquiring/learning cycle of SSM

teaching purposes, has a rather mechanistic flavour and can give the false impression that SSM is a prescriptive process which has to be followed systematically, hence its fall from favour.

These three key thoughts capture succinctly the learning which accumulated with experience of using SSM, and they make sense of its development. The fourth such thought, that models of purposeful activity can provide an entry to work on information systems (which are less than ideal in virtually every real-world situation) is not our concern here, this aspect of SSM's use being the detailed subject of *ISIS*.

Hard and Soft Systems Thinking

Our final concern in this section is the major thought which came from these particular experiences of relating systems thinking to systems practice: the 'hard'–'soft' distinction. This was first sharply expressed in a paper written two years after the publication of *STSP* in 1981 (Checkland 1983). It took some time for this idea to sink in!

In systems engineering (and also similar approaches based on the same fundamental ideas, such as RAND Corporation systems analysis and classic OR) the word 'system' is used simply as a label for something taken to exist in the world outside ourselves. The taken-as-given assumption is that the world can be taken to be a set of interacting systems, some of which do not work very well and can be engineered to work better. In the thinking embodied in SSM the taken-as-given assumptions are quite different. The world is taken to be very complex, problematical, mysterious. However, our coping with it, the process of inquiry into it, it is assumed, can itself be organized as a learning *system*. Thus the use of the word 'system' is no longer applied to the world, it is instead applied to the process of our dealing with the world. It is this shift of systemicity (or 'system-ness') from the world to the process of inquiry into the world which is the crucial intellectual distinction between the two fundamental forms of systems thinking, 'hard' and 'soft'.

In the literature it is often stated that 'hard' systems thinking is appropriate in well-defined technical problems and that 'soft' systems thinking is more appropriate in fuzzy ill-defined situations involving human beings and cultural considerations. This is not untrue, but it does not *define* the difference between 'hard' and 'soft' thinking. The definition stems from how the word 'system' is used, that is from the attribution of systemicity.

Experience shows that this distinction is a slippery concept which many people find it very hard to grasp; or, grasped one week it is gone the next. Probably this is because very deeply embedded within our habits is the way we use the word 'system' in everyday language. In everyday talk we constantly use it as if it were simply a label-word for a part of the world, as when we talk about the legal system, health care systems, the education system, the transport system, etc. even though many of these things named as systems do not in fact exhibit the characteristics associated with the word 'system' when it is used properly. This day-by-day use unconsciously but steadily reinforces the assumptions of the 'hard' systems paradigm; and the speaking habits of a lifetime are hard to break!

As the thinking about SSM gradually evolved, the formulation of this precise definition of 'hard' and 'soft' systems thinking did not arrive in the dramatic way events unfold in adventure stories for children ('With one bound, Jack was free!'). Rather the ultimate definition is the result of our feeling our way to the difference between 'hard' and 'soft', as experience accumulated, via a number of different formulations. These have been spotted and extracted by Holwell (1997, Table 4.2, p. 126) who collects eight different ways of discussing the hard/soft distinction between 1971 and 1990. These begin unpromisingly—judged by today's criteria—by assuming that 'hard' and 'soft' systems (roughly, determinate and indeterminate respectively) exist in the world. The shift in thinking comes between the publication of *STSP* and *SSMA*, its very first explicit appearance being in Checkland (1983), a paper which can now also be seen as part of the developments which have made the phrase 'soft OR' meaningful.

The eventual definition of the hard–soft distinction is succinctly expressed in Figure 2.3 of *SSMA* (p. 23), but this diagram is over-rich for many, and so here it is supplemented by Figure A2, a further attempt to make clear the difference between hard systems thinking and soft systems thinking. Understanding this idea is the crucial step in understanding SSM.

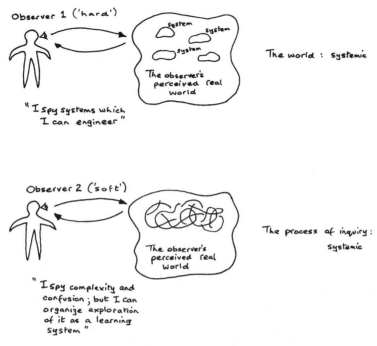

Figure A2   The hard and soft systems stances

SOFT SYSTEMS METHODOLOGY—THE WHOLE

Learning from books or lectures is relatively easy, at least for those with an academic bent, but learning from experience is difficult for everyone. Everyday life develops in all of us trusted intellectual structures which to us seem good enough to make sense of our experiences, and in general we are reluctant to abandon or modify them even when new experience implies that they are shaky. Even professional researchers, who ought to be ready to welcome change in taken-as-given structures of thinking, show the same tendency to distort perceptions of the world rather than change the mental structures we use to give us our bearings. So we were lucky in our research programme that the failure of classic systems engineering in rich 'management' problem situations, broadly defined, was dramatic enough to send us scurrying to examine the adequacy of the systems thinking upon which systems engineering was based. (The early experiences are described in *STSP*, Chapter 7.) But in spite of this it is still the case that the story of our learning is also the story of our gradually managing to shed the blinkered thinking which we started out with as a result of taking classic systems engineering as given.

Holwell (1997) has an appendix to her thesis which collects four different representations of SSM between 1972 and 1990 and correctly suggests that these 'show how the methodology has become less structured and broader as it has developed' (p. 450). It is useful briefly to review this changing perception of the methodology as a whole before moving on to a consideration of its parts.

1972—Blocks and Arrows

The first studies in the research programme were carried out in 1969, and the first account of what became SSM (though that phrase was not used at the time) was published three years later in a paper: 'Towards a systems-based methodology for real-world problem solving' (Checkland 1972). The paper argues the need for methodology 'of practical use in real-world problems' [sic](p. 88), reviews the context provided by the systems movement, introduces the case for action research as the research method, describes three projects in detail, refers to six others, and describes the emerging methodology. It finishes with the very important argument that any methodology which will be used by human beings cannot, as methodology, be *proved* to be useful:

> Thus, if a reader tells the author 'I have used your methodology and it works', the author will have to reply 'How do you know that better results might not have been obtained by an *ad hoc* approach?' If the assertion is: 'The methodology does not work' the author may reply, ungraciously but with logic, 'How do you know the poor results were not due simply to your incompetence in using the methodology?' (p. 114)

With reference to human situations, neither of these questions can be answered. Methodology, as such, remains undecidable.

Nearly 30 years later the paper has a somewhat quaint air, though not embarrassingly so. Apart from the reference noted above to 'real-world problems', rather than problem *situations*, the main inadequacy now is in the legacy of hard systems thinking which leads to reference being made to both 'hard systems' and 'soft systems' as existing in the real world; thus we find a few remarks of the kind: 'In soft systems like those of the three studies under discussion . . .' (p. 96). Such statements would not have been made a few years later. Also the methodology is presented as a sequence of stages with iteration back to previous stages, the sequence being: analysis; root definition of relevant systems; conceptualization; comparison and definition of changes; selection of change to implement; design of change and implementation; appraisal.

The focus on implementing *change* rather than introducing or improving *a system* is a signal that the thinking was on the move as a result of these early experiences, even if the straight arrows in the diagrams and the rectangular blocks in some of the models do now cause a little pain!

1981—Seven Stages

By the time the first book about SSM was written (*STSP*, 1981) the engineering-like sequence of the 1972 paper was being presented as a cluster of seven activities in a circular learning process: the 'seven-stage model', versions of which are Figure 6 in *STSP* (p. 163) and Figure 2.5 in *SSMA* (p. 7). In this model the first two stages entail entering the problem situation, finding out about it and expressing its nature. Enough of this has to be done to enable some first choices to be made of relevant activity systems. These are expressed as root definitions in stage three and modelled in stage four. The next stages use the models to structure the further questioning of the situation (the stage five 'comparison') and to seek to define the changes which could improve the situation, the changes meeting the two criteria of 'desirable in principle' and 'feasible to implement' (stage six). Stage seven then takes the action to improve the problem situation, so changing it and enabling the cycle to begin again.

The arrows which link the seven stages simply show the logical structure of the mosaic of actions which make up the overall process; it has always been emphasized that the work done in a real study will not slavishly follow the sequence from stage one to stage seven in a flat-footed or dogged way. Thus, to give one example, the stage five 'comparison' cannot but enhance the finding out about the situation, leading to new ideas for 'more relevant' systems to model. Similarly, the process can take a real-world change being implemented to be an example of stage seven; you can then work backwards to construct the notional 'comparison' which would lead to *this* change being selected, thus teasing out what world-views are being taken as given by people in the situation.

The seven-stage model of SSM has proved resilient, not least because it is easy to understand as a sequence which unfolds logically. This makes it easy to teach, and that too helps explain its resilience. Certainly it has three virtues worth noting before we begin to undermine it in what follows.

Firstly—an intangible, aesthetic point, but an important one—its fried-egg shapes and curved arrows begin to undermine the apparent *certainty* conveyed by straight arrows and rectangular boxes. These are typical of work in science and engineering, and the style conveys the implication: 'this is the case'. The more organic style of the seven-stage model (and of the rich pictures and hand-drawn models in *SSMA*) is meant to indicate that the status of all these artefacts is that they are working models, currently relevant *now* in *this* study, not claiming permanent ontological status. They are also meant to look more human, more natural than the ruled lines and right angles of science and engineering.

Secondly, it is a happy chance that the learning cycle of this model of the process has seven stages. Miller's well-known account of laboratory experiments on perception (1956) suggests that the channel capacity of our brains is such that we can cope with about seven items or concepts at once, hence the title of his famous paper: 'The magical number seven, plus or minus two: some limits on our capacity for processing information'. (He reminds us that there are seven days of the week, seven wonders of the world, seven ages of man, seven levels of hell, seven notes on the musical scale, seven primary colours . . .) Irrespective of whether or not seven is truly a crucial number in human culture, the comfortable size of the model of the SSM process does mean that you can easily retain it in your mind. You do not have to look it up in a book, and this is very useful when using it flexibly in practice.

Another feature of the seven-stage model worthy of note is that the stages of forming most definitions and building models from them (stages three and four) were separated from the other stages by a line which separates the 'systems thinking world' below the line from the everyday world of the problem situation above the line. This distinguishing between the everyday world and the systems thinking *about* it was intended to draw attention to the conscious use of systems language in developing the intellectual devices (the activity models) which are consciously used to structure debate. The purpose of the line was essentially heuristic, and its elimination from the 1990 model of SSM will be discussed later in this chapter.

Finally, as far as the 1981 model is concerned, it was important at that stage of development to think about what it was you had to do in a systems study if you wished to claim to be using SSM. This problem was first addressed by Naughton (1977). He was tackling the problem of teaching SSM to Open University students, and for the sake of clarity in teaching, distinguished between 'Constitutive Rules' which had to be obeyed if the SSM claim was to be made, and 'Strategic Rules' which allowed a number of options among which the user could choose. Versions of these rules endorsed in *STSP* are given in Table 6 (p. 253). This was a very useful development in its time, though this is another area which will be further discussed in the light of current thinking.

In summary, formulation of SSM in the 1981 book was at least rich enough to enable it to be taught and used; accounts began to appear of uses of SSM by people other than its early developers. See, for example, Watson and Smith (1988) for an account of 18 studies carried out in Australia between 1977 and 1987.

1988—Two Streams

All of the action research which developed and used SSM was carried out in the spirit of Gwilym Jenkins' remark quoted earlier, that 'Analysis is not enough'. The overall aim in all the projects undertaken was to facilitate action, and it was always apparent that making things happen in real situations is a complex and subtle process, something which will not happen simply because some good ideas have been generated or a sophisticated analysis developed. Ideas are not usually enough to trigger action and that is why industrial companies value highly their 'shakers and movers': they are a much rarer breed than intelligent analysts. So, although a debate structured by questioning perceptions of the real situation by means of purposeful activity models was always insightful, moving on to action entailed broader considerations.

In the very first research in the programme, for example, in the failing textile company described in Checkland and Griffin (1970) and in *STSP* (p. 156), we were brought into the situation by a recently appointed marketing director. He had been brought into the company because the crisis due to falling revenues and disappearing profitability had at last been recognized by a relatively unsophisticated and rather inbred group of managers. This was the first instance in that company's history of appointing a senior manager from outside. The newcomer was thus not part of what had become a closed tribe, and though his previous experience gave him many ideas relevant to improving company performance, his effectiveness was profoundly affected by suspicion of the 'off-comer'. Understanding that, and taking it into account in influencing thinking in the company was crucial to initiating action.

It was thus important always to gain an understanding of the culture of the situations in which our work was done. For some years this was done informally, but—we hoped—with insights from experience, since all the original action researchers developing SSM were ex-managers rather than career academics—who are often naïve about life in unsubsidized organizations.

During those years much reflection went on concerning how we went about 'reading' situations culturally and politically, and it was a significant step forward when SSM was presented as an approach embodying not only a logic-based stream of analysis (via activity models) but also a cultural and political stream which enabled judgements to be made about the accommodations between conflicting interests which might be reachable by the people concerned and which would enable action to be taken. This two-stream model of SSM (*SSMA*, Figure 2.6, p. 29) was first expounded at a plenary session of the Annual Meeting of the International Society for General Systems Research in 1987, and was published the following year (Checkland 1988).

This version of SSM as a whole recognizes the crucially important role of history in human affairs. It is their history which determines, for a given group of people, both what will be noticed as significant and how what is noticed will be judged. It reminds us that in working in real situations we are dealing with something which is both perceived differently by different people and is continually changing.

Also, it is worth noting that this particular expression of SSM as a whole omits the dividing line between the world of the problem situation and the systems thinking world. It had served its heuristic purpose.

1990—Four Main Activities

Published in 1981, *STSP* covered broadly the first decade of development of SSM. The seven-stage model gave a version of the approach which was by then sufficiently well founded to be applied in new real-world situations, large and small, in both the public and the private sector. That was what happened during the second decade of development, some of those experiences being described in *SSMA*. They cover action research in different organizational settings (industry, the Civil Service, the NHS) and include involvements which took from a few hours (ICL, Chapter 6, pp. 164–171) to more than a year (Shell, Chapter 9).

When it came to expressing the shape of the methodology in the 1990 book, the seven-stage model was no longer felt able to capture the now more flexible use of SSM; and even the two-streams model was felt to carry a more formal air than mature practice was now suggesting characterized SSM use, at least by those who had internalized it. The version presented was the four-activities model (*SSMA*, Figure 1.3, p. 7) of which Figure A1 in this chapter is a contemporary form. This is iconic rather than descriptive, and subsumes the cultural stream of analysis in the four activities, which it implies rather than declares.

The four activities are, however, capable of sharp definition:

1. Finding out about a problem situation, including culturally/politically;
2. Formulating some relevant purposeful activity models;
3. Debating the situation, using the models, seeking from that debate both
   (a) changes which would improve the situation and are regarded as both desirable and (culturally) feasible, *and*
   (b) the accommodations between conflicting interests which will enable action-to-improve to be taken;
4. Taking action in the situation to bring about improvement.

((a) and (b) of course are intimately connected and will gradually create each other.)

A decade after *SSMA* was published this iconic model of SSM is still relevant. Why that is so will be discussed when we return to discussing the methodology as a whole. But first it is useful to review the evolving thinking about the parts which make up the whole.

SOFT SYSTEMS METHODOLOGY—THE PARTS

The gradual change in the way SSM as a whole has been thought about, described above, has been paralleled by more substantive changes to some of the separate parts which make up the whole. Many of these represent conscious attempts to improve and enrich such things as model building, or the uses to which rich pictures are put; some have entailed dropping earlier ways of doing things, for example the shift away from using 'structure/process/climate' as a framework for initial finding out about a situation (*STSP*, pp. 163, 164, 166), or the deliberate dropping of the 'formal system model' (*STSP*, Figure 9; *SSMA* pp. 41, 42). But whether the changes to the parts were additions or deletions, they were never made by sitting at desks being 'academic'. They have always been made as a result of experiences in using the approach in a complex world, and they have played their part in changing perceptions of SSM as a whole. This section will review the changes to the parts of SSM, the review being structured by the four activities which underpin the mature icon for SSM which is Figure A1 here.

Finding Out about a Problem Situation

Rich Picture Building

Making drawings to indicate the many elements in any human situation is something which has characterized SSM from the start. Its rationale lies in the fact that the complexity of human affairs is always a complexity of multiple interacting relationships; and pictures are a better medium than linear prose for expressing relationships. Pictures can be taken in as a whole and help to encourage holistic rather than reductionist thinking about a situation.

Producing such graphics is very natural for some people, very difficult for others. If it does not come naturally to you, it is a skill worth cultivating, but experience suggests that its *formalization* via use of ready-made fragments, such as is advocated by Waring (1989) is not usually a good idea, except perhaps as a way of making a start. Users need to develop skill in making 'rich pictures' in ways they are comfortable with, ways which are as natural as possible for them as individuals.

As far as use of such pictures is concerned, we have found them invaluable as an item which can be tabled as the starting point of exploratory discussion with people in a problem situation. In doing so we are saying, in effect 'This is how we see this situation at present, its main stakeholders and issues. Have we got it right from your perspective?' For example, when researching the subtle relationship between a health authority and one of its acute hospitals a few years ago (during the short-lived experiment with 'contracting' in the NHS) we assembled from a great many semi-structured interviews a somewhat large and complicated picture—though even very elaborate pictures are of course selections. (Bryant (1989) is correct to emphasize that 'Selection of the key features of a situation is a crucial skill in developing a picture' (p. 260).) The picture in question became known as 'the briar patch', since that was the impression it gave at first glance! Nevertheless it was found extremely useful, in a second round of interviews, to talk people through it and ask them for both their comments about things we had got wrong, as they saw it, and for their views on what were the main issues concerning contracting (Duxbury 1994). Their responses not only improved the picture, and hence our holistic view of the situation, but also contributed to our understanding of the social and cultural features of the situation—the subject, in SSM of Analyses One, Two and Three (discussed below).

In recent work in the Health Service a new role for rich-picture-like illustrations has emerged. In December 1997 the Government White Paper *The New NHS* (HMSO 1997) described a new concept of the NHS, which was to exhibit such features as: led from the front line of health care ('primary care' by family doctors and other local services); founded on evidence-based medicine, with national standards and guidelines; and supported by modern information systems. Achieving this, according to the Minister of Health responsible for it, involved 'a demanding ten year programme' of development (p. 5). In 1998 the necessary information strategy to support this vision was published, the two documents being coherently linked (Burns 1998). Together, these two publications represent the best conceptual thinking about the NHS for 20 years, though realizing the vision will be an immense and difficult task for medics who are usually not very interested in thinking deeply about *managing* their work (as opposed to its professional execution) and for an organization in which sophisticated 'informatics' skills are scarce.

The White Paper and the information strategy are documents of 86 and 123 pages respectively; absorbing their message is not an easy task for people as busy as health care professionals and Health Service managers. We have found it exceptionally useful, in work commissioned by the centre of the NHS on the information system implications of the new concept for acute hospitals, to turn these excellent but overwhelming documents into picture form. (The documents themselves, being products of a Government service in which prose rules, contain only a handful of rather unadventurous diagrams.) For the White Paper, Figure A3 gives the basic shape of the concept, while Figure A4 adds much more detail to this simple picture. The information strategy, more complicated at a detailed level than the White Paper,

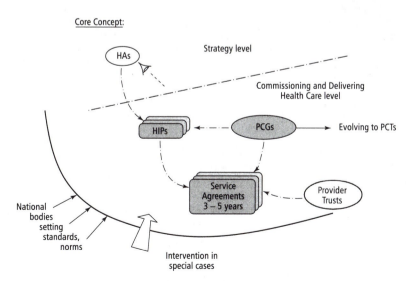

Model of the White Paper – The New NHS – Modern : Dependable

Core Concept:

HAs

Strategy level

Commissioning and Delivering
Health Care level

HIPs

PCGs

Evolving to PCTs

Service
Agreements
3 – 5 years

Provider
Trusts

National
bodies
setting
standards,
norms

Intervention in
special cases

Figure A3 The core concept of the NHS White Paper 1997 (HA = health authority; HIP = health improvement plan; PCG = primary care group; PCT = primary care trust)

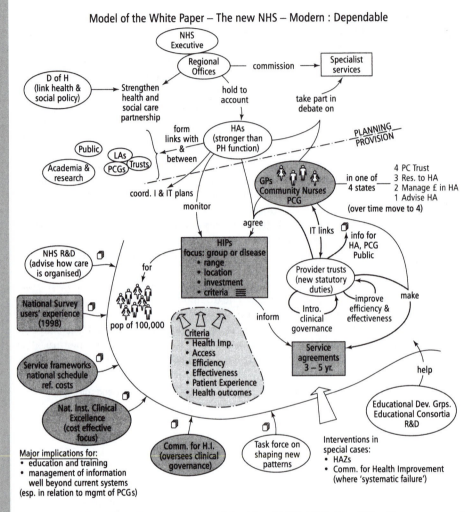

Figure A4 The White Paper concept of the New NHS 1997 (D of H = Department of Health; HAZ = health action zone; HI = health improvement; PH = public health)

was converted into a suite of eight pictures covering its core processes and structures, as well as the intended technical solution: electronic patient records which gradually evolve into each person's lifelong electronic health record. These picture versions of long documents have been very useful in conceptualizing our work, and no NHS audience sees them without asking for copies. This experience does suggest that there is a useful role for pictures of this kind wherever there is detailed written exposition of plans and strategies—at least until the happy times when such documents will themselves use seriously the medium of pictures as well as words.

Figures A3 and A4 can be seen as representations of combined structures and processes which enable the relation between the two elements to be debated. But the use of 'structure', 'process' and 'the relation between them' as a *formal* framework for 'finding out' in SSM, emphasized in the 1972 paper and in *STSP* (pp. 163, 164, 166), has not survived. I believe personally that I still use that framework *mentally*, without giving it much focused thought, but its more formal use, as described in 1972, has fallen into disuse. This seems to be because when you are faced with the energy and confusion which greet you whenever you enter any human problem situation, that particular framework seems highly abstract, a long way away from enabling you to grapple with pressing issues. However, as always with methodology, if it seems useful to you, then use it!

Analyses One, Two and Three

In addition to rich picture building, other frameworks which help to make the grasp of the problem situation as rich as possible are provided by Analyses One, Two and Three (*STSP*, pp. 194–198, 229–233; *SSMA*, pp. 45–53). Analysis One is an examination of the intervention itself, and its development was a direct result of our experience of research for the late Kenneth Wardell, a respected mining engineering consultant in that industry. (He is the 'Mr Cliff' of *STSP* (pp. 194–198).) This analysis is now a deeply embedded part of the thinking. The rich pictures will draw attention to the (usually) many people or groups who could be seen as stake-holders in any human situation, and Analysis One's list of possible, plausible 'problem owners', selected by the 'problem solver', is always a main source of ideas for 'relevant systems' which might usefully be modelled.

The freedom of the person or group intervening in a problem situation to answer the question: 'Who could I/we take the problem owner to be?' is important in achieving a grasp of the situation which is as holistic as possible. Thus in work which helped a community centre in Liverpool to rethink its role in a run-down part of that city, it was relevant to consider Liverpool Social Services Department as one among many possible problem owners, even though at the time the relationship between the centre and the department had not surfaced as an issue for anyone in the department. This kind of choice is what trying to be 'as holistic as possible' entails— even though *the* whole will always remain an unreachable grail. To adopt the coun-ter-view suggested by Bryant (1989) that to be a problem owner you have to be *aware* of owning the problem, would put a completely unnecessary constraint on interventions founded on soft systems thinking.

Analyses Two and Three, comprising a framework for the social and political analyses, are also now thoroughly embedded in praxis. Some commentators have suggested that they are less highly developed than some of the other parts of SSM, such as model building, but that is to misunderstand them. The roles/norms/values framework and the ongoing analysis of 'commodities which embody power' are certainly simply expressed. That is the point of them. You can keep them in your head, and they can constantly guide all of the thinking which goes on throughout an intervention. But though they are simple in expression they reflect one of the main underlying conclusions from the whole 30 years of SSM development: that to make sense of it you have to adopt the view argued in Chapter 8 of *STSP* (pp. 264–285), namely that social reality is no reified entity 'out there', waiting to be investigated. Rather, it is to be seen as continuously socially constructed and reconstructed by individuals and groups (the latter never perfectly coherent). This represents an intellectual stance defined by such features as: deriving from the work of Max Weber; articulated, for example, in the sociology of Alfred Schutz; and underpinned by the philosophy of Edmund Husserl (Luckmann 1978, pp. 7–13). In practical terms, the usable framework which underpins Analyses Two and Three was found in the autopoietic model teased out of the work of Vickers on 'appreciative systems' (Checkland and Casar 1986). That will be discussed further towards the end of this chapter.

Analysis Three moves beyond the model of an appreciative system but is compatible with it. (The appreciative system model describes a social process; Analysis Three covers one of the main determinants of the *outcomes* of that process: the distribution of power in the social situation.) This analysis is avowedly practical, a highly significant contribution to the development of SSM from the action research carried out by Stowell in a light engineering company and in an educational publisher (Stowell 1989). He reviews the extensive social science literature on 'power', but his main aim is not to add to that literature—which is strong on words, less interested in action—but to find practical ways of enabling open discussions to take place on topics which are usually taboo, or emerge only obliquely in the local organizational jokes. These are discussions focused on power, its manifestations and the pattern of its distribution.

Analysis Three is not based on an answer to the question: What is power? It works with the fact that everyone who participates in the life of any social grouping quickly acquires a sense of what you have to do to influence people, to cause things to happen, to stop possible courses of action, to significantly affect the actions the group or members of it take. The metaphor of the 'commodities' which embody power is used to encourage discussion of these matters. Views can be elicited on what you have to possess to *be* powerful in this group or this organization. Is it knowledge, a particular role, skills, charisma, experience, clubbability, impudence, commitment, insouciance … etc? Recent history of the organization or group can be questioned and/or illustrated in these terms, all with the aim of finding out as deeply as possible how this particular culture 'works', what change might be feasible and what difficulties would attend that change. Stowell (1989) describes the use of the metaphor 'commodity' thus:

> 'Commodity', then, is the proposed means of providing organisational members involved in change with a practical means of addressing power. Acknowledging, with Giddens, 'that speech and language provide us with useful clues as to how to conceptualise processes of social production and re-production', what has been suggested within this thesis is an idea by which the notion of power can be articulated in terms which are appropriate to a given organizational culture and which can be understood by those most affected (p. 246).

The aspiration of openness here is admirable, but do not be surprised if Analysis Three has to be carried out with great sensitivity and tact. In many human situations there is not the confidence necessary for open discussion of issues hinging on power.

Before moving on from the 'finding out' activity, it is worth reiterating that 'finding out' is never finished; it goes on *throughout* a study, and must never be thought of as a preliminary task which can be completed before modelling starts.

Building Purposeful Activity Models

The Role of Modelling in SSM

The purposeful activity models used in SSM are *devices*—intellectual devices—whose role is to help structure an exploration of the problem situation being addressed. This is not an easy thought to absorb for many people, since the normal connotation of the word 'model', in a culture drenched in scientific and technological thinking, is that it refers to some representation of some part of the world outside ourselves. This is the case, for example, for models as used within classic operational research. If an operational researcher builds a model of a production facility, then there is a need, before experimenting on the model to obtain results which can be used to improve the real-world performance, to first show that the model is a 'valid' representation. This might be done by showing that the model, fed with the last six months' input, can generate something which is close to the actual output produced over that period. But models in SSM are not at all like this. They do not purport to be representations of anything in the real situation. They are accounts of concepts of pure purposeful activity, based on declared world-views, which can be used to stimulate cogent questions in debate about the real situation and the desirable changes to it. They are thus not models *of . . .* anything; they are models *relevant to debate* about the situation perceived as problematical. They are simply devices to stimulate, feed and structure that debate.

In the early stages of SSM's development, devices of only one type were built. Blinkered by our starting position in systems engineering, we tended only to make models whose (systems) boundaries corresponded to real-world organizational boundaries. This self-imposed limitation derived, we can now see, from the systems engineers' view that the world consists of interacting systems. Thus, working in, say, a manufacturing company with a conventional functional organization structure, we would make models of a production system, an R&D system, or a marketing system. These would map on to Production, R&D and Marketing departments. But organizations have to carry out, corporately, many more purposeful activities than the handful which can be institutionalized in an organization structure. For example, suppose the manufacturing company in question to be in the petrochemicals business. Such companies, in order to survive in a science-based international business full of very smart competitors, have to be technological innovators. In a systems study carried out in just such a company (the study being concerned with improving relations between R&D and other functions) it was found very useful to make a model based on the core idea of *innovating* in that industry. That model a . . . system . . . to innovate . . . in the petrochemical industry . . .) had a boundary which did not coincide with the organizational boundaries of the (functionally defined) existing departments. Not surprisingly, many of the activities in that model were actually taking place in the company: some in R&D, some in Production, some in Marketing. Also, some of the activities in the model were missing in the real situation. The great value of the model was that its boundary *cut across* the organizational boundaries of the actual departments. This was very helpful in stimulating discussion and debate within the company, when the model was used to question the existing situation.

Models which map existing organization structures (such as 'a system to carry out R&D' in this example) are thought of as 'primary task' models; models like that of the innovation system are 'issue-based'—the notional issue here being that some-how or other this particular company has to ensure that it has the ongoing capability to innovate. This primary task/issue-based distinction (*SSMA*, p. 31) has been found to be a source of confusion for many. This is probably because the distinction is not absolute. The petrochemical company, if its thinking had been a little differ-ent, *might* well have brought together people with the appropriate skills and exper-tise to staff an Innovation Department. Had they done so the issue-based model here would then have been a primary task model. Pragmatically, to make sure that the useful provocation provided by models whose boundaries cut across existing orga-nization boundaries is not neglected, the rule from experience is simple: make sure that you do not think only in terms of models which map existing structures. This will help ensure that the modelling fulfils its intended role in SSM: to lift the thinking in the situation out of its normal, unnoticed, comfortable grooves.

Root Definitions, CATWOE and Multi-level Thinking

To build a model of a concept of a complex purposeful activity for use in a study using SSM, you require a clear definition of the purposeful activity to be modelled. These definitional statements, SSM's 'root definitions', are constructed around an expression of a purposeful activity as a transformation process T. *Any* purposeful activity can be expressed in this form, in which an entity, the input to the transform-ing process, is changed into a different state or form, so becoming the output of the process. A bold sparse statement of T could stand as a root definition, for example 'a system to make electric toasters', but this would necessarily yield a very general model. Greater specificity leads to more useful models in most situations, so the T is elaborated by defining the other elements which make up the mnemonic CATWOE, as described in *STSP* (Chapters 6, 7, Appendix 1) and *SSMA* (Chapter 2; illustrated *passim*).

These are not abstruse ideas; the skills required for model building are not arcane: logical thought and an ability to see the wood *and* the trees; also, any model should be built in about 20 minutes. Nevertheless there are classic errors which recur time and time again. The most common error, often found in the literature, is to confuse the input which gets transformed into the output with the resources needed to carry out the transformation process. This conflates two different ideas: input and resources, which coherency requires be kept separate. Also, when people realize that there is a formula (an abstract one) which will always produce a formulation which is at least technically correct, namely: 'need for X' transformed into 'need for X met', they seize on this with glee. Unfortunately, they then often slip into writing down such transformations as 'need for food' transformed into 'food'. What a fortune you could make in the catering industry if you knew how to bring off that remarkable transformation! It is evidently not easy to remember that in a transfor-mation what comes out is the same as what went in, but in a changed (transformed) state.

In recent years experience has shown the value of not only including CATWOE elements in definitions but also casting root definitions in the form: do P by Q in order to contribute to achieving R, which answers the three questions: What to do (P), How to do it (Q) and Why do it (R)? [This formulation was, alas, initially given in terms of XYZ rather than PQR (*SSMA*, p. 36). Using P, Q and R avoids the chance that Y may be confused with why?] The simplest possible definition is of 'a system to do P'. 'Do P by Q' is richer, answering the question: how? And also forcing the model builder to be sure that there is a plausible theory as to why Q is an appropriate means of doing P. For example: 'communicate (P) by letter writing (Q)' is certainly plausible, but would provoke examination of the reasons for doing this communication (i.e. the R question) by this chosen means. In this particular case, the question of required timing would have to be thought about. This could lead to examining, for example, whether there was a case for replacing the cultural resonance which goes with writing a letter to someone by the more brutal but quicker e-mail.

The formal aim of this kind of thinking prior to building the model is to ensure that there is clarity of thought about the purposeful activity which is regarded as relevant to the particular problem situation addressed. The idea of levels, or layers (or 'hierarchy', though that word tends to carry connotations of authoritarianism which are not relevant here) is absolutely fundamental to systems thinking. Much human conversation is dogged by the confusion which follows from the common inability to organize thoughts and expression consciously in several layers. Thus, the Chair of the Tennis Club Social Committee opens a meeting by saying that the committee needs to decide whether or not to organize the club fête this year, given the wet day last July and the unfortunate arrival of a gang of unruly bikers! As you begin to think about this, sitting in committee, you are surprised (but should not be) that the first member to speak says: 'My sister and I will do the cake stall as usual'. Systems thinkers are adept at consciously separating 'whether' from 'what' and 'how'.

In selecting some hopefully relevant systems to model, there are in principle always a number of levels available, and it is necessary to decide for each root definition which level will be that of 'the system', the level at which will sit the T of CATWOE. This makes the next lower level the 'sub-system' level: that of the individual activities which, linked together, meet the requirements of the definition. The next higher level is then defined automatically as that of the 'wider system': the system of which the system defined by T is itself only a sub-system. In SSM this higher level is the level at which a decision to stop the system operating would be taken: it is the level of the system 'owner', i.e. the O of CATWOE. Thus, this intellectual apparatus of T, CATWOE, root definition and PQR, ensures that the thinking being done covers at least three levels, those of system, sub-systems and wider system. It prevents the thinking from being too narrow, and stimulates thoughts about whether or not to build other models. For example it might be decided also to model at the wider-system level, or to expand some of the individual activities in the initial model by making them sources of further root definitions. (Figure 8.14 in *SSMA*, p. 231, for example, shows a model in which activity 1 has been expanded into four more detailed activities. Similar structuring is shown in Figure 5.6, p. 136, and Figure 7.8 of *ISIS*, p. 209 shows a simple model in which most of the activities have been expanded in this way.) Figure A5 summarizes the importance of thinking consciously at several different levels, and also makes the point that different people might well make different judgements about which level to take as that of 'the system'. 'What' and 'how', 'system' and 'sub-system' are relative, not absolute concepts.

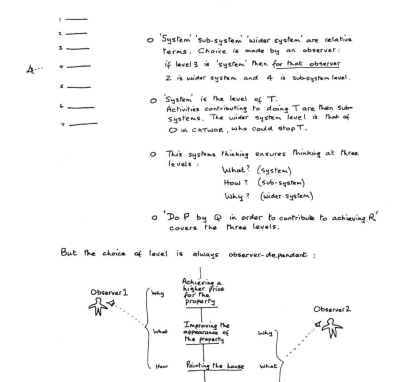

Figure A5   Systems thinking entails thinking in layers defined by an observer

Measures of Performance

It is obvious from the form of SSM (as in Figure A1) that it would be possible to use the approach without creating *systems* models as the devices used to shape the exploration of the situation addressed. It would be possible to use instead models based on theories of, say, aesthetics, psychology, religion, or even, if you were foolish enough to abandon rationality completely, astrology! We use systems models because our focus is on coping with the complexity in everyday life, and that complexity is always, at least in part, a complexity of interacting and overlapping relationships. Systems ideas are intrinsically concerned with relationships, and so systems models seem a sensible choice; and since they have been found, time after time, to lead to insights, they have not been abandoned.

Now, the core systems image is that of the whole entity which can adapt and survive in a changing environment. So our models, to use systems insights, need to be cast in a form which in principle allows the system to adapt in the light of changing circumstances. That is why models of purposeful activity are built as sets of linked activities (an operational system to carry out the T of CATWOE) together with another set of activities which monitor the operational system and take control action if necessary. Since there is no such thing as completely neutral monitoring, it is necessary to define the criteria by which the performance of the system as a whole will be judged. Hence the core structure of the monitoring and control sub-system is always the same: a 'monitor' activity contingent upon definition of the criteria by which system performance will be judged, and an activity rendered as 'take control action' which is contingent upon the monitoring. This can of course be augmented if justified in particular cases—as in the model in *STSP*, p. 291. The basic structure is seen in many of the models in *SSMA* and *ISIS*.

For many years the concept of 'measures of performance' was felt to be sufficient for use in models, but was then enriched by an analysis which flows from the consideration that SSM's models are simply logical machines for carrying out a purposeful transformation process expressed in a root definition. Measuring the performance of a logical machine can be expressed through an instrumental logic which focuses on three issues: checking that the output is produced; checking whether minimum resources are used to obtain it; and checking, at a higher level, that this transformation is worth doing because it makes a contribution to some higher level or longer-term aim. This gives definition of the '3Es' which will be relevant for every model: the criteria of efficacy (E_1), efficiency (E_2), and effectiveness (E_3), first developed in 1987 (Forbes and Checkland 1987; Checkland *et al* 1990; *SSMA*, pp. 38, 39). This core set of criteria can be extended in particular cases—for example by adding E_4 for ethicality (is this transformation morally correct?) and E_5 for elegance (is this an aesthetically pleasing transformation?). Since it will not be possible to name the criteria for effectiveness without thinking about the aspirations of the notional system owner (O in CATWOE), this analysis is another contribution which prevents the modeller's thinking being restricted only to one level, that of the system itself.

Model Building

Given the preliminary thinking expressed in root definition, CATWOE, the three Es and PQR, assembling an activity model ought not to be too difficult: simply a matter of assembling the activities required to obtain the input to T, transform it, and dispose of the output, ensuring that activities required by the other CATWOE elements are also covered; then link the activities according to whether or not they are dependent upon other activities. And the task ought not to be an elaborate one either, given the oft-proved value of the heuristic rule that the overall activity of the operational sub-system should be captured in Miller's (1968) 'magical number' 7 ± 2 individual activities (any of which can if necessary be made the source of a more detailed model). Nevertheless some people manage to make model building a task fraught with difficulty. This is probably because there are in fact subtle features of the process which are masked in the simple account just given.

These subtleties are illustrated by the fact that, for example, the distribution manager of a manufacturing company is probably not the person who will find it very easy to build a model from a root definition of a system to distribute manufactured products. The difficulty for such a person is to focus only on unpacking and displaying the *concept in the root definition*; the tendency will be to slip into describing the real-world arrangements for distributing products in his or her own company. Equally, inexperienced users, fresh-from-school undergraduates especially, find all such models difficult to build because they know so little about real-world arrangements. The fact is—and this is where the unobvious difficulties of modelling lie—it is not usually possible to construct a model *exclusively* on the basis of root definition, CATWOE, three Es and PQR; real-world knowledge *does* inform model building, but, crucially, must not dominate it. The craft skill is to build a model using a background of real-world knowledge without including features of typical practice which are not justified by the root definition, CATWOE, 3Es and PQR. As always with craft skills, practice, practice, practice is the watchword.

Because most practitioners initially 'feel their way' to a method of modelling comfortable for them, it may be helpful to provide some templates which derive purely from the logic of the process and which may provide help for those just starting to use the process of SSM. Two such templates are provided here; they are meant to be abandoned as experience grows.

Figure A6 sets out a logical procedure for modelling purposeful activity systems in a series of steps; Figure A7 expresses the process in Figure A6 as a partial activity model. These are self-explanatory, though it may be remarked with reference to Figure A6 that although the stages can be carried out on a computer screen, there is a good case, as long as you can manage it in good visual style, for producing the final model in hand-drawn form. The reason for this is psychological, and is the

Given : definition of T, E 1,2,3 , CATWOE, Root Definition (PQR)

(1) Using verbs in the imperative ('obtain raw material X') write down activities necessary to carry out T (obtain I, transform it, dispose of Output). Aim for 7±2 activities.

(2) Select activities which could be done at once (ie not dependent on others) :

(3) Write these out on a line, then those dependent on these first activities on a line below ; continue in this fashion until all activities are accounted for.

Indicate the dependencies :

(4) Redraw to avoid overlapping arrows where possible and add monitoring and control

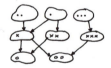

Figure A6   A logical procedure for building activity models

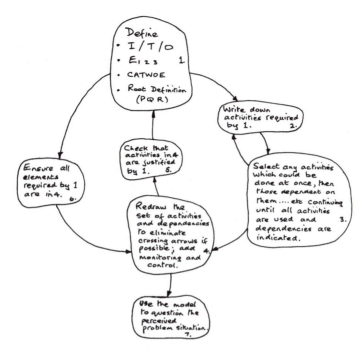

Figure A7   The process of modelling in SSM, embodying the logic of Figure A6

same as that for drawing egg or cloud shapes rather than rectangular boxes: it acknowledges the models' role as pragmatic devices, not definitive once-and-for-all statements. In Figure A7 the process form emphasizes the exercise of judgement during modelling. Iteration around activities 2, 3, 4 continues until it is felt that the minimum but necessary cluster of activities has been assembled; the wider iterations around activities 1 to 6, and around 1–6–4–5 represent the checks that the model is *defensible* in relation to the concept being expressed.

Once a model is constructed by such a process, the golden rule for 'reading' a model—something which the many people unconsciously straitjacketed in linear thinking find difficult—is always to start from the activities which are not dependent upon other activities but have others dependent upon them, i.e. those which have arrows from them but none to them.

Finally, on modelling, a few remarks about the formal system model are in order (*STSP*, pp. 173–177; *SSMA*, pp. 41–42). As formulated in Figure 9 of *STSP* (p. 175) this was useful when we were acquiring a sense of what is meant to treat purposeful activity seriously as a systems concept. In *SSMA* it is said that it can now be 'cheerfully dropped' (p. 92). Its language was the problem. Since it was built using concepts such as boundary, sub-systems, decision-takers, resources, etc. the unfortunate effect of its use was to reinforce the wrong impression that the devices called 'human activity systems' are in some way to be thought of as would-be descriptions of real-world purposeful action. Since that is a main source of misunderstanding about SSM, and since what it offers conceptually is captured in CATWOE, the 3Es and PQR, it can indeed be 'cheerfully dropped'. The same argument speaks against the phrase 'human activity system', but that is probably too deeply embedded to be prised out of SSM and ditched. The best antidote to these dangerous phrases is undoubtedly to encourage the use of Arthur Koestler's neologism for the abstract concept of a whole, namely 'holon' (Checkland 1988). That is what models in SSM are: holons for use in structuring debate.

Exploring the Situation and Taking Action

As human beings experience the unrolling flux of happenings and thoughts which make up day-to-day life, both professional and private, they are all the time likely to see parts of that flux as 'situations', and certain features of it as 'problems', or 'issues'. These concepts and this kind of language—of 'situations', 'issues', 'problems'—are very commonly used in everyday talk, but they are subtle concepts, and we need to beware of giving them a status they do not deserve. We must not reify them; they do not exist 'out there', beyond ourselves, as we can assume 'that beech tree' and 'that dog scratching itself' do. 'This situation', and 'this problem' indicate dispositions to think about (parts of) the flux in particular ways, and they are themselves *generated by human beings*; also, no two people will see them in *exactly* the same way. If, for example, the senior managers of a company all agree in discussion that they have a problem due to the failure of a new product to build up sales following its launch, no two of them will have precisely the same perception of this situation and/or this problem. What is more, some among those who 'agree' about the situation/problem may privately be seeking to ensure the failure of the new product in order that more resources can then come their way! (Remember, we can never know for sure what is going on inside the head of another person; and we cannot assume that their words necessarily reveal it.)

These are bleak thoughts, but necessary ones if applied social science is to be pursued with adequate intellectual rigour. They mean that neither problem situations nor problem types can be classified and made the basis of pigeon holes into which particular examples may be slotted, for one person's 'major issue' or 'serious problem' may well be another's unruffled normality. Both the existence of a problem situation and its interpretation are human judgements, and human beings are not like-thinking automata.

A result of this is that the later stages of a study using SSM cannot be pinned down and as sharply defined as the early stages, in which a situation can be tentatively defined and explored, plausible 'problem owners' named, 'relevant' systems selected and models built. The many uses of SSM described in *STSP*, *SSMA*, *SCMA* and *ISIS*, as well as the many accounts in the literature from people outside the Lancaster group, reveal the variety which is possible. This ranges from quick, short, tactical studies to much longer ones oriented to strategy. Because of this, comments on the later parts of SSM are bound to be generalizations from experiences which are very diverse, those generalizations being themselves subject to change as the flux of experience rolls on. Nevertheless a few very broad generalizations from a rich mass of experience can be entertained.

The initial ways of using models, described in *STSP* (pp. 177–180) and *SSMA* (pp. 42–44 and *passim* in the cases described), are still the most common way of initiating the 'comparison' stage of SSM, in which well-structured debate about possible change is sought. Most common, at least as an *initiator* of debate, is the completion of charts in which questions derived from the models brought to the debate are answered from perceptions of current reality on the part of people in the situation addressed. But do not expect the debate to be tidy or predictable; be deft, light on your feet, ready to follow where the debate leads, unready to follow any dogmatic line.

Looking back over experiences in the last decade, an emerging pattern can be discerned in which there are two common foci of the later stages of SSM, during which the driving principle is to bring the study to some sort of conclusion. The first of these is the original one: SSM as an action-oriented approach, seeking the accommodations which enable 'action to improve' to be taken. This is exemplified in the work in 'Index Publishing and Printing Company', described in *STSP*, pp. 183–189. Here action was taken to improve the working relationship between publishers and printers, who represent two very different cultures. A new process to deal with issues surrounding the decision 'where to print' was established, and a new unit to carry out the work was set up. The second focus, very prevalent in the great complexity which characterizes the public sector, is on SSM as a sense-making approach. This is exemplified in recent work in the NHS, and is discussed below.

In the first (action-oriented) case the change sought can usefully be thought about in terms of structural change, process change and changes of outlook or attitude. Normally in human affairs any explicitly organized change will entail all three, and the relationship and interactions between the three need careful thought. Of course the easy option to take—in the public sector for Government or, in other organizations, for senior managers—is to impose *structural* change; and that is often done without serious attention to the other two dimensions: process and attitude. The long series of changes imposed by the UK Government on the Health Service, for example, give a good illustration of imposed *structural* change with relatively little attention to the process and attitudinal change also required (Ham 1992; Rivett 1998; Webster 1998). [It has been significant, recently, that an experienced commentator on the Health Service, Chris Ham, has detected that 'the obsession with structural change that has dominated health policy in recent years has given way to a focus on how staff and services can be developed...' (Ham, 1996). That is a much-needed change.]

In general, thinking about desirable and feasible change can initially be structured in the way shown in Figure A8. A most important feature of this is the need in human affairs to think not only about the substance of the intended change itself but also about the additional things you normally have to do in human situations to *enable* change to occur. (In introducing a clinical information system in a big acute hospital, for example, a project described in *ISIS* (pp. 192–198), its instigator Peter Wood, Chairman of the District Health Authority had already spent several years preparing the ground with hospital consultants inevitably suspicious that such systems could lead to greater control of their clinical activity by hospital managers.)

The second broad category of use to which SSM-style activity models can be put is to use them to make sense of complex situations (though that sense making may of course also lead on to action being taken). It is significant that this category of use has grown markedly in the last decade of SSM development, as concepts such as 'organization', 'function', 'profession' and 'career' have all become more fluid.

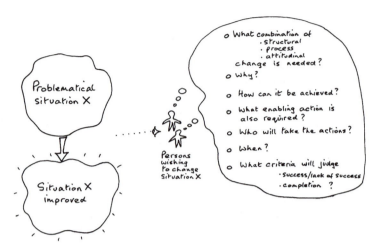

Figure A8   Thinking about desirable and feasible change

Sense-making use of models is well illustrated in recent research in the NHS. The work has been described in some detail in Checkland (1997) and in *ISIS*, pp. 165–172 and will only be sketched in here. Setting out to research the new 'contract'-based relationship between purchasers of health care for a given population and providers of that care (such as acute hospitals, for example) a research team from Lancaster University Management School, using SSM, first built an activity model of the contracting process (Figure 6.2 in *ISIS*). The concept expressed in this model did not rely at all upon observation of the NHS. It reflected simply the interests of the multi-disciplinary research team: information support, organization change, etc. This model was used as a source of structure for open-ended interviews with more than 60 NHS professionals. This produced a daunting mass of interview material. This was analysed by extracting from it the nouns and verbs used by NHS professionals in describing the contracting process and their expectations of it during its first year. These nouns and verbs were fashioned into the elements of an activity model, and these elements were combined to make an activity model relevant to the contracting process as it was initially being interpreted by both purchasers and providers of health care. (This 'backwards' modelling—not based on a root definition but teased out of the interview material—represented an innovation within SSM. See *ISIS*, pp. 165–172.) The difference between the first model (based on the researchers' world-views) and the second one (based on the world-views of NHS professionals) defined the learning achieved in this first phase of the research. This led to 10 pieces of action research in the NHS, and eventually to another sense-making model which helped to unpack and illuminate the purchaser–provider relationship.

This second sense-making model sought to flesh out coherently the complex interactions between a particular purchaser (a health authority) and a particular provider (an acute hospital), interactions to which we had had access over a two-year period. In order to find our way to a model which would richly express all we had observed, 47 previous models relevant to NHS purchasing/providing were first examined (Duxbury 1994). (These came from earlier SSM-based work in the NHS.) This established what language had been found relevant to describing purchasing or providing. This language, together with the recorded observations of what had happened in the present experience between the collaborating hospital and their local health authority, yielded an activity model which makes sense of all that had been observed. The derivation of the model [Figure 8.4 in the book about research in the NHS edited by Flynn and Williams (1997)] was a subtle process. The guide to that process was the question: What activity model could generate all the happenings observed over the two-year period? Its role was to provide a coherent frame for the 10 further pieces of action research at NHS sites.

This completes the necessarily tentative discussion of the variegated later stages of SSM-based studies or projects. Enough has been said to illustrate that the just described sense-making use of activity models calls for rather more than a slavish adherence to the apparently prescriptive seven-stage model of SSM! It also illustrates the fact that the role of methodology, properly interpreted as a set of guiding principles, is not to produce 'answers': that it can never do on its own; it is to enable you, the user, to produce better outcomes than you could without it.

This examination of the parts of SSM is now complete, and we can return to a re-examination of the methodology as a whole, a re-examination which we may hope is made richer by this examination of the parts.

SOFT SYSTEMS METHODOLOGY—THE WHOLE REVISITED

In the earlier section which examined SSM as a whole the focus was on the way in which its representation changed as experience of use accumulated and the different parts of it gradually became more sophisticated. This indicated a shift from the rather biff-bang 'engineering' atmosphere of the 1982 paper to the 'four-activities' model of Figure A1, with its deliberate reticence about the 'hows' and its avoidance of any implication of a prescription to be followed. Having now examined the parts of SSM in their developed form, a re-examination of the whole can try to address the question of what it is which characterizes the approach, making it more than the sum of its parts. This requires an examination of three things: the fundamental notion of *methodology*, as opposed to *method*; the question of what *constitutes* SSM (what you must do if you wish to claim to be guided by it in a particular study); and what happens to SSM when it is internalized in the practice of experienced users—at which point it is apparently a world away from the original formulation in the 1972 paper.

Methodology and Method: the LUMAS Model

The word 'methodology' was originally used to mean 'the science of method', which technically makes the concept of '*a* methodology' meaningless. I remember clearly the day in the early 1970s when my colleague, the late Ron Anderton, said to me: 'You're misusing the word "methodology"; you can't have *a* methodology, the word refers to the whole body of knowledge about method', to which I replied: 'We'll have to change the way the word is used, then.' The deplorable arrogance of that reply stemmed from the fact that I was at that time just becoming aware that, outside the study of social facts, as Durkheim (1895) advocated, the normal scientific method is inadequate as a way of inquiring into human situations; and I was starting to see systems thinking as a holistic reaction against the reductionism of natural science. This meant that the principles of scientific investigation, as used to underpin investigation of natural phenomena, would not adequately support our work. We needed a different methodology, that is a different set of principles. Happily for me, the way that the word 'methodology' is now used has indeed changed, and in the late 1990s Oxford dictionaries of current English now define it not only as 'the science of method' but also as 'a body of methods used in a particular activity' (*Concise Oxford Dictionary of Current English*, 1996). This latter definition makes the crucial distinction between 'methodology' and 'method', and it is the failure to understand this which characterizes much of the secondary literature on SSM.

As the structure of the word indicates, methodology, properly considered, is 'the logos of method', the *principles of method*. When those principles are used to underlie, justify and inform the things which are actually done in response to a particular human problem situation, those actions are at a different level from the overarching principles. Methodology in that situation leads to 'method', in the form of the specific approach adopted, the specific things the methodology user chooses to do in that particular situation. If the user is competent then it will be possible to relate the approach adopted, the specific 'method', to the general framework which is the methodology. And if the methodological principles are well thought out and clearly expressed, then a repertoire of regularly used methods which are found to work will emerge over time as experience is gained. (And of course some methods, over time, in some fields of study, acquire the status that they can—if skilfully employed—guarantee a particular result; they become *techniques*. Examples are the simple algebraic technique which enables you to solve any pair of simultaneous equations, or the physical technique which will cause a cricket ball to 'swing' (move sideways) in mid-air as it is bowled, this latter being a rather more difficult technique to master! Given the multiple perceptions which define and characterize human situations, it is extremely unlikely that any of the methods used within a methodology like SSM could become techniques in the sense used here.)

Since methodology is at a meta level with respect to method (i.e. *about* method) this argument means that no generalizations about methodology-in-use can ever be taken seriously. Thus to read commentators who declare that SSM is 'managerialist', or 'radical', or 'conservative', or 'emancipatory', or 'authoritarian' tells you something about the writer—that they have confused methodology with method—but it tells you nothing about SSM. SSM may exhibit any of these characteristics, as *method*, when it is used by particular users in particular situations. In fact whenever a user knowledgeable about a methodology perceives a problem situation, and uses the methodology to try to improve it, the three elements in Figure A9 are intimately linked: user; methodology as words on paper, and situation as perceived by the user. Any analysis of what happens, carried out by an outsider, would have to embrace all three elements and the interactions between them. This would include converting the methodology (as a set of principles) into a specific approach or 'method' which the user felt was appropriate for *this* particular situation at a particular moment in its history. What happens whenever a methodology is

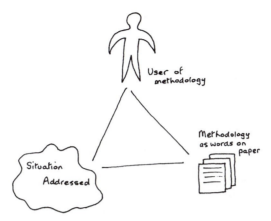

Figure A9   Three interacting elements always present in methodology use

used is shown in the LUMAS model which is Figure A10. Here a user, U, appreciating a methodology M as a coherent set of principles, and perceiving a problem situation S, asks himself (or herself): *What can I do?* He or she then tailors from M a specific approach, A, regarded as appropriate for S, and uses it to improve the situation. This generates learning L, which may both change U and his or her appreciations of the methodology: future versions of all the elements LUMAS may be different as a result of each enactment of the process shown. All the systems studies described in *STSP*, *SSMA* and *ISIS* can be seen as enactments of this process, which accepts that what the user can do depends upon the nexus consisting of U, U's perceptions of M, and U's perceptions of S (Tsouvalis 1995). Never imagine that any methodology can itself lead to 'improvement'. It may, though,

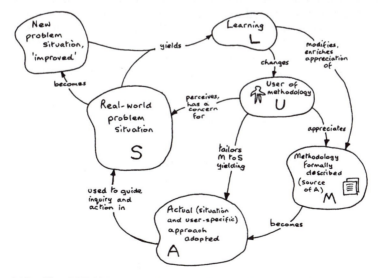

Figure A10   The LUMAS model: Learning for a User by a Methodology-informed Approach to a problem Situation

help *you* to achieve better 'improvement' than you would without its guidelines. But different users tackling the same situation would achieve different outcomes, and an outside observer can form sensible judgements not about M, as if it could be isolated and judged on its own, but about LUMAS as a whole. The model in fact pictures the process by which SSM was developed.

SSM's Constitutive Rules

In the early 1980s Atkinson researched SSM in use. His work included a very detailed examination of three completed systems studies in which different people had made use of SSM as their guiding methodology. He found their uses to show interesting differences. His shorthand summary for the three modes of use he observed were: 'liberal' (eclectic, problem-oriented); 'professional' (SSM as a management consultant's expertise, not necessarily shared with clients); and 'ideological' (the work dominated by an ideological commitment to help cooperatives become more effective). This kind of observation supports the argument developed in the previous section, that methodology use will always be user-dependent. But at the same time that he is noting these differences Atkinson (1984) also observes that the studies all show 'a family resemblance', which raises the questions: What then *is* SSM, the source of this resemblance? and What must a user do if he or she wishes to claim to be 'using SSM'? In *SSMA* the statement is made (p. 58) that

> . . . *mouldability* by a *particular user* in a *particular situation* is the point of methodology. . .

which prompts us to ask what it is that gets shaped into the different forms which different users and different situations evoke.

This question had been addressed before Atkinson did his research, being raised initially by Naughton (1977) in the context of *teaching* SSM. In his commentary on SSM written for Open University students, Naughton argued that there were 'Constitutive Rules' which had to be followed if a claim to be using SSM was to be accepted as valid, and 'Strategic Rules' which 'help one to select among the basic moves'; for example the user might choose (or not) to use the structure/process/climate model in doing the initial exploration of the problem situation. These rules, deriving from the seven-stage model of SSM, were very helpful at the time, and were endorsed in *STSP* (pp. 252, 253). By the time that SSMA was written, however, the seven-stage model was no longer the preferred expression of SSM as a whole, and a new set of constitutive rules were proposed (*SSMA*, pp. 284–289). These defined five characteristics of uses of SSM and set out its epistemology (rich pictures, CATWOE, etc.). A use of SSM was one which could be described using these concepts and language.

In 1997, in the most cogent exegesis of SSM carried out so far, Holwell found these 1990 rules to be at the same time 'both too loose and not extensive enough' (p. 398). They are too loose because they allow people who have done no more than draw a rich picture to claim they are using SSM (the literature contains such examples!). And they are too restrictive, in the sense of being not extensive enough, because they are silent on some basic assumptions which SSM always takes as given. To correct this, Holwell (1997) argues that the answer to the question:

what is SSM? has to be made at three levels: the taken-as-given assumptions; the process of inquiry; and the elements used within that process. She writes:

> ... there are three necessary statements of principle or assumption:
>
> (1) you must accept and act according to the assumption that social reality is socially constructed, continuously;
> (2) you must use explicit intellectual devices consciously to explore, understand and act in the situation in question; and
> (3) you must include in the intellectual devices 'holons' in the form of systems models of purposeful activity built on the basis of declared worldviews.
>
> Then there are the necessary elements of process. The activity models ... are used in a process informed by an understanding of the history of the situation, the cultural, social and political dimensions of it ... (the process being) about learning a way, through discourse and debate, to accommodations in the light of which either 'action to improve' or 'sense making' is possible . Such a process is necessarily cyclical and iterative. Finally, while not limited to this pool ... a selection from Rich Picture, Root Definition, CATWOE ... etc may be used in the process.

These arguments are well made, and this work gives us a solid basis for definitive constitutive rules for SSM. We need rules which are oriented to practice rather than teaching, and which can encompass the wide range of sophistication brought to the use of SSM. At one end of the spectrum is a naïve following of the seven stages in sequence. This is not necessarily wrong, simply something users quickly grow out of as the ideas take root in their thinking. Once internalized, SSM's concepts lead to the deft, light-footed and flexible use which characterizes the other end of the spectrum of sophistication. The two 'ideal types' of SSM use which define the spectrum are termed Mode 1 and Mode 2 in *SSMA* (pp. 280–284). The difference between them is very relevant to the question of SSM as a whole, and is discussed in the next section.

Prescriptive and Internalized SSM: Mode 1 and Mode 2

SSM grew out of the failure of systems engineering—excellent in technically defined problem situations—to cope with the complexities of human affairs, including management situations. As systems engineering failed we were naturally interested in discovering what kind of approach *could* cope with problems of managing. So the research programme which yielded SSM was initially rather methodology-oriented. Then what happened was that as the shape of SSM emerged, as its assumptions became clearer, and its process and elements became firm, so the whole methodology became, for its pioneers, internalized. SSM became the way we thought about coping with complexity in real situations, and the research itself could become more problem-oriented. The process of internalization is a very real one for those for whom it is happening, but it is not an easy process to describe, certainly not as a series of steps recognized at the time they occur, for the steps are often not so recognized. The descriptions of the two ideal types of SSM use in *SSMA* enabled the 10 studies described in the book to be (subjectively) placed relative to each other on the spectrum between Mode 1 and Mode 2 (Figure 10.3 in *SSMA*; see also *ISIS*, pp. 163–172). This implicitly invited the reader to get a feel for what internalizing the methodology means, and to see whether he or she agrees with the placings.

Certain dimensions may be used to differentiate the two ideal types, recognizing that actual studies will never exactly match either of the two idealized concepts, but will reflect elements of both. Such dimensions are:

Mode 1		Mode 2
Methodology-driven	versus	situation-driven
Intervention	versus	interaction
Sometimes sequential	versus	always iterative
SSM an external recipe	versus	SSM an internalized model

and it follows from these that there will never be a generic version of what happens in 'near-Mode 2' studies precisely because they are situation-driven. Perhaps the best approach to understanding internalized SSM in action is through examples. One was given in the previous section (in which an activity model was teased out from the nouns and verbs used by Health Service professionals in talking about the then mandatory contracting process between purchasers and providers of health care). Another is now briefly described.

This example of near-Mode 2 use of SSM occurred at a one-day conference on 'Mergers in the NHS'. This was a topic of interest because the Health Service has seen many mergers in recent years—between district health authorities joining to form bigger purchasers of health care, between hospitals, and more recently between health authorities; and ministers have indicated that more such mergers will occur. In the morning the conference heard a number of talks from people who had been involved in mergers, in industry as well as the NHS, including in the case of the latter, examples from both a health authority and a hospital perspective. After lunch the participants split into small groups for discussion, this to be followed by a final plenary session to summarize the day. The organizers were anxious to avoid the usual problem in such circumstances: small-group discussions generate flip charts containing long unstructured lists of points made, usually covering several different (unstated) levels; and so everyone ends up unable to see any patterns which would help the audience to see and retain important lessons. To do better than this the people chairing the small groups were asked to structure the discussion by following an explicit agenda written out for them. Three of us spent the discussion period touring the various groups, trying to get a feel for the content and tone of the group discussions.

Alas for the well-laid plans, and in spite of the best efforts of those in the chair, what happened was what always happens when health professionals meet on occasions like this: uncontrollable discussion broke out and anecdotes were exchanged! The problem now to be solved during the afternoon tea-break was to prepare for the final plenary presentation and discussion in the absence of the hoped-for coherent responses from the groups. This is where SSM was helpful.

To provide a recognizable context for talk of mergers, a simple model relevant to the Health Service was jotted down, as shown in Figure A11. Here the public (who are occasionally patients of the NHS) both elect a Government and—in the UK— provide resources through direct taxation. Those resources are disbursed via NHS structures so that appropriate configurations of services can be made available to the

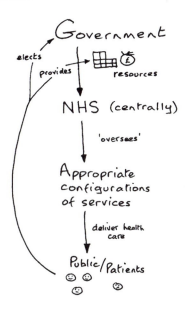

Figure A11   The simple model built to explore mergers in the NHS

public. Talk of 'mergers' can be thought about as talk about changes which will affect those configurations of services, changes which will involve any or all of: Health Authorities, hospitals, community service providers, family doctors and local authorities. The three of us who had spent the small-group discussion period touring the groups now annotated the model with our generic impressions of either the issues which were being discussed, or the issues which underlay the stories being told. These consolidated into five main points, and the final plenary session was opened by my displaying the model of Figure A11 and then adding the five main discussion points, as shown in Figure A12. This served to structure the final discussions. Feedback from delegates about the coherence of the day was good.

I can guarantee that this near-Mode 2 use of SSM was problem-oriented, not methodology-oriented. The fact that we had only the half-hour tea-break to prepare for the final discussion session concentrated the mind. Figures A11 and A12 represent the only explicit output from the work done in the tea-break, but I could retrospectively produce a conventional SSM-style model, together with root definition, CATWOE, E_1, E_2, E_3, etc., which would map Figure A11, as well as an issue-based root definition and model relevant to 'talk of mergers' (a system to decide the structural and service entailments of a configuration of health services considered desirable for population x in area y . . . etc.). None of that work was done at the time— or has been done since, for that matter; the internalization of SSM enabled the practical response to the 'tea-break problem' to be generated. Reconstructing this Mode 2-like use of SSM after the event, we can see that the small-group discussions three of us had dipped into were the source of a holistic impression of the work done in the small groups. We then made sense of that overall impression—for the purpose, on the day, of exposition—by means of the models in Figures A11 and A12.

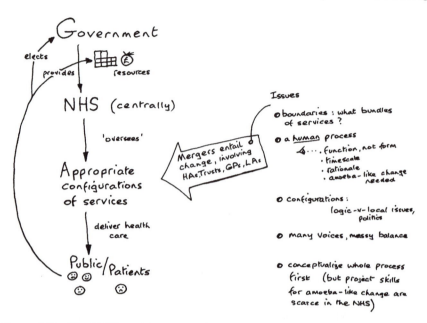

Figure A12   The NHS model annotated to structure presentation of merger issues

It is inevitable that users of SSM will internalize its guidelines and use them in an increasingly sophisticated way. This is akin to learning physical skills: beginners at rock climbing treat each hold as a new problem, appearing clumsy as they make their jerky progress up a rock face. Experienced climbers who have internalized their skill, at whatever level they have attained, put together sequences of moves and appear to 'flow' up a climb. They are likely to believe that you cannot be said to be *truly* rock climbing until this internalization has occurred, and so it is with use of SSM. The more subtle nature of human situations will be revealed to sophisticated users while the novice is struggling to remember what Analysis Two is, and what CATWOE and E_1, E_2, E_3 mean. So the disappearance of near-Mode 1 use is to be welcomed, apart, that is, from the fact that it has its virtues for initial teaching purposes. Just as novice climbers need to be taken up easy climbs, and to have the next hold, and how to use it, pointed out to them, so people coming to SSM for the first time *need* to treat it as a series of stages, each with a definite output, just as Naughton declared in the original constitutive rules.

Finally, though, we cannot advise inexperienced users simply to seek out straight-forward problem situations to tackle, since *all* human situations have their subtleties!

SOFT SYSTEMS METHODOLOGY—THE CONTEXT

Before concluding, two aspects of the context of SSM's development are worth attention, since they have emerged as virtually inseparable from SSM as a way of conducting inquiry in human affairs. The two are: the 'action research' mode; and the assumptions about the nature of the social process which underpin SSM as a whole.

Action Research

The fact that the research which produced SSM started out from a base in systems *engineering* indicates that it was part of the strand of research which concentrates on situations in which people are trying to take action. From the start the researchers tried not simply to observe the action as external watchers but to *take part* in the change process which the action entailed; this made change, and how to achieve it, the object upon which research attention fastened. This puts the research into the 'action research' tradition stemming from Kurt Lewin's views, developed in the 1940s, that real social events could not be studied in a laboratory. This mode of research is discussed in *STSP*, pp. 146–154 and illustrated throughout *SSMA*. Here I shall focus only briefly on what experience and reflection have shown to be an important requirement of this kind of research, a requirement which is, surprisingly, almost completely neglected in the literature of action research. (It is discussed in *ISIS*, pp. 18–28, and Checkland and Holwell 1998a).

The point is this. For findings to be accepted as part of the body of 'scientific knowledge' they have to be repeatable, time and again, by scientists other than those who first discovered them. If you announce that you have discovered the 'inverse square law of magnetism', working in Berlin, then that finding has to be repeatable in Brazil, Barnsley, Brisbane and Bournemouth if the happenings in your experiments are to be accepted as 'scientific knowledge'. Apparent findings in human situations, however, no one of which is ever either static or exactly the same as any other human situation, cannot match this strong criterion. It is the public testability which makes 'scientific knowledge' different from other kinds of knowledge; though do not expect unanimity on any *interpretation* of the findings, since the interpreting is a human act, and can in principle be as various as the people who make the interpretations.

In the human domain, in the province of 'social science', the findings are of a different nature, as are the criteria by which they can be judged. Emile Durkheim (1895), who made up the word 'sociology', suggested that the concern of this new 'science of society' should be 'social facts'. 'Treat social facts as things' is his best-known dictum. Social facts refer to aggregates, and are defined by an observer: for example, the fraction of marriages which end in divorce in a given society, or the rate at which people commit suicide—which was the subject of a famous study by Durkheim himself. But action research in local situations is concerned not with social facts but with study of the myths and meanings which individuals and groups attribute to their world and so make sense of it. This is part of that other great strand within sociology, the interpretive tradition stemming from Max Weber (1904). This is relevant to SSM since the meaning attribution by individuals and groups leads to their forming particular intentions and undertaking particular purposeful action.

The question of the criteria by which findings of this kind can be judged is obviously a tricky one. I have heard sociologists argue that the criterion by which their findings can be judged can be no more than mere plausibility: do these findings make a believable story? But if this weak criterion is accepted there seems to be virtually no difference between writing novels and doing social research. Surely we can do better than that?

In between the strong criterion of repeatability (of the happenings) and the weak criterion of plausibility, we argue (Checkland and Holwell 1998a) that action research should be conducted in such a way that the whole process is subsequently *recoverable* by anyone interested in critically scrutinizing the research. This means declaring explicitly, at the start of the research, the intellectual frameworks and the process of using them which will be used to define what counts as knowledge in this piece of research. By declaring the epistemology of their research process in this way, the researchers make it possible for outsiders to follow the research and see whether they agree or disagree with the findings. If they disagree, well-informed discussion and debate can follow. Also, the learning gained in a piece of organization-based action research may concern any or all of: the area focused on in the research; the methodology used; or the framework of ideas embodied in the methodology. SSM is itself the result of 30 years of this kind of learning in real-world situations.

The Social Process: Appreciative Systems

Once a systems thinker has taken on board the idea of conceptualizing the world and its structures in terms of a series of layers, with any layer being justified by definable emergent properties at that level (see *STSP*, Chapter 3), it is always appropriate to think at more than one level. As discussed earlier, the 'apparatus' of SSM ensures that whatever level is taken by an observer or researcher to be that of 'system', the level above ('wider system') and that below ('sub-system') will always be taken into account, as Figure A5 illustrates. But the systems thinker also accepts that an observer, investigator or researcher will not only select the level which is to be that of 'system' but will also interpret the nature of 'system' according to his or her own *Weltanschauung* or world-view (or, in SSM, deliberately select multiple world-views whose adoption might yield insights into the problem situation). These ideas of 'layers' and 'world-views' mean that developers of SSM could not avoid taking a position on both the nature of the methodology and the higher-level assumptions which it takes as given.

The methodology is taken to be a process of social inquiry which aims to bring about improvement in areas of concern by articulating a learning cycle (based on systems concepts) which can lead to action. This raises the question of what higher-level assumptions about the nature of social reality SSM implicitly takes as given: hence the discussion in Chapter 9 of *STSP*. The conclusion there is that in order to make sense of the research experiences it is necessary to take 'social reality' to be

the ever-changing outcome of the social process in which human beings, the product of their genetic inheritance and previous experiences, continually negotiate and re-negotiate with others their perceptions and interpretations of the world outside themselves (pp. 283, 284).

This makes SSM in harmony with the sociology of Alfred Schutz and the philosophy of Edmund Husserl; but in practical terms it was Geoffrey Vickers' work on what he calls 'appreciative systems' which mapped most completely our experiences.

Vickers' theoretical work was done in his retirement after 40 years in what he always referred to as 'the world of affairs'. (He was City lawyer, a civil servant, a member of the National Coal Board responsible for management, training and personnel issues, and a member of many public bodies—as well as a young subaltern who won the Victoria Cross on the day after his twenty-first birthday at the Battle of Loos in 1915.) In his retirement he set himself the task of making sense of all his experience, and wrote a series of books in which he developed his account of 'the social process' in terms of his theory of 'appreciation'. SSM's debt to Vickers is recorded in *STSP* (Chapter 8) but more work has been done since then, and is here summarized in the Appendix.

In essence: Vickers discovered systems thinking in his retirement, found it very helpful for sense-making purposes, and was amazed that the greatest use of systems ideas seemed to be made in a technical context, whereas he saw them as richly relevant to 'human systems'. In a taped interview at the Open University in 1982 he said:

> While I was pursuing these thoughts, everyone else who was responding at all was busy with man-made systems for guided missiles and getting to the moon or forcing the most analogic mental activities into forms which would go on digital computers. 'Systems' had become embedded in faculties of technology and the very word had become dehumanized (quoted in Blunden 1984, p. 21).

In his thinking he rejected first the 'goal-seeking' model of human life (the core of Simon's great contribution to management science) and then the cybernetic model because in it the course to which the Steersman steers is a 'given' from outside the system whereas in human affairs the course being followed is continuously generated and regenerated from inside the system. This led him to his notion of 'appreciation' in which, both individually and in groups, we all do the following: selectively perceive our world; make judgements about it, judgements of both fact (what is the case?) and value (is this good or bad, acceptable or unacceptable?); envisage acceptable forms of the many relationships we have to maintain over time; and act to balance those relationships in line with our judgements. [The Appendix contains our model of what Vickers meant by 'an appreciative system' (Checkland and Casar 1986), and links his work to SSM.]

In summary, SSM can be seen as a systemic learning process which articulates the working of 'appreciative systems' in Vickers' sense.

CONCLUSION

The saxophonist John Coltrane was the greatest innovator in the jazz idiom since Charlie Parker reminted the coinage of jazz expression in the mid-1940s. Playing with the Miles Davis Quintet, Coltrane took to playing long long solos which might last for 20 minutes or more. On one occasion at the Apollo in Harlem, when he eventually finished a very lengthy solo he was asked why he had gone on so. He replied 'I couldn't find nothing good to stop on', whereupon Davis said, 'You only have to take the horn out of your mouth.' Authors too face the problem of finding 'something good to stop on', and obviously all they have to do is lift the pen from the page. But that would not satisfy a systems thinker, who would want to effect some kind of closure. Hence this conclusion, which adds some final comments on what has been an enthralling 30-year research experience for this writer.

SSM has been ill-served by its commentators, many of whom demonstrably write on the basis of only a cursory knowledge of the primary literature. However, both life and this chapter are too short to expend time and energy on correcting these nonsenses; but it is probably worth illustrating the size of the problem by recording the spectacular example which Holwell found during her masterly exegesis of the secondary literature (1997, p. 335). It is from a book on information systems published in 1995. The authors refer to *STSP* but—all too typically—do not mention *SSMA*, even though it had been published for five years when they wrote their book.

This methodology stems from the work of Checkland (1981) who took a radically different approach to the *analysis and design of information human activity systems.* Starting from the premise that *organizations* (and therefore their *subsystem information systems*) are *open systems that interact with their environment,* he *includes the human activity subsystems as part of* his modelling process. The methodology starts by taking a particular view *of the system* and incorporating subjective and objective impressions into a 'rich picture' *of the system* that includes the people involved, the problem areas, sources of conflict and other 'soft' aspects *of* the overall system. *A* 'root definition' is then formed *about the system* which *proposes improvements* to *the system* to *tackle the problems identified in the rich picture.* Using *the* root definition, *various conceptual models of the new system* can be built, compared and evaluated against the *problems in the rich picture.* A *set of recommendations* is then suggested to deal with the specific changes that are necessary to *solve the problems.*

The italics here are used to highlight fundamental errors: nearly 20 in less than 200 words! Cheerful stoicism seems to be the necessary response to a lack of understanding as profound as this. Pity the poor students.

Although the secondary literature often creates a barrier, it is not the only reason that *teaching* SSM is not straightforward. In teaching such a methodology you are teaching not what to think, but a *way of thinking* which the user can consciously reflect upon. Many people coming to SSM for the first time in a classroom have never before consciously thought about their own thinking, and there is some re-arrangement of mental furniture entailed in this which many find strange. Certainly the biggest difficulty in understanding SSM is to absorb its shift from assuming 'systems' exist in the world (as in everyday language) to assuming that the process of inquiry into the world can be a consciously organized learning system.

This is to say that *process thinking* is very unfamiliar for many people, and there is no doubt that teaching a way of thinking is harder than teaching substantive factual material—which is why many MBA courses, which ought to focus on teaching 'how to think in problem situations', instead opt for current factual material about marketing, finance, and other common organizational functions. How strange process thinking is for some people was illustrated recently by a journalist, Matthew Parris, who described in *Literary Review* (December 1998) how much he hated a training week in Brussels to which he was sent as a junior Foreign Office employee. He found a suffocating respect for questions of process combined with a carefree disregard for questions of substance. They kept telling me how a policy was steered into being. I kept wanting to know whether the policy was any good. They looked at me as though I was missing the point.

Of course, he *was* missing the point. A systems thinker would know that the process of policy creation and policy content are entirely complementary, the process itself conditioning what might emerge as content. Both need to be thought about together; but this is not yet a familiar concept.

The other difficulty faced by teachers of SSM is overcoming the shock some people feel when they discover the rigour involved in building purposeful models, thinking out their measures of performance, and so on. (Perhaps there is a tendency for newcomers to equate 'soft' with sloppy or casual: as if anything will do.) But the rigour helps clear the mind, as well as ensuring that the devices which will structure debate are themselves defensible.

In the first heady days of the Gorbachev reforms in the USSR the Institute for Systems Analysis in Moscow wanted to hear about SSM, since the Institute's researchers were intrigued by the idea of undertaking action research projects in Soviet industry. At the end of a week of lectures and seminars, the Director of the Institute, J. M. Gvishiani, said to me that he saw SSM as 'a rigorous approach to the subjective'. This struck me as a very insightful phrase. Both the primacy of the subjective in human affairs *and* the rigour in the thinking about process are important.

Oddly enough, the difficulties of teaching a systemic way of thinking in a classroom disappear when people learn it by using SSM in a real situation. But the situation has to be real for this to happen. There is a huge gap between real uses of SSM and 'pretend' uses on case studies in the classroom. Pretending to invest £10m., or deciding who to make redundant, in a case study, costs you nothing; doing it in real life is a world away from the pretence. But, when the use is real, our experience is that SSM is quickly grasped, and seems 'natural' to those using it. This adds weight to the argument in *SSMA* (p. 300) that the process of SSM reflects the everyday process we all engage in whenever we form sentences and entertain alternative predicates, comparing them with each other and with the perceived world in order to make judgements about action. SSM simply makes a special kind of predicate, in the form of models of purposeful activity, each of which expresses a pure world-view. It is a more organized, more holistic form of what we do when we engage in serious conversation.

But in observing that SSM, in use, seems natural, we need also to remind ourselves that its concern is with would-be purposeful action; and we should never forget how easy it is to overestimate the role of the purposeful in human life. Being able to act with intention, purposefully, is an important part of what makes us human. But it is only a part, and maybe not the most important part.

It has been argued above (and that argument is extended in the Appendix) that SSM can be seen as articulating 'the social process', in the form of what Vickers calls an 'appreciative system'. If, thinking systemically, we ask: what is the level above that of 'the social process'? then we are moving into very abstract realms indeed: in this case into the level at which the concern could be defined as 'being human'. This is two levels above that at which the concern is 'use of SSM', but it provides the ultimate context in which SSM is used.

This suggests several self-admonitory instructions for the user of SSM. We should be rigorous in thinking but circumspect in action. We should remember that many people painfully find their way unconsciously to world-views which enable them to be comfortable in their perceived world. Coming along with a process which challenges world-views and shifts previously taken-as-given assumptions, we should remember that this can hurt. So what right do we have to cause such pain? None at all unless we do it with respect and in the right spirit: no lofty hauteur. And we must remember, feet on ground, that all we can do with our 'natural' but intellectually sophisticated process of inquiry is learn our way to improved purposeful action, which is a ubiquitous part of human life but only a limited part of it, not the whole.

And so, to complete this chapter, let us remind ourselves—using a true story—of what it means to be *fully* human, and end with that image.

In 1993 in south London a black teenager, Stephen Lawrence, was fatally stabbed in a racist attack by a group of white youths. Six years later, with no one found guilty of the murder, Sir William Macpherson delivered to the Home Secretary his report on the incompetence of the criminal investigation, precipitating national soul searching and debate about institutionalized racism in British society. A writer, Richard Norton-Taylor, brilliantly crafted a play—*The Colour of Justice*—from the transcript of the Lawrence tribunal; this was shown on BBC television in February 1999. The production contained one of those moments, exceptionally rare on television, when the viewer is transfixed and transformed. A witness described how he and his wife, returning home from a church meeting, came upon Stephen as he lay bleeding on the pavement. The wife cradled Stephen, as the young man's life ebbed away. Knowing that hearing is the last sense to go, she whispered in his ear 'You are loved'. When he got home, the man washed his blood-ied hands into a container and poured the water on to the roots of a favourite rosebush. He said that he supposed that in some way Stephen lived on.

We should never entertain the idea, even for a moment, that a mere 'systems approach', or any 'systems methodology' could ensure that we behave as Louise and Conor Taaffe did on that April night in Eltham in 1993.

APPENDIX: SYSTEMS THEORY AND MANAGEMENT THINKING

Two inquiring systems developed since the 1960s—Vickers's concept of the appreciative system and the soft systems methodology, are highly relevant to the problems of the 2lst century. Both assume that organizations ane more than rational goal-seeking machines and address the relationship-maintaining and Gemeinschaft aspects of organizations, characteristically obscured by functionalist and goal-seeking models of organization and management. Appreciative systems theory and soft systems methodology enrich rather than replace these approaches.

Two rich metaphors provide a useful frame within which any consideration of the problems facing us in the late 20th century can, with advantage, be placed. As a result of the first industrial revolution, based on energy, and the current second one, based on information, the world is increasingly Marshall McLuhan's 'global village'. More and more problems need to be examined in a global rather than a local context and, as we do so, we need to remember that we are all of us, in, Buckminster Fuller's great phrase, 'the crew of Spaceship Earth'.

Thanks to the material successes of the two industrial revolutions we are a crew with rising expectations of high living standards. But we are increasingly aware that the wealth-generating machine may not be able to meet those expectations without doing unacceptable damage to Spaceship Earth, which, together with the free supply of energy from our sun, is the only given resource we have.

This triangle—of expectations, wealth generation, and protection of the planet— will have to be managed with great care at many different levels as we enter the 2lst century if major disasters are to be avoided. Unfortunately, our current ideas on *management* are rather primitive and are probably not up to the task. They stem from the technologically oriented thinking of the 1960s, and they now need to be enlarged and enriched. This may well be possible from the systems thinking of the 1970s and 1980s, which has placed that body of thought more firmly within the arena of human affairs.

This article will examine the legacy of thinking about management and organizations that we get from the 1960s and develop a richer view that stems from more recent systems thinking, especially Vickers's work on the theory of *appreciative systems* and work on soft systems methodology, which can be seen as a way of making practical use of Vickers's concepts. This, it is argued, is. more relevant than the current conventional wisdom to managing the problems of the new century.

Checkland, P. B., AMERICAN BEHAVIORAL SCIENTIST **38**(1) pp. 75–91, copyright © 1994 by Sage Publications, Inc. Reprinted by Permission of Sage Publications, Inc.

MANAGEMENT AND ORGANIZATION

In spite of a huge literature—some of it serious, much of it at the level of airport paperbacks—and courses in colleges and universities worldwide, the role of the manager and the nature of the process of managing remain problematic, whether we are concerned with trying to manage global, institutional, or personal affairs. Anyone who has been a professional manager in an organization knows that it is a complex role, one that engages the whole person. It requires not only the ability to analyse problems and work out rational responses but also, if the mysterious quality of leadership is to be provided, the ability to respond to situations on the basis of feelings and emotion.

One of the reasons the manager's role remains obstinately problematic stems from our less than adequate thinking about the context in which managers perform, namely the organization. Some basic systems thinking indicates that if we adopt a limited view of organization then the conceptualization of the manager's role will inevitably also be rather threadbare. Thus a manager at any level occupies a role within a structure of roles that constitutes an organization. The activity undertaken by managers can be seen as a system of activity that serves and supports and makes its contribution to the overall aims of the organization as a purposeful whole. Now, if one system serves another, it is a basic tenet of systems thinking that the system that serves can be conceptualized only after prior conceptualization of the system served (Checkland 1981, p. 237). This is so because the form of a serving system, if it is truly to serve, will be dictated by the nature of the system served: That will dictate the necessary form of any system that aspires to serve and support it.

Now there is a conventional wisdom about the nature of organization that persists in spite of the fact that anyone who has worked within an organization knows that this image conveys only part of the story. The conventional model is that an organization is a social collectivity that arranges itself so that it can pursue declared aims and objectives that individuals could not achieve on their own. Given this view of organization, the manager's role is to help achieve the corporate goals, and it follows that the manager's activity is essentially rational decision making in pursuit of declared aims. This is the conventional wisdom even though intuitively we all have a rich sense that organizations in which we have worked are more than rational goal-seeking machines. The experienced day-to-day reality of organizations is that they have some of the characteristics of the tribe and the family as well as the characteristics necessary if they are to order what they do rationally so as to achieve desired objectives such as, in the case of industrial companies, survival and growth. In spite of this folk knowledge, the orthodoxy has been very strong, and we can see this both in the literature of organization theory and in that of management science.

Organization Theory

This is not the place to discuss the development of organization theory in any detail, but it is useful for present purposes to mark the general shape of this field as it emerges in such wide-ranging studies as Reed's (1985) *Redirections in Organisational Analysis*. The general shape is that of the establishment of an orthodoxy (the systems/contingency model that held sway from the 1930s to the 1960s) and the challenge to that orthodoxy since then, with no single dominant alternative.

Nevertheless, the challenging models do, in general, have in common the fact that they see organizations not as reified objects independent of organizational members, as in the orthodox systems model, but as the continually changing product of a human process in which social reality is socially constructed: the title of Berger and Luckmann's (1966) well-known book—*The Social Construction of Reality*—neatly captures this alternative strand of thinking.

At a broad level of generalization, we can see the two major approaches as reflecting the two main categories of thinking about organizations on which a pioneering sociologist, Ferdinand Tönnies, built his account. In his major work *Gemeinschaft und Gesellschaft* (1887) (translated as *Community and Association* by Loomis, 1955), Tönnies constructed models of two types of society or organization. There is the natural living community into which a person is born, the family or the tribe (Gemeinschaft), and there are the formally created groupings (Gesellschaft) that a person joins in some contractual sense, as when he or she becomes an employee of a company or joins a climbing club.

In general, the orthodox view of organizations emphasizes their Gesellschaft nature, that they are created to do things collectively (*achieve goals* is the usual language) that would be beyond the reach of individuals. The alternatives emphasize rather that all social groupings take on some flavour of Gemeinschaft: being in an organization is something like being part of a family. Intuitively, the lived experience of organizations that we all gradually acquire gives us the folk knowledge that organizations exhibit some of the characteristics of both models.

That the orthodox view has been dominant can be seen by perusing college textbooks, which present students with the conventional wisdom. For example, in Khandwalla's (1979) *The Design of Organizations*, the view of organizations as open systems devoted to achieving corporate objectives is described as 'the most powerful orientation in organization theory today' (p. 251). Much attention is paid to well-known work aimed at correlating an organization's structure with its core tasks carried out in an environment with which it interacts (Lawrence & Lorsch, 1967; Pugh & Hickson, 1976; Woodward, 1965; etc.). Reed's (1985) survey argues that 'systems theorists . . . had dominated organizational analysis since the 1930s' (p. 35) but that by the 1960s there was no common history or intellectual heritage. By the 1970s, a systems-derived approach was 'struggling to retain its grip on organizational studies' (p. 106). This does not mean that the orthodoxy has lost its adherents, however. In the same year that Reed's book was published, Donaldson (1985) brought out his *In Defence of Organization Theory*, the defence being of the 'relatively accepted contingency-systems paradigm' (p. ix).

Both Reed and Donaldson make much reference to a book that marks as much as any other the challenge to the orthodox systems view: Silverman's (1970) *The Theory of Organizations*. Silverman contrasts the systems view from the 1950s and 1960s with what he calls 'the Action frame of reference' in which action results from the meanings that members of organizations attribute to their own and each other's acts. Organizational life becomes a collective process of meaning attribution; attention is displaced away from the apparently impersonal processes by means of which, in the conventional model, a reified organization as an open system responds to a changing environment. Some of Silverman's subheadings convey the nature of his argument: Action not behaviour, Action arises from meanings, Meanings as social facts, Meanings are socially sustained, Meanings are socially changed.

This important work opens the way to various alternatives to the systems orthodoxy. Donaldson's discussion, for example, includes social action theory, the sociology of organizations, and the strategic choice thesis. Just as the orthodoxy draws on a positivist philosophy and a functionalist sociology, the alternatives are underpinned philosophically by phenomenology, and sociologically by an interpretive approach derived from Weber and Schutz.

It has to be said that the orthodox view provides a much clearer model of organization, and hence the manager's role, than is provided by the alternatives. Concentrating on the Gesellschaft aspects of an organization, the conventional view sees it as an open system seeking to achieve corporate objectives in an environment to which it has to adapt. Its tasks are analysed and assigned to groups within a functionalist structure, and the managers' role is essentially that of decision making in pursuit of corporate aims that also provide the standards against which progress will be judged. No similarly clear picture is provided by the alternatives, beyond the notion that organizations are characterized essentially by discourse that establishes the meanings that will underpin action by individuals and groups.

It is not at all surprising that that section of the management literature most concerned with intervening in, in order to influence and shape, real-world situations, namely management science, should itself focus on the orthodox systems model.

Management Science

In examining briefly the state of thinking in management science, it is useful to focus on the work of Herbert Simon. There are two reasons for this. First, it has been a dominating contribution in the field; second, in developing an approach based on the work of Vickers, we find that he explicitly contrasted his approach with that of Simon, drawing attention to the reliance of Simon on a goal-seeking model of human action that he himself was deliberately trying to transcend.

In the period after the Second World War, strenuous efforts were made to apply the lessons from wartime operations research to industrial companies and government agencies. In doing this, a powerful strand of systems thinking was developed— it would now be thought of as 'hard' systems thinking—concerned broadly with engineering a system to achieve its objectives. Systems were here assumed to exist in the world; it was assumed that they could be defined as goal seeking; and ideas of system control were generalized in cybernetics. These ideas mapped the orthodox stance of organization theory discussed in the previous section, and they conceptualized the manager's task as being to solve problems and take decisions in pursuit of declared goals. Indeed, this paradigm is succinctly expressed in Ackoff's (1957) assumption that problems ultimately reduce to the evaluation of the efficiency of alternative means for a designated set of objectives.

This is the field to which Simon has made such a significant and influential contribution, the flavour of which is captured in the title of his 1960 book: *The New Science of Management Decision*.

At a round table devoted to his work, Zannetos (1984) summarized Simon's legacy as 'a theory of problem solving, programs and processes for developing intelligent machines, and approaches to the design of organizational structures for managing complex systems' (p. 75).

Overall, Simon sought a science of administrative behaviour and executive decision making. In an intellectually shrewd move that has no doubt helped to make this body of work so influential, Simon wisely abandoned the notion that managers and administrators seek to *optimize*, replacing it with the idea of *satisficing*: the idea that the search is for solutions that are *good enough* in the perceived circumstances, rather than optimal (March & Simon, 1958). Nevertheless, the flavour of hard systems thinking is retained in the claim that the search is 'motivated by the existence of problems as indicated by gaps between performance and goals' (p. 73).

Similarly in another of Simon's major contributions, the development with Newall of GPS (general problem solver), a heuristic computer program that seeks to simulate human problem solving, the whole work is built on the concept of problem solving as a search for a means to an end that is already declared to be desirable (Newall & Simon, 1972). Simon (1960) stated,

> Problem solving proceeds by erecting goals, detecting differences between present situation and goal, finding in memory or by search tools or processes that are relevant to reducing differences of these particular kinds, and applying these tools or processes. Each problem generates subproblems until we find a subproblem we can solve—for which we have a program stored in memory. We proceed until, by successive solution of such subproblems, we eventually achieve our overall goal—or give up. (p. 27)

This is an especially clear statement of the thinking, derived from the systems theory of the 1950s, that has dominated management science and that underlies organization theory's orthodox model of what an organization is.

It is the argument here that this goal-seeking model, largely adequate though it was in the management science that contributed to post-Second World War industrial development, is not rich enough to support and sustain the management thinking now needed by the crew of Spaceship Earth, that spaceship having become akin to a global village.

An alternative, richer perspective is provided by the systems thinking of the 1970s and 1980s, and in particular by Vickers's development of appreciative systems theory and by an approach to intervention in human affairs that can be seen as making practical use of that theory, namely, soft systems methodology.

These are discussed in the next section, but it may be useful to point out at once that these are developments in what is now known as 'soft' systems thinking, as opposed to the hard systems thinking of the 1950s and 1960s that permeates both orthodox organization theory and Simonian management science. The usual distinction made between the two is that the hard systems thinking tackles well-defined problems (such as optimizing the output of a chemical plant), whereas the soft approach is more suitable for ill-defined, messy, or wicked problems (such as deciding on health care policy in a resource-constrained situation). This is not untrue, but it fails to make an intellectual distinction between the two. The real distinction lies in the attribution of systemicity (having the property of system-like characteristics). Hard systems thinking assumes that the world is a set of systems (i.e. is *systemic*) and that these can be *systematically* engineered to achieve objectives. In the soft tradition, the world is assumed to be problematic, but it is also assumed that *the process of inquiry* into the problematic situations that make up the world can be organized as a system. In other words, assumed systemicity is shifted: from taking the world to be systemic to taking the process of inquiry to be systemic (Checkland, 1983, 1985b).

Thus in the following section both appreciative systems theory and soft systems methodology describe inquiring processes—the former with a view to understanding, the latter with a view to taking action to improve real-world problem situations.

Finally, we may note that soft systems thinking can be seen as representing the introduction of systems thinking into Silverman's action frame of reference, although the organization theory literature is apparently at present innocent of any knowledge of post-1960s developments in systems thinking (Checkland, 1994).

APPRECIATIVE SYSTEMS THEORY

The Nature of an Appreciative System

The task that Vickers set himself in his 'retirement' after 40 years in the world of affairs was to make sense of that experience. In the books and articles that he then wrote he constructed

> an epistemology which will account for what we manifestly do when we sit round board tables or in committee rooms (and equally though less explicitly when we try, personally, for example, to decide whether or not to accept the offer of a new job). (G. Vickers, personal communication, July 1974)

In his thinking as this project developed, Vickers first rejected the ubiquitous goal-seeking model of human activity; then he found systems thinking relevant to his task; but he also rejected the cybernetic model of the steersman (whose course is defined from outside the system), replacing it by his more subtle notion of 'appreciation' (Vickers, 1965, is the basic reference). He expressed his intellectual history in the following terms in a letter to the present writer in 1974:

> It seems to me in retrospect that for the last twenty years I have been contributing to the general debate the following neglected ideas:
> (1) In describing human activity, institutional or personal, the goal-seeking paradigm is inadequate. Regulatory activity, in government, management or private life consists in attaining or maintaining desired relationships through time or in changing and eluding undesired ones.
> (2) But the cybernetic paradigm is equally inadequate, because the helmsman has a single course given from outside the system, whilst the human regulator, personal or collective, controls a system which generates multiple and mutually inconsistent courses. The function of the regulator is to choose and realise one of many possible mixes, none fully attainable. In doing so it also becomes a major influence in the process of generating courses.
> (3) From 1 and 2 flows a body of analysis which examines the 'course-generating' function, distinguishes between 'metabolic' and functional relations, the first being those which serve the stability of the system (e.g. budgeting to preserve solvency and liquidity), the second being those which serve to bring the achievements of the system into line with its multiple and changing standards of success. This leads me to explore the nature and origin of these standards of success and thus to distinguish between norms or standards, usually tacit and known by the mismatch signals which they generate in specific situations, and values, those explicit general concepts of what is humanly good and bad which we invoke in the debate about standards, a debate which changes both. (G. Vickers, personal, communication, 1974)

In developing the theory of appreciative systems and relating it to real-world experience, Vickers never expressed the ideas pictorially, in the form of a model, although this seems a desirable form in which to express a system. (His explanation for this lack was disarming: 'You must remember,' he said, 'that I am the product of an English classical education' [G. Vickers, personal communication, 1979]). What follows is an account of the model of an appreciative system developed by Checkland and Casar (1986) from the whole corpus of Vickers's writings.

From those writings we may highlight some major themes that recur:

- A rich concept of day-to-day experienced life (compare Schutz's [1967] *Lebenswelt*)
- A separation of judgments about what is the case, *reality judgments*, and judgments about what is humanly good or bad, *value judgments*
- An insistence on *relationship maintaining* as a richer concept of human action than the popular but poverty-stricken notion of goal seeking
- A concept of action judgments stemming from reality and value judgments
- A notion that the cycle of judgments and actions is organized as a system

The starting point for the model is the Lebenswelt, the interacting flux of events and ideas unfolding through time. This is Vickers's 'two-stranded rope', the strands inseparable and continuously affecting each other. Appreciation is occasioned by our ability to select, to choose. Appreciation perceives (some of) reality, makes judgments about it, contributes to the ideas stream, and leads to actions that become part of the events stream. Thus the basic form of the model is that shown in Figure A13. There is a recursive loop in which the flux of events and ideas generates appreciation, and appreciation itself contributes to the flux. Appreciation also leads to action that itself contributes to the flux.

It is now necessary to unpack the process of appreciation. From Vickers's writings we take the notion of perceiving reality selectively and making judgments about it. The epistemology of the judgment making will be one of relationship managing rather than goal seeking, the latter being an occasional special case of the former. And both reality and value judgments stem from standards of both fact and value: standards of what *is*, and standards of what is good or bad, acceptable or unacceptable. The very act of using the standards may itself modify them.

These activities lead to a view on how to act to maintain, to modify, or to elude certain forms of relevant relationships. Action follows from this, as in Figure A13.

The model also tries to capture Vickers's most important point and greatest insight, namely, that there is normally no ultimate source for the standards by

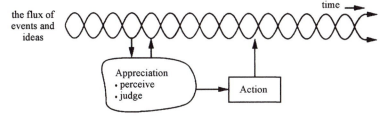

Figure A13 The structure of an appreciative system
SOURCE: Checkland and Casar (1986).

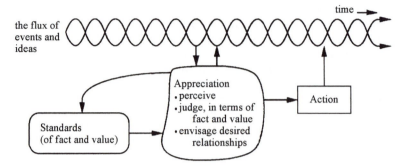

Figure A14 The structure of an appreciative system expanded
SOURCE: Checkland and Casar (1986).

means of which what is noticed is deemed good or bad, important or unimportant, relevant or irrelevant, and so on. The source of the standards is *the previous history of the system itself*. In addition, the present operation of the system may modify its present and future operation through its effect on the standards. These considerations, together with those already discussed, yield Figure A14 as a model of an appreciative system. The most difficult aspect to model is the dynamic one, but it should be clear from Figure A14 that the dynamics of the system will be as shown in Figure A15. The form of the appreciative system remains the same, whereas its contents (its *setting*) continually (but not necessarily continuously) change. An appreciative *system* is a process whose products—cultural manifestations—condition the process itself. But the system is not operationally closed in a conventional sense. It is operationally closed via a structural component (the flux of events and ideas) that ensures that it does not, through its actions, reproduce exactly itself. It reproduces a continually changed self, by a process that Varela (1984) called the 'natural drift' of 'autopoietic systems' (Maturana & Varela, 1980), systems whose component elements create the system itself. Through its (changing) filters the appreciative system is always open to new inputs from the flux of events and ideas, a

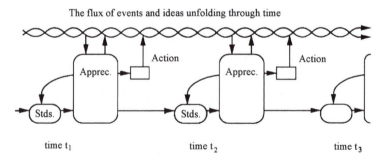

Figure A15 The dynamics of an appreciative system
SOURCE: Checkland and Casar (1986).

characteristic that seems essential if the model is to map our everyday experience of the shifting perceptions, judgments, and structures of the world of culture.

Vickers's claim was that he had constructed an epistemology that can provide convincing accounts of the process by which human beings and human groups deliberate and act. The model in Figures A14 and A15 is a systemic version of the epistemology.

Checkland and Casar (1986) used it to give an account of the learning in a systems study of the Information and Library Services Department of what was then ICI Organics (a manufacturer of fine chemicals within the ICI Group), a study that has been described in detail elsewhere (Checkland, 1985a; Checkland & Scholes, 1990). This study was carried out by a group of managers in the function with some outside help in the use of soft systems methodology (SSM), which was the methodology used. It is a way of making practical use of the notion of an appreciative system, and it will be discussed briefly in the next section. It entails structuring a debate about change by building models of purposeful activity systems and comparing them against perceptions of the real world as a means of examining what the appreciative settings are in the situation in question and how they and the norms or standards are changing. In the study in question, there were three cycles of this learning process.

In the first cycle, the study team's interest and concern were to rethink the role of their function in a changing situation. They perceived many facts relevant to this, which resulted in 26 relevant systems. They selected and judged these facts in terms of a conception of a particular relationship and standards relevant to it: they accepted the relevance of a simple model that took as given that their function was a support to the wealth-generating operations of their company, and they implicitly made use of standards according to which a good version of this relationship would be to make efficient, effective, and timely provision of information to other parts of the company.

These considerations contributed to the ideas stream of the Lebenswelt and led to the action of exploring several perceptions of the relationship between the function and the rest of the company in greater depth. In this second methodological cycle, the focus was still on the relationship between function and company but the appreciative settings began to change. This can be expressed as a change in standards resulting from the first cycle of appreciation. The shift was in the concept of what would constitute a good relationship:

> The focus shifted from ILSD (Information and Library Services Department) as a reactive function responding quickly and competently to user requests and having the expertise to do it, to ILSD as a proactive function, one which could on occasion tell actual and potential users what they *ought* to know. (Checkland 1985a, p. 826)

In the third cycle, the new concept of ILSD was developed and, in the language of Figure A14, several hypothetical forms of relevant relationships were considered. This led to attention being given both to internal relationships within the function (How different would they have to be to sustain a proactive role?) and to the relationship between the function and the company. These considerations led to decisions on actions necessary to broaden the appreciative process. The actions taken were to make both internal (within ILSD) and external presentations of the results of the study. These events entered the company's Lebenswelt and had the effect of starting to bring about the change in the company's appreciative system, as evidenced by the remark made by the research manager at the external presentation, namely, that 'I have known and worked with ILSD for 20 years and I came along this morning out of a sense of duty. To my amazement I find I now have a new perception of ILSD' (Checkland, 1985a, p. 830).

Finally, the company's subsequent allocation of significant new resources to ILSD can be described as illustrating its implicit adoption of new standards with respect to the Information and Library Services function, standards whose change stems from the recent history of the company's appreciative system, involving input of ideas and events from the systems study itself.

The Appreciative Process in Action: Soft Systems Methodology

It is not appropriate here to give a detailed account of SSM, which is described in numerous books and articles since the early 1970s. (The basic books describing its development are Checkland, 1981; Checkland & Scholes, 1990; and Wilson, 1984; a burgeoning secondary literature may be sampled via, for example, Avison & Wood-Harper,1990; Davies & Ledington, 1991; Hicks, 1991; Patching, 1990; and Waring, 1989.)

SSM was not an attempt to operationalize the concept of an appreciative system; rather, after SSM had emerged in an action research programme at Lancaster University, it was discovered that its process mapped to a remarkable degree the ideas Vickers had been developing in his books and articles (Checkland, 1981, chap. 8).

The Lancaster programme began by setting out to explore whether or not, in real-world managerial rather than technical problem situations, it was possible to use the approach of systems engineering. It was found to be too naïve in its questions (What is the system? What are its objectives? etc.) to cope with managerial complexity, which, we could now say, was always characterized by conflicting appreciative settings and norms. Systems engineering as developed for technical (well-defined) problem situations had to be abandoned, and SSM emerged in its place.

The development of SSM has been characterized by four points in time at which what can now be seen, with hindsight, as crucial ideas moved the project forward (Checkland & Haynes, 1994). The first was the realization that all real-world problem situations are characterized by the fact that they reveal human beings seeking or wishing to take *purposeful action*. This led to purposeful action being treated seriously as a systems concept. Ways of building models of human activity systems were developed. Then it was realized that there can never be a single account of purposeful activity, because one observer's terrorism is another's freedom fighting. Models of purposeful activity could only be built on the basis of a declared Weltanschauung. This meant that such models were never models *of* real-world action; they were models *relevant to* discourse and argument about real-world action; they were epistemological devices that could be used in such discourse and debate; they were best thought of as *holons*, using Koestler's (1967) useful neologism, which could structure debate about different ways of seeing the situation. This led to the third crucial idea, that the problem-solving process that was emerging would inevitably consist of a learning cycle in which models of human activity systems could be used to structure a debate about change. The structure was

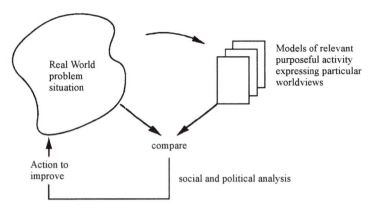

Figure A16   Soft systems methodology as a learning system
SOURCE: Checkland and Scholes (1990).

provided by carrying out an organized comparison between models and perceived real situations in which accommodations between conflicting perspectives could be sought, enabling action to be taken that was both arguably desirable—in terms of the comparisons between models and perceived situation—and culturally feasible for a particular group of people in a particular situation with its own particular history. (The fourth crucial idea, not relevant here, was the realization that models of human activity systems could be used to explore issues concerning what information systems would best be created to support real-world action—which took SSM into the field of information systems and information strategy.)

Given these considerations, SSM emerged as the process summarized in Figure A16. This is a picture of a *learning system* in which the appreciative settings of people in a problem situation—and the standards according to which they make judgments—are teased out and debated. Finally, the influence of Vickers on those who developed SSM means that the action to improve the problem situation is always thought about in terms of managing relationships—of which the simple case of seeking a defined goal is the occasional special case.

CONCLUSION: THE RELEVANCE OF APPRECIATIVE SYSTEMS THEORY AND SSM TO MODERN MANAGEMENT

It is not difficult to envisage the situations in both industry and the public sector in which the thinking about problems and problem solving would be significantly helped by the models underpinned by hard systems thinking, namely the models that see organizations as coordinated functional task systems seeking to achieve declared goals and that see the task of management as decision making in support of goal seeking. These models would be useful in situations in which goals and measures of performance were clear-cut, communications between people were limited and prescribed, and in which the people in question were deferential toward the authority that laid down the goals and the ways in which they were to be achieved. But this image has never accurately described life in most organizations as most people experience it, and it has become less and less true since the end of the Second World War. Since that time the trends have been toward much increased capacity for communication, greater complexity of goals as economic interdependence has increased, much reduced deference toward authority of any kind, and the dismantling of monolithic institutionalized power structures. The dethronement of the mainframe computer by the now ubiquitous personal computer is at once both a metaphor for these changes and one of the catalysts for their occurrence.

In such a situation richer models of organization and management will be helpful, and it has been argued that those based on Vickers's appreciative systems theory and SSM have a role to play here. More important, they do not replace the older models but rather subsume and enhance them. In SSM, focusing on a unitary goal is the occasional special case of debating multiple perceptions and proceeding on the basis of *accommodations* between different interests. For Vickers, managing relationships is the general case of human action, the pursuit of a goal the occasional special case.

Vickers himself has usefully differentiated his stance from that of Simon in remarks that relate to the latter's *Administrative Behaviour* (Simon, 1957):

> The most interesting differences between the classic analyses of this book and my own seem to be the following:
>
> (1) I adopt a more explicitly dynamic conceptual model of an organisation and of the relations, internal and external, of which it consists, a model which applies equally to all its constituent sub-systems and to the larger systems of which it is itself a part.
>
> (2) This model enables me to represent its 'policy makers' as regulators, setting and resetting courses or standards, rather than objectives, and thus in my view to simplify some of the difficulties inherent in descriptions in terms of 'means' and 'ends'.
>
> (3) I lay more emphasis on the necessary mutual inconsistency of the norms seeking realisation in *every* deliberation and at *every* level of organisation and hence on the ubiquitous interaction of priority, value and cost.
>
> (4) In my psychological analysis linking judgments of fact and value by the concept of appreciation, I stress the importance of the underlying appreciative *system* in determining how situations will be seen and valued. I therefore reject 'weighing' (an energy concept) as an adequate description of the way criteria are compared and insist on the reality of a prior and equally important process of 'matching' (an information concept).
>
> (5) I am particularly concerned with the reciprocal process by which the setting of the appreciative system is itself changed by every exercise of appreciative judgment.
> (Vickers, 1965, p. 22)

As an example of the relevance of SSM to current problems of managing complexity, we offer recent work done within the National Health Service (NHS) in the United Kingdom. (Some of this is described in Checkland & Holwell, 1993; Checkland & Poulter, 1994; and Checkland &. Scholes, 1990, chap. 4.)

In recent years the NHS has been subjected to several waves of government-imposed change. First there was the imposition of a system of accountable management, replacing the previous consensus management of teams of professionals. This had hardly settled down before it was replaced by an internal quasi-market. In this development the old district health authorities (into which the previous change had introduced district general managers) became purchasers of health care for a defined population, whereas hospitals and some general practitioners became providers of health care, the two being linked by contracts (although not legally binding ones) for particular services at a negotiated price. All these changes have entailed a considerable shift in appreciative settings for health professionals, and the NHS has been experiencing a period of considerable turmoil.

In the study described in Checkland and Scholes (1990), the problem was addressed of how a Department of Community Medicine in what was then a district health authority could evaluate its performance. Clearly the evaluation standards would depend completely on this department's image of itself and its role within the district. This is not a casual consideration, because concepts of community medicine range from *providing epidemiological data* to *managing the delivery of health care*. In this work, SSM-type models of purposeful activity relating to concepts of community medicine were built, with participation of members of the department, and eventually an evaluation methodology was developed. This was based on a structured set of questions derived from models that members of the department felt expressed their shared appreciative settings with regard to their image of the role of community medicine, which in their case was a very proactive interventionist one.

More recently, much work has been done in NHS hospitals and purchasing authorities as they assimilate and adapt to the purchaser–provider split (Checkland & Holwell, 1993). The new appreciative settings have been explored with participants via models of notional systems to enact the purchaser and provider roles. These have served to structure coherent debate concerning the requirements of the new roles.

In a recent study in a large teaching hospital, the work was part of a project to re-create an information strategy for the hospital suitable to cope with the new arrangements (Checkland & Poulter, 1994). In this work half a dozen teams of hospital workers representing the different professions were set up; members included clinicians, nurses, professionals from the finance and estates offices, and so on. Over a period of about 6 months, with a plenary meeting of team leaders every month, the teams discussed their activity and its contribution to meeting the requirements of the contracts for providing particular health care services that the hospital would in future negotiate with purchasers. Activity models were built and then used to structure analysis of required information support. This was related to existing information systems, and the information gaps identified helped in the formulation of the new information strategy.

One incident that occurred during this process may be recounted. It illustrates, in microcosm, the change of appreciative settings that can occur in the process of using SSM. It concerns a working group made up of nurses in the teaching hospital, led by a senior nurse. The group was building activity models relevant to providing nursing care, before using them to examine required information support.

Within SSM, when would-be relevant activity models are built, careful concise accounts of them as transformation processes are formulated (so-called Root Definitions). Various questions are asked in clarifying these definitions, one of which is 'If this notional system were to exist, who would be its victims or benefi-ciaries?' Nurses asking this question naturally wish to answer, 'The patients'. That is what their whole ethos, education, and professionalism tell them. That illustrates why they are in the profession. It was therefore something of a shock to this group—brought to their attention by the structured requirements of the SSM process—to realize in discussion that under the new arrangements the *technically* correct answer is nearer to being 'The hospital contracts manager'. This is because, under the so-called internal market, each contract for a health care service that involves nursing care ought technically to include the cost of providing a certain level (and quality) of nursing care. The nurses' task is then to provide what the contract calls for. Beyond this, of course, there is a theory according to which the interests of patients will, in fact, best be met by the new purchaser–provider contracts.

But it is not easy for nurses to accept this. The senior nurse who described this incident at one of the plenary discussions said that this question, and the issue it exposed, occupied the team for much of one of their meetings. It gave her insight into the NHS changes and helped her to understand her own misgivings about a supposed internal market in health care. Geoffrey Vickers would have appreciated this story.

Given our self-consciousness and the degree of mental autonomy that we seem to possess as human beings, that part of our thinking that is beyond the unreflecting stream-of-consciousness involvement in everyday life can itself be thought about. This can be done by examining the mental models that we use to make sense of our worlds. It is entirely plausible that our perceptions will be coloured by those mental models. And it follows that they need both to be better than primitive and to change as our human and social world changes.

It has been argued here that the models of organization and management that have been useful since the 1950s need to be enriched. It has then been argued that appreciative systems theory and SSM can help to provide such enrichment. They do not replace the earlier functionalist and goal-seeking models: They enclose and enhance them in ways more appropriate to institutional life at the end of the century.

NOTE

This article appeared in a special issue of *American Behavioral Scientist* **38**(1) September/October, 1994 devoted to: Rethinking Public Policy-Making: questioning assumptions, challenging beliefs. Essays in Honour of Sir Geoffrey Vickers on his Centenary. Edited by Margaret Blunden and Malcolm Dando.

The whole issue was republished as a book in 1995: *Re-thinking Public Policy Making* Blunden, M. and Dando, M. (Eds) Sage Publications, London.

The author is grateful to Sage Publications for permission to reprint the article here.

REFERENCES

Ackoff, R. L. (1957). Towards a behavioural theory of communication. In W. Buckley (Ed.), *Modern systems research for the behavioural scientist* (pp. 209–218). Chicago, IL: Aldine.

Avison, D. E., & Wood-Harper, A. T. (1990). *Multiview: An exploration in information systems development.* Oxford: Blackwell.

Berger, P. & Luckmann, T. (1966). *The social construction of reality.* Harmondsworth: Penguin.

Checkland, P. (1981). *Systems thinking, systems practice.* Chichester: Wiley.

Checkland, P. (1983). OR and the systems movement. *Journal of the Operational Research Society,* *34,* 661–675.

Checkland, P. (1985a). Achieving desirable and feasible change: An application of soft systems methodology. *Journal of the Operational Research Society, 36,* 821–831.

Checkland, P. (1985b). From optimizing to learning: A development of systems thinking for the 1990s. *Journal of the Operational Research Society, 36,* 757–767.

Checkland, P. (1994). Conventional wisdom and conventional ignorance: The revolution organization theory missed. *Organization, 1*(1), 29–34.

Checkland, P., & Casar, A. (1986). Vickers' concept of an appreciative system: A systemic account. *Journal of Applied Systems Analysis, 13*, 3–17.

Checkland, P., & Haynes, M. G. (1994). Varieties of systems thinking: The case of soft systems methodology. *System Dynamics Review, 10*, 189–197.

Checkland, P., & Holwell, S. (1993). Information management and organizational processes: An approach through soft systems methodology. *Journal of Information Systems, 3*, 3–16.

Checkland, P., & Poulter, J. (1994). *Application of soft systems methodology to the production of a hospital information and systems strategy.* United Kingdom: HISS Central Team of the NHS Management Executive.

Checkland, P., & Scholes, J. (1990). *Soft systems methodology in action.* Chichester: Wiley.

Davies, L., & Ledington, P. (1991). *Information in action: Soft systems methodology.* London: Macmillan.

Donaldson, L. (1985). *In defence of organization theory.* Cambridge: Cambridge University Press.

Hicks, M. J. (1991). *Problem solving in business and management.* London: Chapman & Hall.

Khandwalla, P. N. (1979). *The design of organizations.* New York: Harcourt Brace.

Koestler, A. (1967). *The ghost in the machine.* London: Hutchinson.

Lawrence, P. R., & Lorsch, J. W. (1967). *Organization and environment.* Cambridge, MA: Harvard University Press.

March, J. G., & Simon, H. A. (1958). *Organizations.* New York: Wiley.

Maturana, H. R., & Varela, F. J. (1980). *Autopoiesis and cognition.* Dordrecht: Reidel.

Newall, A., & Simon, H. A. (1972). *Human problem solving.* Englewood Cliffs, NJ: Prentice-Hall.

Patching, D. (1990). *Practical soft systems analysis.* London: Pitman.

Pugh, D. S., & Hickson, D. J. (1976). *Organization structure in its context.* Farnborough: Saxon House.

Reed, M. I. (1985). *Redirections in organisational analysis.* London: Tavistock.

Schutz, A. (1967). *The phenomenology of the social world.* Evanston, IL: Northwestern University.

Silverman, D. (1970). *The theory of organizations.* London: Heinemann.

Simon, H. A. (1957). *Administrative behaviour* (2nd ed.). New York: Macmillan.

Simon, H. A. (1960). *The new science of management decision.* New York: Harper & Row.

Tönnies, F. (1955). *Community and association [Gemeinschaft und Gesellschaft]* (C. P Loomis, Trans.). London: Routledge & Kegan Paul. (Original work published 1887)

Varela, F. J. (1984). Two principles of self-organization. In H. Ulrich & G. J. Probst (Eds), *Self-organisation and management of social systems* (pp. 25–32). Berlin: Springer-Verlag.

Vickers, G. (1965). *The art of judgment.* London: Chapman & Hall.

Waring, A. (1989). *Systems methods for managers.* Oxford: Blackwell.

Wilson, B. (1984). *Systems: Concepts, methodologies and applications.* Chichester: Wiley.

Woodward, J. (1965): *Industrial organization: Theory and practice.* London: Oxford University Press.

Zannetos, Z. S. (1984). Decision sciences and management expectations. In J. P. Brans (Ed.), *Operational research '84* (pp. 69–76). Amsterdam: North Holland.

BIBLIOGRAPHY

Atkinson, C. J. (1984) Metaphor and systemic praxis, Ph.D. Dissertation, Lancaster University, UK.

Blunden, M. (Ed.) (1984) *The Vickers Papers*, Harper and Row, London.

Bryant, J. (1989) *Problem Management*, John Wiley, Chichester.

Burns, F. (1998) *Information for Health: an Information Strategy for the Modern NHS 1998–2005*, NHS Executive, Department of Health Publications, Wetherby, UK.

Checkland, P. (1972) Towards a systems-based methodology for real-world problem solving, *Journal of Systems Engineering*, **3**(2), 87–116.

Checkland, P. (1981) *Systems Thinking, Systems Practice*, John Wiley, Chichester.

Checkland, P. (1983) OR and the systems movement: mappings and conflicts, *Journal of the Operational Research Society*, **34**(8), 661–675.

Checkland, P. (1988) The case for 'holon', *Systems Practice*, **1**(3), 235–238.

Checkland, P. (1995) Soft Systems Methodology and its relevance to the development of information systems. In Stowell, F.A. (Ed.), *Information Systems Provision: the Contribution of Soft
Systems Methodology*, McGraw-Hill, London.

Checkland, P. (1997) Rhetoric and reality in contracting: research in and on the National Health
Service. In Flynn, R. and Williams, G. (Eds), *Contracting for Health*, Oxford University Press,
Oxford.

Checkland, P. and Casar, A. (1986) Vickers' concept of an appreciative system, *Journal of Applied
Systems Analysis*, **13**, 3–17.

Checkland, P., Forbes, P. and Martin, S. (1990) Techniques in soft systems practice: Part 3:
monitoring and control in conceptual models and in evaluation studies, *Journal of Applied
Systems Analysis*, **17**, 29–37.

Checkland, P. and Griffin, R. (1970) Management information systems: a systems view, *Journal of
Systems Engineering*, **1**(2), 29–42.

Checkland, P. and Holwell, S. (1998) *Information, Systems and Information Systems*, John Wiley,
Chichester.

Checkland, P. and Holwell, S. (1998a) Action research: its nature and validity, *Systemic Practice and
Action Research*, **11**(1), 9–21.

Checkland, P. and Scholes, J. (1990) *Soft Systems Methodology in Action*, John Wiley, Chichester.

Chorley, R. J. and Kennedy, B. A. (1971) *Physical Geography: a Systems Approach*, Prentice-Hall
International, London.

Durkheim, E. (1895) *The Rules of Sociological Method* (Catlin, G. E. G., Ed.) 1964, The Free Press,
New York.

Duxbury, J. (1994) The development and testing of a model relevant to the decision making process
concerning the provision of health care in Morecambe Bay, M.Sc. Dissertation, Lancaster
University.

Flynn, R. and Williams, G. (Eds) (1997) *Contracting for Health*, Oxford University Press, Oxford.

Forbes, P. and Checkland, P. (1987) Monitoring and control in systems models, Internal Discussion
Paper 3/87, Department of Systems and Information Management, Lancaster University.

Ham, C. (1992) *Health Policy in Britain* (3rd edn), Macmillan, London.

Ham, C. (1996) The future of the NHS, *British Medical Journal*, **313**, 1277–1278 (23rd November).

HMSO (1997) *The New NHS*, Cm 3807.

Holwell, S. E. (1997) Soft systems methodology and its role in information systems, Ph.D.
Dissertation, Lancaster University.

Luckmann, T. (Ed.) (1978) *Phenomenology and Sociology*, Penguin Books, Harmondsworth, UK.

Maturana, H. R. and Varela, F. J. (1980) *Autopoesis and Cognition*, D. Reidel, Dortrecht.

Miller, G. A. (1956) The magical number seven plus or minus two: some limits on our capacity for
processing information, *Psychological Review*, **63**(2), 81–96.

Miller, G. A. (1968) *The Psychology of Communication*, Allen Lane, The Penguin Press, London.

Miller, J. G. (1978) *Living Systems*, McGraw-Hill, New York.

Morse, J. M. (Ed.) (1994) *Critical Issues in Qualitative Research Methods*, Sage, Thousand Oaks
(Calif.).

Mueller-Vollmer, K. (Ed.) (1986) *The Hermeneutics Reader: Texts of the German Tradition from the
Enlightenment to the Present*, Basil Blackwell, Oxford.

Naughton, J. (1977) *The Checkland Methodology: a Reader's Guide* (2nd edn), Open University
Systems Group, Milton Keynes.

Optner, S. L. (1965) *Systems Analysis for Business and Industrial Problem-solving*, Prentice-Hall,
Englewood Cliffs, NJ.

Rivett, G. (1998) *From Cradle to Grave: Fifty Years of the NHS*, King's Fund Publishing, London.

Simon, H. A. (1960) *The New Science of Management Decision*, Harper and Row, New York.

Simon, H. A. (1977) *The New Science of Management Decision* (revised edn), Prentice-Hall,
Englewood Cliffs, NJ.

Stowell, F. A. (1989) Change, organizational power and the metaphor 'commodity', Ph.D.
Dissertation, Lancaster University, UK.

Stowell, F. A. (Ed.) (1995) *Information Systems Provision: the Contribution of Soft Systems
Methodology*, McGraw-Hill, London.

Tsouvalis, C. N. (1995) Agonistic thinking in problem-solving: the case of soft systems methodology,
Ph.D. Dissertation, Lancaster University.

Waring, A. (1989) *Systems Methods for Managers*, Blackwell Scientific Publications, Oxford.

Watson, R. and Smith, R. (1988) Applications of the Lancaster soft systems methodology in
Australia, *Journal of Applied Systems Analysis*, **15**, 3–26.

Weber, M. (1904) 'Objectivity' in social science and social policy. In Shils, E. A. and Finch, H. A. (Eds), Max Weber's *Methodology of the Social Sciences*, Free Press, New York.

Webster, C. (1998) *The National Health Service: a Political History*, Oxford University Press, Oxford.

Wilson, A. (1973) *The Concorde Fiasco*, Penguin, Harmondsworth.

Wilson, B. (1984) *Systems: Concepts, Methodologies and Applications*, John Wiley, Chichester, 2nd edn 1990.

Author Index

Subject Index

Systems Thinking,
Systems Practice

Systems Thinking, Systems Practice

Peter Checkland
Department of Systems
University of Lancaster

JOHN WILEY & SONS

Chichester · New York · Brisbane · Toronto

Copyright © 1993 by John Wiley & Sons Ltd,
Baffins Lane, Chichester,
West Sussex PO19 1UD, England
National   Chichester (01243) 779777
International   +44 1243 779777

Reprinted with corrections February 1984
Reprinted February 1985, July 1986, April and November 1988, October 1989, August 1990,
September 1991, March 1994, June 1995, November 1996, March and September 1998.

Other Wiley Editorial Offices

John Wiley & Sons, Inc., 605 Third Avenue,
New York, NY 10158-0012, USA

Jacaranda Wiley Ltd, 33 Park Road, Milton,
Queensland 4064, Australia

John Wiley & Sons (Canada) Ltd, 22 Worcester Road,
Rexdale, Ontario M9W 1LI, Canada

John Wiley & Sons (SEA) Pte Ltd, 37 Jalan Pemimpin #05-04,
Block B, Union Industrial Building, Singapore 129809

British Library Cataloguing in Publication Data:

Checkland, P. B.
    Systems thinking, systems practice.
    1. Systems theory
    I. Title
    003   Q295   80-41381

Typeset by Preface Ltd., Salisbury, Wilts.
Printed and bound in Great Britain by TJ International Ltd, Padstow, Cornwall

To

Philip and Margaret Youle

*who   encouraged   me   in   a
change   of   career   I   have
had   no   cause   to   regret*

Contents

PART 3: CONCLUSION

The cradle rocks above an abyss, and common sense tells us that our existence is but a brief crack of light between two eternities of darkness.

VLADIMIR NABOKOV

...no intention of any kind can be discerned in nature, no concerted action by environment on heredity that might direct variation into predetermined paths ...

FRANCOIS JACOB

It is very nice to have feet on the ground if you are a feet-on-the-ground person. I have nothing against feet-on-the-ground people at all. And it is very nice to have feet off the ground if you are a feet-off-the-ground person. I have nothing against feet-off-the-ground people. They are all aspects of the truth, or motes in the coloured rays that come from the coloured glass that stains the white radiance of eternity.

STEVIE SMITH

Preface

A research student recently suggested to me that no book ought to be published unless it makes at least two other books obsolete! This is an austere doctrine, but an understandable one, given the exponential growth of the literature, and the fact that so many books are produced to meet an identified market opportunity rather than written because an argument presses to be expounded. But a literature which would slowly converge on a single priceless tome is a daunting prospect, and I hope my student Graham Moss would allow that there is a place for a book which, while not rendering other books obsolete, tries to break a measure of new ground by relating some systems *theory* to some systems *practice*; for that is the intention of this book.

Its origin lies in the belief that, in the words of West Churchman (1968a) 'The systems approach is not a bad idea'.

This gnomic statement is a very cautious one if we read it colloquially ('not bad, eh?'); but it is a very bold statement if we read it literally ('The systems approach is not an evil idea'), I believe that both the cautious and the bold versions are true, and the argument of this book can be seen as an attempt both to establish this in theory and to demonstrate it in practice, relating how theory has led to practice and how the practice has modified theory.

The situation in which the argument is presented seems to me a strange one. For some years now 'a systems approach' has been a modish phrase. Few are prepared publicly to proclaim that they do not adopt it in their work, and it would be an unwise author of a management science text, for example, who failed to subtitle his book: 'a systems approach'. And yet, in spite of this fashionable lip-service to systems ideas, and in spite of frequent exhortations to use a systems approach, we are rarely told what it consists of, or exactly how we might use it. There has been a notable lack of determined persistent efforts, first to define what 'a systems approach' means and then to go out and use it in tackling problems, in order to experience that interaction between theory and practice which is the best recipe for intellectual progress.

The work described here has been an attempt *first*, to develop an explicit account of the systems outlook, the 'systems view of the world'; *second*—based on that view—to develop ways of using systems ideas in practical problem situations; *third*, to modify both the systems outlook and the ways of using systems ideas as experience was gained, as mistakes were made, as lessons were learned; *fourth*, to reflect on the interaction between systems thinking and systems practice in order to draw conclusions which will allow future theory to benefit from practice and future practice from theory.

The area selected for the experiments in problem-solving which are described in Part 2 was that of 'real-world problems'. By 'real world' is meant the interacting human activity which makes up the business of living, as opposed to the 'artificial' world of the laboratory experiment, in which the researcher is free to decide what to vary and what to keep constant. That the real world is the arena in which a systems approach must prove itself will emerge from the argument of this book. By 'problem' is meant not the puzzle, paradox or conundrum which exercises the philosopher, but simply any situation in which there is perceived to be a mismatch between 'what is' and what might or could or should be.

Obviously the work is not finished, and can never be finished. There are no absolute positions to be reached in the attempt by men to understand the world in which they find themselves: new experience may in the future refute present conjectures. So the work itself must be regarded as an on-going system of a particular kind: a learning system which will continue to develop ideas, to test them out in practice, and to learn from the experience gained. But after ten years something of a pattern can be discerned in the work, and it seemed worth while attempting to give an account of the position reached.

The project of which this book is the realization has been carried out at the University of Lancaster from 1969. The work has been done in the Department of Systems and I am much indebted to my colleagues in the Department, to successive generations of Master's students and to the staff of ISCOL Ltd, the University-owned consultancy company associated with the Department.

A very special debt is owed to Gwilym Jenkins who founded the Department (as 'Systems Engineering') in 1965, and to Philip Youle, then of ICI Ltd, who helped considerably in getting the Department going. They were bold enough both to envisage and then to realize a Master's course which attracted mainly students in their late twenties or early thirties and included as part of the MA Course a five-month involvement in a systems study for a client who wanted something done about an actual problem. The resulting programme of action research was a main vehicle for the 'systems practice' described here, and without it progress would have been slow indeed. The advantage of a research vehicle of this kind—part of a one-year course given each year, so that there is never a course-free gap—is that there can be a very rapid modification of ideas and concepts as a result of pressing experience. It provides an opportunity to do experiments in the real world which are closer than is often the case in social science to the kind of experiments which natural scientists carry out in laboratories. The disadvantage of the arrangement is that as one course ends on 30 September and another begins on 1 October each year, experience is piled up quicker than it can be digested, quicker than lessons can be extracted. Hopefully this book does something to make up for that.

In the Acknowledgements section of a paper published in 1975 I thanked the students, colleagues and managers 'with whom I have struggled' on the studies which were the source of that paper. The irony was intended, but was not meant in any critical sense: an intellectual struggle which does not slip into personal acrimony is the best present that any collaborator can give you. Although I have gained much from arguing about the ideas in this book with each student class since 1969 and with many individual students during the course of their projects, I am especially grateful for their ideas, industry, and resilience, to those students with whom I have worked directly on the projects which have led to the formulation and modification of the methodology for 'soft' systems studies which is described and discussed in Parts 2 and 3: D. G. W. Allen, C. J. Atkinson, S. Cornock, R. Fuenmayor, W. Gunawardena, R. Griffin, G. R. Hewitt, M. C. Jackson, R. W. Keen, G. Kirwan, Su Lau, G. L. Moss, M. Mouthon, W. P. Murray, A. Orr, C. W. Pritchard, M. Rafat, R. Salinas, G. Severn, D. S. Smyth, D. I. Thomas, A. M. Waugh. I have gained much also from working and arguing with ISCOL consultants David Brown, Brian D'Arcy, Bob Galliers, Rod Griffiths, Iain Perring, Chris Pogson, and Chris Woods on projects which were even more demanding than those which involved post-experience postgraduate students.

It will be apparent that I am intellectually in debt not only to Gwilym Jenkins who founded the Department of Systems, but also to many other members of staff since 1969 including Ron Anderton, Ross Barnett, John Collins, John Denmead, Mike Jackson, Brendan Stannard, Lewis Watson, Brian Wilson, and Ian Woodburn. My special thanks for a long continuing discussion about systems methodology and how to teach it go to Brian Wilson. The members of the Systems Group of the Open University, too, have been a welcome source of intellectual stimulation; invidiously, I mention John Naughton in particular. Many members of client organizations have also made a valued contribution to the work, as have those managers who have attended ISCOL's short courses. For the typing of the manuscript I am grateful to Pauline Cookson, Helen Jump, Jeanette Davies, and Karen Grey, both for their fine professional skill and their good humour.

Figure 18 has been redrawn from Figures in *Social Paradigms and Organisational Analysis* by G. Burrell and G. Morgan, 1979, published by Heinemann Educational Books, and is reproduced by permission of the publishers.

The quotation on p. 262 is reprinted by permission of Penguin Books Ltd. from *Freedom in a Rocking Boat* by G. Vickers, published by Pelican Books, 1972, p. 128. © Sir Geoffrey Vickers, 1970.

The quotation on p. 82 is reprinted by permission of Editions Gallimard from *Logic of Living Systems* by F. Jacob. © Editions Gallimard 1970.

INTRODUCTION

Chapter 1

The Subject of Systems

We are all of us aware of ourselves as beings in the world, and we are aware also of a very complex world outside ourselves, of which we are a part. Perhaps the most surprising thing about this mysterious world in which we find ourselves is that it is in fact intelligible. The world shows regularities and continuities upon which we may rely. We may not be able to prove in advance that the sun will rise tomorrow, or that the inverse square law of magnetism will continue to hold, but coherent life on this planet is possible for us because the world outside ourselves does appear to be regular, not capricious. What is more, the process of evolution has provided our species with a brain mechanism with which we can comprehend the regularities of the world we inhabit. We can observe the world, and think about our observations, and so gain knowledge of the world. Western civilization, in particular, has in the last 300 years provided us with a powerful means of observing the world, and thinking about it, and acquiring well-tested knowledge of its regularities: I refer to the method of science.

This book is about a particular way of thinking about the world, one which although broadly a part of the science movement, uses some concepts which are complementary to those of classical natural science. This book is about *systems thinking*, and about the use of a particular set of ideas, systems ideas, in trying to understand the world's complexity. The central concept 'system' embodies the idea of a set of elements connected together which form a whole, this showing properties which are properties of the whole, rather than properties of its component parts. (The taste of water, for example, is a property of the substance water, not of the hydrogen and oxygen which combine to form it.) The phrase 'systems thinking' implies thinking about the world outside ourselves, and doing so by means of the concept 'system', very much in the way envisaged in the following passage by Einstein:

> What, precisely, is 'thinking'? When at the reception of sense impressions, memory pictures emerge, this is not yet 'thinking'. And when such pictures form series, each member of which calls forth another, this too is not yet 'thinking'. When, however, a certain picture turns up in many such series, then—precisely through such

3

return—it becomes an ordering element for such series. . . . Such an element becomes an instrument, a concept. I think that the transition from free association or 'dreaming' to thinking is characterised by the more or less dominating role which the 'concept' plays in it (Einstein, in Schilpp, 1949).

Systems thinking, then, makes conscious use of the particular concept of wholeness captured in the word 'system', to order our thoughts. 'Systems practice' then implies using the product of this thinking to initiate and guide actions we take in the world. This book is about both systems thinking and systems practice, and about the relation between the two.

The Subject: 'Systems'

Memory is a curious faculty, constantly surprising us with its achievements and its lapses. Although I usually have difficulty remembering what I did last Tuesday, I have a clear memory of a school science lesson in the 1940s when the chemistry master put into my astonished mind the idea that Nature did not consist of physics, chemistry, and biology: these were *arbitrary* divisions, man-made, merely a convenient way of carving up the task of investigating Nature's mysteries. This surprising thought, the correctness of which immediately impressed me, gave me a new picture of the universe, and, incidentally, made it clear to me that I had to try and become a scientist. More than thirty years later I think I have not lost the scepticism about subjects of study and their boundaries which began during that chemistry lesson. History shows us that subjects rise and fall in importance; intellectual fashions change, new concerns emerge, or old situations are newly perceived as a set of problems worthy of study. In the 1880s, for example, developments in experimental physiology and in 'mental philosophy' led to the new science of 'psychology'. Earlier in the 19th century, Auguste Comte, apostle of 'positive science', conceived of a new science of society as the apex of a hierarchy of sciences; he called it first 'social physics' and then 'sociology', a new subject from the time Comte christened it, and one which is still with us under that name, although its themes have been written about from the time of the ancient Greeks. Discoveries in electricity and the realization of their practical value make 'electrical engineering' a subject separable from the rest of engineering. More recently, 'ecology' emerges. The examples could be multiplied: the esoteric subject of 'linguistics' lies apparently fallow for many years and then is revolutionized by Chomsky's *Syntactic Structures* and becomes widely recognized as extremely important for the impact it can have on many other disciplines; and so on.

This book concerns a new subject, one only thirty years old, at least in its self-conscious form; and I shall argue that it has a special importance which will give it a long and active life. But I shall have to argue both that it is in fact a new subject, a collection of concerns and methods recognizable as an entity, and that it is important, so little recognized is it at the present time. I refer to the subject: 'systems'.

The reason why systems is not readily recognized as a legitimate subject is that it is different in kind from most other disciplines. Its concern is not a particular set of phenomena, as is the case with chemistry and physics, for example; neither is it, like biochemistry, a subject which has arisen at the overlapping of existing subjects. Nor is it a subject which exists because a particular problem area is recognized as important, and requires the bringing together of a number of different streams of knowledge—as do town planning or social administration, for example. What distinguishes systems is that it is a subject which can talk *about* the other subjects. It is not a discipline to be put in the same set as the others, it is a meta-discipline whose subject matter can be applied within virtually any other discipline. This is a point which is not readily understood; but it is reflected in a phrase capturing the idea which most people who care about the matter at all would associate with the subject of systems: the phrase 'a systems approach'.

An *approach* is a way of going about tackling a problem, and obviously a particular approach may be relevant to more than one subject, just as 'an experimental approach' might be taken to the problems of physics, psychology, agriculture, and many other subjects. 'A systems approach', however, although it conveys the idea of a method of attack, does not readily convey to most people much idea about the content of the method. In the early 1970s when I asked managers on short courses in systems ideas to write down what the phrase 'a systems approach' conveyed to them, I used to find that many imagined that the approach necessarily involved the use of computers in problem-solving. It is no doubt the case that the spread of computer systems has led to an increasing understanding of the concept 'system'. In the late 1970s, however, I have found that computers are now (very properly) regarded only as a means of achieving various ends, as a particular kind of tool—even though one which might cause you to contemplate an objective you would not previously have considered. Answers to the question: What is a systems approach? tend now to be of the kind: an approach to a problem which takes a broad view, which tries to take all aspects into account, which concentrates on interactions between the different parts of the problem. For the infant discipline, this is progress!

In the early days of the new subject the enthusiasts who sought to establish its legitimacy frequently used the words 'organised complexity' to describe its subject matter (Rapoport and Horvath, 1959; Weaver, 1948), this being a region between 'organised simplicity' and 'chaotic complexity'. The concern was to be *organization* as such, the principles underlying the existence of any whole entity, whether a soap bubble, a slow-worm or a social system. This is a rather abstract subject matter, and for that reason is more elusive than the content of subjects more obviously rooted in everyday experience—which is not to say that systems is not also a highly practical subject, or can be made to be.

In order to grasp the significance of a subject whose concern is 'organised complexity', it is useful to ask what other subjects are comparable with it. One subject of the same kind as 'systems' is so well established, and its findings are so well accepted that it is now studied only by a relatively small group of philosophers; most of the work done on the subject is now done in the more detailed areas into which it has split. I refer to the subject 'science'. Only the philosophers of science are now concerned with science as a whole. Professional scientists work in one of the many branches of the subject.

Now, just as we may adopt 'a systems approach' to some subject area, or not, so we may use, or not use, 'a scientific approach' in many different disciplines. We may, for example, adopt a scientific approach to the subject matter of biology (most biologists do!) but there is also the descriptive semi-mystical approach found in the works of Teilhard de Chardin. English literature, to take another example, is frequently studied via subjective critical appraisal, but it is possible to adopt a thorough-going scientific approach even here, for example by analysing and collating the content of metaphors in early and late Shakespeare. Science provides us with the phrase 'a scientific approach' just as systems provides 'a systems approach'. Both are meta-disciplines, and both embody a particular way of regarding the world. The scientific outlook assumes that the world is characterized by natural phenomena which are ordered and regular, not capricious, and this has led to an effective way of finding out about the regularities—the so-called 'laws of Nature'. The systems outlook, accepting the basic propositions of science, for it is a part of the scientific tradition, assumes that the world contains structured wholes (which include soap bubbles, slow-worms and social systems) which can maintain their identity under a certain range of conditions and which exhibit certain general principles of 'wholeness'. Systems thinkers are interested in elucidating these principles, believing that this will contribute usefully to our knowledge of the world.

The best understanding of the new subject comes from examining the history of the scientific method. There we observe that the idea of connected wholes emerges as something worth studying as a result of some intractable problems which defeat the classical scientific method, with its emphasis on reducing the situation observed in order to increase the chance that experimentally reproducible observations will be obtained. The emergence of systems thinking as a response to some problems in science is the content of the next chapter. Meanwhile we may observe that in practical terms there is a good deal of circumstantial evidence that the subject of systems does exist. At least the trappings of a subject exist! There is a Society for General Systems Research; there are conferences and journals; there are a few university chairs; and there are bold claims that we are now living in 'the Systems Age' (Ackoff, 1974), that systems theory 'heralds a new world-view of considerable impact' (Bertalanffy, 1968), and that 'we need nowadays to be able to think not just about simple processes but about complex systems' (Waddington, 1977). More important, a systems orientation is recognizable in the work of some practitioners in many different disciplines, including biology, geography, economics, anthropology, sociology, psychology, political science, social administration, and management science, thus confirming the status of systems as a *meta*-discipline.

A Prescription for a Subject

I assume that the subject 'systems' aspires to the status of a serious academic discipline. I hesitate before using the word 'academic' because the way the word is used in a world I hope to influence, namely that of management, is almost always in the sense that to be academic is to be unpractical. I use the word in the sense 'scholarly', and assume that scholarship involves not only speculation, but also the establishment of knowledge which can withstand serious tests, if possible practical ones. In this sense of the phrase, chemistry is a serious academic discipline; phrenology and flower arranging are not. If systems is to be taken seriously it will have to show that within the subject there is a cycle of interaction between the formulation of theory relevant to serious problems or concerns, and the testing of that theory by the application of methodology appropriate to the subject matter. Figure 1 (Anderton and Checkland, 1977) shows the picture of this cyclic interaction which underlies everything in this book. (Figure 1 is, in fact, a *systems model* of any developing discipline.) It assumes that the focus of interest is a set of concerns, issues or problems perceived in the real world, or something there about which we have aspirations. For chemistry, one problem is to discover the laws which govern the transformation of one substance into another. For town planning the focus is a particular set of issues, while engineering science serves the aspiration of establishing the general scientific principles which will enable us to make efficient machines. Whatever the focus, it will lead to ideas from which we can formulate two kinds of theory, *substantive* theories about the subject matter (for example a theory concerning catalysis in chemistry) and *methodological* theories concerning how to go about investigating the subject matter. Once such theories exist, it is possible to state problems, not merely as problems existing in the world, but as *problems within the discipline*. All the resources of the discipline—previous results within it, its paradigms, models, and techniques—can then be used in an appropriate methodology to test the theory. The results from this test, which will itself involve action in the real world (intervention, influence, observation) will provide what in Figure 1 are called 'case records', records of happenings under certain

8

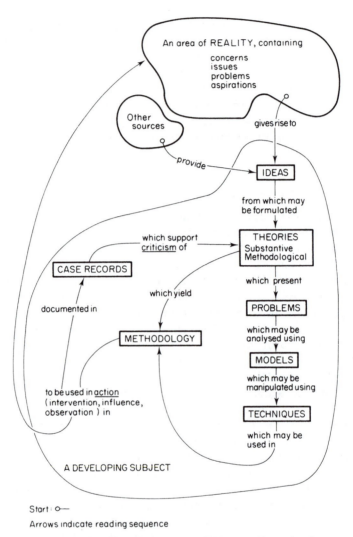

An area of REALITY, containing

concerns
issues
problems
aspirations

Other sources

gives rise to

provide

IDEAS

from which may
be formulated

which support
criticism of

THEORIES
Substantive
Methodological

CASE RECORDS

which yield

which present

documented in

PROBLEMS

which may be
analysed using

METHODOLOGY

MODELS

which may be
manipulated using

to be used in action
(intervention, influence,
observation) in

TECHNIQUES

which may be
used in

A DEVELOPING SUBJECT

Start : o—

Arrows indicate reading sequence

Figure 1. Relationships between activities and results in a
developing subject

conditions. These provide the crucial source of *criticism* which enables
better theories to be formulated, better models, techniques, and
methodology to be developed.

The application of this cycle of interaction between theory and practice
is perhaps easiest in the case of the natural sciences. For a would-be
meta-discipline like systems (or, as we shall see later, for the social
sciences) there are some additional problems.

For systems as a meta-discipline with a highly abstract subject matter
('organized complexity') there are two obvious possibilities for work within

it. There will be work on the general principles of 'wholeness', applicable, hopefully, to any perceived 'whole', and there will be work which brings systems ideas to bear on problems within other disciplines. Work of the first kind, for example, might be the attempt to formulate in general terms the requirements for any controller which governs the overall behaviour of a system in an environment with which it interacts (Ashby, 1956); work of the second kind covers the attempt to picture the professional activity of a social worker as an interaction between systems of different kinds (Pincus and Minahan, 1973).

Work of the first kind, abstract work on general systems principles, is obviously in danger of losing contact with reality. Its theory will be highly abstract, and it will be difficult to translate it into testable propositions. Writing about high-level 'grand theory' of this kind in the context of sociology, Mills (1959) observes, with reference to theorists like Talcott Parsons (1951; see also Black, 1963)

> The basic cause of grand theory is the initial choice of a level of thinking so general that its practitioners cannot logically get down to observation. They never, as grand theorists, get down from the higher generalities to problems in their historical and structural contexts. This absence of a firm sense of genuine problems, in turn, makes for the unreality so noticeable in their pages.

Early systems thinkers were aware of this danger. In an influential paper, Boulding (1956) tries to steer a difficult course between the highly abstract and the unduly specific:

> [General Systems Theory] does not seek, of course, to establish a single, self-contained 'general theory' of practically everything' which will replace all the special theories of particular disciplines. Such a theory would be almost without content, for we always pay for generality by sacrificing content, and all we can say about practically everything is almost nothing. Somewhere however between the specific that has no meaning and the general that has no content there must be, for each purpose and at each level of abstraction, an optimum degree of generality. It is the contention of the General Systems Theorists that this optimum degree of generality is not always reached by the particular sciences.

Trying to avoid content-free abstraction leads to systems work of the second kind, in which systems models, techniques, and methodology are applied to a problem defined within another discipline. Here difficulties may arise as a result of using one (meta) discipline which is itself still developing to tackle issues which will be part of the continuing development of another discipline. Systems thinkers need to be mindful of the distinction between using systems ideas to obtain 'case records' within another discipline, and using that discipline simply as a vehicle for further developing systems ideas themselves. Any actual systems study may well provide lessons of both kinds, of course, which makes the appraisal of results a complex business and the need to make the distinction crucial.

When I came to start the particular work described in this book I had completed a career of fifteen years in which I had been first a scientist, then a technologist, and then a manager. These experiences had given me a taste for rubbing ideas against reality, so I was not inclined to be satisfied with the austere pleasures of General Systems Theory. I believed that systems ideas ought to be relevant to tackling the kind of problem I had faced as manager of a group developing new products in a science-based industry; I was aware that practising managers had a poor opinion of so-called management science, which seemed to consist of a set of techniques relevant to a small number of textbook problems which never seemed to match the idiosyncrasies of real life; and I had come to the conclusion that even highly sophisticated industrial firms were rather bad at learning from their experiences. I joined a university department of systems engineering which intended the word 'engineering' to be read in its general sense: you can engineer a meeting or a political agreement. The task was to apply systems ideas to the (soft) engineering of systems relevant to real-world problems, including, especially, problems of managing. In terms of Figure 1, the concerns or problems were those of 'managing', defining the word broadly and not in a class sense. This means that the work to be described is neither an exercise in General Systems Theory nor, exactly, the application of systems ideas and methodology within another discipline. (It *would* be the latter if 'real-world problem-solving' were a recognized discipline.) It is simultaneously the application of systems ideas to real-world situations in which problems are perceived, and the use of such experience to modify the systems ideas and the methodology for their use, on the pattern of Figure 1. Again referring to the figure, 'reality' is taken to contain 'real-world problems', situations about which we have, in the happy phrase of Miller and Starr (1967), not idle, but *busy* curiosity', 'busy' signifying a readiness to take action, to change the situation and hopefully to secure improvement. These problems located in reality may then be formulated, by means of systems theories, as problems in the discipline of systems, and tackled by means of systems methodology which incorporates systems models and techniques. The systems methodology involves intervention in 'reality', the intervention being a source of case records of systems studies. These in turn are a source of criticism of systems theory as such and of the systems methodology used. Both can thus be improved on the basis of experience, and over a period of time the interventions in reality should become more successful if, that is, systems concepts provide a reasonable mapping of the real world, which is the conjecture underlying the whole programme. If this conjecture survives the testing of it, the work will help to establish the use of systems ideas as a fruitful way of gaining knowledge of the world. Hopefully it may provide not only knowledge which is relevant to General Systems Theory but also knowledge which is relevant to any discipline concerned with real-world problems; this embraces a very large number of disciplines including all the social sciences. All of the procedure outlined should help to establish systems as a legitimate subject of study and do so by means of a method which retains scepticism and ensures critical debate.

Difficulties for the Reader

Before summarizing the argument of the book as a whole, it seems wise to point out some of the difficulties the reader may have with it. When Hayek reviewed Keynes's *Treatise on Money*, Keynes wrote on his copy of the review

> Hayek has not read my book with that measure of 'goodwill' which an author is entitled to expect of a reader. Until he does so he will not know what I mean or whether I am right (Moggridge, 1976).

It is of course easy for an author to sympathize with these thoughts! But there is a serious sense in which it is worth while pointing out what kind of goodwill is needed from a reader anxious to understand the argument developed here, so that he may agree with it or, more important, fruitfully disagree with it. I have tried to make my meaning clear, believing that clarity is a better basis for critical debate than obscurity. The argument of the book moves from the kind of thinking embodied in the practice of science and engineering, via the experience of a large number of systems studies, to a stance in agreement with some of the positions argued by the Frankfurt School of 'Critical Sociology'. But I find the apparently wilful obscurity of much of the writing of that School self-defeating, in that it demands more goodwill from the reader than an author ought to expect. I have tried to avoid their kind of obscurity, but nevertheless I need the reader's goodwill on three counts.

First, I need the reader's goodwill in accepting the need for some systems language which could, and no doubt will, be called 'jargon'. It is very noticeable that the word 'jargon' is hurled as a missile at any attempt to use technical terms in discussing human activity or social systems. It is certainly the case that much writing in sociology, in particular, is characterized, in Mill's (1959) words by

> a seemingly arbitrary and certainly endless elaboration of distinctions which neither enlarge our understanding nor make our experience more sensible.

and this has led to the cry of 'jargon!' greeting any use of technical terms. I would restrict the word 'jargon' to *unnecessary* technical terms of this kind, but not deny the need, in any argument which is intellectually serious, for a number of precisely defined terms. This is accepted in the natural sciences. No doubt physicists could by a circumlocution avoid ever using the word 'entropy'. It is not the name of any object in the real world, for which a name is essential, it is the physicists' concept of a measure of the unavailability of a system's thermal energy for conversion into mechanical work. The word is a useful label for that concept, and when a physicist uses it we know exactly what he is talking about, which must help the clarity of argument. My argument will similarly require the use of a few technical terms—such as 'root definition' and 'human activity system'. I have tried to keep their number small, and to choose words which in their everyday language meaning convey the right flavour. The Glossary lists and defines them.

Second, I need the reader's goodwill in being ready to follow an argument which extends throughout the book. I hope the individual chapters are individually intelligible, but their role is to contribute to the argument as a whole. The aim is that this argument should be more than an aggregate of its component parts; it should be systemic.

Finally, the book needs the goodwill of the reader because although it is not difficult to understand, it will make a special kind of demand on him. René Descartes taught Western civilization that the thing to do with complexity was to break it up into component parts and tackle them separately. The lesson has been well learned, and the idea is deeply embodied not only in scientists, for whom the idea is central, but in anyone who has a Western-style education. Systems thinking, however, starts from noticing the unquestioned Cartesian assumption: namely, that a component part is the same when separated out as it is when part of a whole. This makes finding out about systems thinking very different from finding out about, say, Renaissance literature, the politics of the Middle East, or natural selection in the fruit fly. The Cartesian legacy provides us with an unnoticed framework—a set of intellectual pigeon-holes—into which we place the new knowledge we acquire. Systems thinking is different because it is *about* the framework itself. Systems thinking does not drop into its pigeon-hole, it changes the shape and structure of the whole framework of pigeon-holes. This questioning of previously unnoticed assumptions can be painful, and many people resist it energetically. Hence I need the reader's goodwill in at least entertaining the idea of it.

The Argument of this Book

The considerations just discussed suggest that it would be useful to set out in summary form the argument to be developed in Chapters 2 to 8.

Chapter 2

1 Science

1.1 We find ourselves for a brief span inhabiting a mysterious universe. We have an itch, a need, to try to understand it, and to preserve and pass on that knowledge.

1.2 One very successful way of finding out about our situation is by the use of the crucial cultural invention made by Western civilization: the method of science.

1.3 We may define that method in terms of the three characteristics: reductionism, repeatability, and refutation. By means of it a continuously refined account of the universe is built up. This account is a successful guide to many kinds of action.

Chapter 3

1.4 But science also has many limitations. Within science there are problems of methodology as we move from the 'restricted' sciences (e.g. physics) to the 'unrestricted' sciences (e.g. biology). Outside science, but within the area of problems we might hope it could help us tackle, we find the method of natural science apparently impotent. Complexity, in general, and social phenomena, in particular both pose difficult problems for science; neither has it been able to tackle what we perceive as 'real-world problems' (as opposed to the scientist-defined problems of the laboratory). These are frequently problems of teleological kind, concerned with ends and means. They are problems of 'managing', defining that term broadly.

1.5 Unfortunately 'management science' has not been able to resolve these problems. Hence there is an incentive to examine alternative paradigms to those of natural science, while continuing to build on the scientific bedrock: rationality applied to the findings of experience.

2 Systems

2.1 Within natural science itself, the existence at certain levels of complexity of properties which are *emergent* at that level, and which cannot be reduced in explanation to lower levels, is an illustration of an alternative paradigm—that of 'systems'. The systems paradigm is concerned with wholes and their properties. It is holistic, but not in the usual (vulgar) sense of taking in the whole; systems concepts are concerned with wholes and their hierarchical arrangement rather than with *the* whole.

2.2 Systems thinking developed in several different disciplines and activities including biology (organismic biology) and control and communications engineering. From the 1940s there is a discernible 'systems movement' with institutions and a growing literature. Its core concerns are the two pairs of ideas: emergence and hierarchy, communication and control.

Chapter 4

2.3 The system concept, the idea of a whole entity which under a range of conditions maintains its identity, provides a way of viewing and interpreting the universe as a hierarchy of such interconnected and interrelated wholes. A number of taxonomies and typologies have been described.

2.4 In a systems typology four kinds of system are required: natural, designed physical, designed abstract, and human activity systems. We may *learn* about systems and systemic properties by observation and experiments with natural systems; we may *use* designed systems, physical or abstract; and we may aspire to design, modify, affect or improve (i.e. 'engineer') what we perceive as human activity systems, i.e. sets of purposeful human activities.

2.5 The concept *human activity system* is crucially different from the concepts of natural and designed systems. These latter, once they are manifest, 'could not be other than they are', but human activity systems can be manifest only as perceptions by human actors who are free to attribute meaning to what they perceive. There will thus never be a single (testable) account of a human activity system, only a set of possible accounts all valid according to particular *Weltanschauungen*.

Chapter 5

3 Use of systems ideas: systems practice

3.1 The idea that real-world problem situations might be improved by using the concept of a human activity system—by seeking to 'engineer' such systems—suggests that we look first to the world of professional engineering for relevant concepts and experience. This was an original growth point for systems thinking (2.2) which arose from the engineers' need to solve the problem of designing and implementing controllable complexes of equipment rather than simply components.

3.2 There are many expositions of ways of creating 'hard' systems, and professional engineers regard doing this as a normal part of their professional activity. In the 1950s in America the RAND Corporation developed 'systems analysis' as a means of rationally appraising the alternatives facing a decision maker.

3.3 The thinking embodied in 'systems engineering' and 'systems analysis' is essentially the same. Analysis of many different accounts of these activities shows that they all assume that problems can be formulated as the making of a choice between alternative means of achieving a known end. The belief that real-world problems can be formulated in this way is the distinguishing characteristic of all 'hard' systems thinking.

3.4 The success of systems engineering and systems analysis led to many attempts to use the concepts in problems of social systems, including those of formulating public policy. It was not always noticed that these problems could not necessarily be formulated as hard problems in the sense defined above (3.3). The results were on the whole disappointing, and this in turn led to polemic criticism of the whole idea of making the transfer.

3.5 The research described in Chapters 6 and 7 assumed that the concept of a human activity system would be relevant to tackling the 'soft' ill-structured problems of the real world, those before which the methodology of natural science is impotent. The idea was to apply one of the versions of hard systems thinking to real-world situations in which the actors perceived they had problems, in order to find out whether, why, and how the hard methodology was inadequate. The intention was to find a systems methodology for tackling problems which defy formulation in the hard sense, and also to enrich the concept human activity system in order to understand better the 'social systems' of the real world.

Chapter 6

3.6 In soft problems the designation of objectives is itself problematic. Not surprisingly, hard systems thinking was not usable in these problems, which were always those of a kind to which the concept human activity system was relevant. It was found impossible to start the studies by naming 'the system' and defining its objectives.

Without this base, hard systems thinking collapses. During nine studies in 1969–72 a different methodology was evolved and has been subsequently tested and modified in more than a hundred studies since then. Such work yields, beyond the action in specific situations, a cumulative account of the nature of the concept *human activity system*.

3.7 In the methodology, the situation in which the perceived problem lies (rather than the problem itself) is 'expressed' and this is done, not in systems terms, but by using the concepts 'structure' and 'process' and the relation between the two. These are gentle guidelines which do actually guide the analyst while not distorting the problem into a preconceived or standard form.

Against the background of this examination there is now assembled a number of possible 'relevant systems', relevant, that is to solving the problem or improving the problem situation. The relevant systems encapsulate various specific undiluted viewpoints, and these are brought out in 'root definitions' of the systems named as relevant. (Ways of ensuring that root definitions are 'well formed' have been developed. See 3.13.)

For each such definition a model is now made of the human activity system which would *be* the notional system in question. The model building language is flexible and powerful: all the verbs in English! Once built, the model may be transformed into (or 'validated' against) other forms considered appropriate, for example Beer's model of organization or the Tavistock socio-technical system model. (And it is of course the case that the activity system conceptualized may require for its internal functioning models of other kinds—linear programming models perhaps, depot location models, or models of operations built using the 'systems dynamics' language.)

The conceptual models of human activity systems built in this phase may now be set alongside what is actually observed in the real-world problem situation. This comparison phase is the formal vehicle for a debate about change which involves both the systems analyst and concerned actors. Its object is to enable potential changes to be defined which meet two criteria: Are they both *desirable* and *feasible*—systemically desirable and culturally feasible?

Assuming that, as a result of debate, concerned actors in the problem situation can agree some changes, then the implementation of those changes becomes the new definition of 'the problem' and the methodology comes full circle. If changes cannot be agreed, then further examination must be made of 'relevant systems', of their root definitions, and of *Weltanschauungen* which make those definitions meaningful to the actors in the situation.

(It is of course the case that the views of the world on which different actors base their actions may well be incompatible. The methodology can orchestrate conflict as well as promote consensus, and it is a rejoinder from the systems movement to those social scientists who claim that a systems approach automatically assumes a consensus model.)

3.8 Numerous practical studies have illuminated these concepts and ensured the evolution of the methodology.

Chapter 7

3.9 Within the group of studies in which the soft methodology has been used, five different study types may, with hindsight, be recognized. These are problems of system design; problems in which the task is to define and implement changes knowing that different actors have different values (most studies have been of this type); studies of which the outcome is a survey of some problem area; studies in which past events have been re-examined in a search for insights useful in the future; and purely theoretical studies to elucidate concepts.

3.10 A number of individual studies have provided particular insights. It is possible, for example, for the systems thinker to tackle a problem of which he is himself the owner. Also, although the methodology is most easily described as a sequence of phases, it is not necessary to move from phase 1 to phase 7: what is important is the content of the individual phases and the relationships between them. With that pattern established, the good systems thinker will use them in any order, will iterate frequently, and may well work simultaneously on more than one phase. This 'out of sequence' usage is particularly important in tackling broad problems which are not owned within a single organization.

3.11 Regarded as a whole, the soft systems methodology is a learning system which uses systems ideas to formulate basic mental acts of four kinds: *perceiving* (stages 1 and 2), *predicating* (stages 3 and 4), *comparing* (stage 5), and *deciding* on action (stage 6). The output of the methodology is thus very different from the output of hard systems engineering: it is learning which leads to a decision to take certain actions, knowing that this will lead not to 'the problem' being now 'solved' but to a changed situation and new learning.

3.12 This is a direct consequence of the nature of the concept 'human activity system'. We *attribute meaning* to human activity and our attributions are meaningful in terms of a particular image of the world, or *Weltanschauung* which in general we take for granted. The methodology teases out such world-images and examines their implications. This concept is the most important one in the methodology.

3.13 Apart from these broad generalizations, there have been a number of other outcomes from the use of the methodology in practice which are general in nature: namely (1) root definitions may be well formulated by ensuring that they include (or consciously exclude) a number of considerations which form the mnemonic CATWOE—the *customers* (beneficiaries or victims of a notional system) the system's *actors*, its *transformation* process, the *Weltanschauung* which gives it meaning, the *owners* who could destroy it and the *environmental constraints* to which it is subject; (2) root definitions may express the 'primary task' of organizational entities or express current issues; (3) systems thinking, to remain coherent, must distinguish between levels of resolution, and this can be aided by a careful distinction between 'whats' and 'hows'; (4) in the initial stages of a study, the examination of elements of 'structure' and 'process' may be enriched by the examination of social systems as sets of *roles* with associated expectations of role holders' behaviour *(norms)* and *values* by which that behaviour is judged; (5) two methodological laws have emerged: any system which serves another cannot be modelled until a definition and model of the system served is available (a rule ignored by much real-world work on information systems!) and models of human activity systems should consist of structured sets of verbs which actors could in principle directly carry out; finally, (6) any example of the use of the methodology may itself be modelled as the operation of a system containing 'problem-solving' and 'problem-content' systems, these containing the roles 'would-be problem solver' and 'problem owner'.

3.14 Overall, the stages of the methodology for work on ill-defined problems (which do not have to be followed in fixed sequence) constitute a *learning* system, a system which finds things out in a situation which at least one person regards as problematic. For ill-structured problems involving a number of people the very idea of 'a problem' which can be 'solved' has to be replaced by the idea of dialectical debate, by the idea of problem-solving as a continuous, never-ending process, but one which can be aided, and orchestrated, by the application of systems ideas, particularly that of a human activity system. The nature of the methodology in fact derives from the special nature of such systems (see 2.5 above).

Chapter 8

4 Implications of systems practice

4.1 Three aspects of the learning which has occurred in the action research programme stand out: first, that the methodology is a mosaic of activities having certain relationships with each other, rather than a required sequence of activities (although most studies will begin with an expression of the problem situation); second, that a crucial distinction is between taking action in the real world (stages 1, 2, 5, 6, 7) and doing some systems thinking *about* the real world (stages 3, 4)—the models of human activity systems in the latter being intellectual constructions, 'ideal types', for use in a debate, *not* attempts to describe reality; third, that the most important systems ideas are emergence, hierarchy, communication, and control.

4.2 The methodology is not a technique which, properly applied, can guarantee a particular kind of result; it leaves room for personal styles and strategies of problem-solving (though it also has some defining constitutive rules). The process of developing it has been the operation of a closed system in which learning from the use of the methodology creates the methodology; initial entry to this closed system was forced by using (unsuitable) hard systems methodology. The success of the methodology, in the real-world actor's terms rather than in the analyst's, makes it useful both to compare it with other work elsewhere and to ask: What does the success imply about the ultimate object of enquiry, namely social reality?

4.3 Regarding work elsewhere, soft systems methodology can be seen to be an operationalization of Churchman's philosophical analysis of enquiry systems, or a means of formally orchestrating the workings of what Vickers terms 'appreciative systems'.

4.4 The nature of social reality implied by the methodology is very different from that implicit in the approach which is usually taken to be the application of systems theory within social science, namely functionalism. Functionalism is a part of the Durkheimian (or positivistic) tradition in sociology. Soft systems methodology implies, rather, a model of social reality such as is found in the alternative (phenomenological) tradition deriving sociologically from Weber and philosophically from Husserl. Ironically, the methodology is also highly compatible with the ideas of the 'Critical Sociology' of the Frankfurt School, although Habermas—its leading theorist—regards himself as opposed to systems theory. Another way to describe the methodology, in fact, is as a formal means of achieving the 'communicative competence' in unrestricted discussion which Habermas seeks.

5 Conclusion

5.1 This work started by trying to use systems ideas in ill-defined problem situations. Its outcome is a systems-based methodology for tackling real-world problems, and incidentally for exploring social reality. It lends support to the view that the latter is not a 'given' but is a process in which an ever-changing social world is continuously re-created by its members. The implicit value system of soft systems methodology is that never-ending learning is a good thing.

Part 1

SYSTEMS THINKING—The Systems Movement in
the Context of Science

Chapter 2

Science as a Human Activity: Its History and its Method

The long story of the development of science in Western civilization from the 6th century B.C. to the Scientific Revolution of the 17th century (which is a very short story in the history of *Homo sapiens*) has been told in many excellent histories of science and cannot be retold here (see, for example, Singer, 1941; Jeans, 1947). What follows is only a brief outline sketch of some of the important features of the history of the Science Movement. Its purpose is to provide an account of science which in subsequent chapters will explain the emergence of the Systems Movement within the broader sweep of science itself.

Clearly it is not possible to write objective history. As Popper (1957) points out the best we can do is to write history which is consistent with a particular point of view. We ought if we can to state that point of view plainly. What follows, then, is a sketch of the development of science which enables us to understand the nature of systems thinking as being complementary to scientific thinking. The assumed problem is that of understanding the nature of systems thinking and explaining why the Systems Movement, conscious of itself as such, emerged in the middle of this century (Checkland, 1976b).

Western civilization is characterized by a particular religion, that of the Judaeo–Christian tradition, by forms of art and crafts which are its own, and by technologies beyond those developed in any previous civilization. Other civilizations have had other religions, other arts, other technologies. But Western civilization is unique in having developed an organized human activity unknown in any previous civilization: the activity we call science. When, in the 1930s, Joseph Needham began his monumental study *Science and Civilisation in China,* he took his essential problem to be the question why modern science (i.e. since the 17th century) had not developed in Chinese, or in Indian civilization but only in Europe (Needham, 1966). He concludes that the answer to his question is a broad one, involving social and economic aspects as well as purely intellectual ones. Science is a product of Western civilization as a whole.

More than being merely a product, however, science is an *invention* of our civilization—a cultural invention—and it is probably the most powerful invention ever made in the whole history of mankind. Our world in the 20th century is essentially the world created by the activity of science, and not only created physically in our cities, our transportation, and our communication systems, but also created institutionally in our political and administrative procedures, in the way we organize our society. Rationalism and empiricism, twin outcomes of the Scientific Revolution of the 17th century, have influenced all of our civilization, not only its more recent technology. The fruits of modern science are now all-pervading in their influence. Science has given us testable knowledge of the way the natural world works, and has provided us with at least the possibility of material well-being, even on a planet with finite resources; and it has also given us the means of destroying all life on our planet.

All this we owe to science; and the initiation of the process which led to the development of modern science we owe to the Greeks (Hutten, 1962).

The impulse behind science (*scientia, epistémé*) is the itch to know things, to find out how and why the world is as it is. This is different from the drive behind technology *(techné)* which is the itch to do things, to achieve practical ends. Of course, once a scientific method exists there can be *applied* science or a *science* of techniques (Korach 1966; Davies 1965) but 'the urge to know' and 'the urge to do' are different motives, and it is not surprising that much technology (for example pottery making, paper making, gunpowder) was developed long before the Greeks provided the society in which the scientific outlook was born, and initiated the rise of science which culminated in the Scientific Revolution of the 17th century.

The urge to know, to find things out, which science satisfies, has at its core the Greeks' greatest legacy to us: the art of rational thinking. The story of the rise of science is the story of the creation of this priceless weapon, of the hiatus in its use in the 'Dark Ages' (5th to 10th centuries) and its recovery in medieval times, when the scholastic philosophers created the medieval picture of the world and brought Aristotle's thought within the orbit of the Christian faith (see Fremantle, 1954). The medieval world-view, with its Aristotelian science, survived until the burst of intellectual energy which we know as the Renaissance of Learning led to its replacement by the new world-view created by Copernicus, Kepler, Galileo, and Newton—the world-view which is still recognizably our own, in spite of some sophisticated modifications in the 20th century (see Santillana, 1956; Hampshire 1956).

Greek Science

Following Singer (1941) we may briefly examine the 900 years of Greek science in terms of three periods. In the first period, 600–400 B.C., the foundations were laid; Singer calls it 'The Rise of Mental Coherence'. In the second period, 400–300 B.C., were established the systems of thought associated with the Athenians Plato and Aristotle, which for 2000 years provided educated men with their picture of the universe. In the third period, 300 B.C.–200 A.D., associated with the city of Alexandria with its library and museum, professional science separated from more general philosophy, and the concepts established in the earlier periods were elaborated and exploited in a range of specialist sciences—geometry, astronomy, mechanics, geography, medicine. Table 1a at the end of this section provides a summary.

The foundations of the art of rational thinking were laid by Asiatic Greeks who had colonized the shores of Asia Minor. Thales, a citizen of the Ionian city of Miletus, speculated on the constancy which might underlie the manifest variety we see in nature. What is crucial is that his speculations were of a different kind to those made in other civilizations. Bertrand Russell (1946) remarks in his impish history of Western philosophy that it is discouraging for respectful young students to be told that philosophy begins with Thales, who said that everything is made of water! This is unfair on the Greeks; it is unfair to view the content of ancient philosophy through the *Weltanschauung* of the 20th century. If, instead of the content, we note the *kind* of statement Thales is making, and the nature of the world in which he says it, we appreciate its importance, for this is the birth of rational speculation in a superstitious myth-laden world. The Babylonians, for example, had a myth that Marduk, a divine being, had caused dry land to appear by placing a rush mat and dirt upon the waters. Now, the significant thing is that *Thales leaves out Marduk* (Farrington, 1944). The myths of Thales and his fellow philosophers were rational myths, required to be logically coherent without recourse to divine beings—through whom anything can be 'explained'.

Thales founded at Miletus a philosophical school, now known as the 'pre-Socratics', which formulated a number of cosmologies as reasoned speculations about the nature of things. For Anaximander, second of the school, water, like earth, or mist, or fire was a form of an indeterminate substance which was the origin of all things. The world consisted of things which were impermanent mixtures of these elements, and so exhibited constant change which proceeded by the combination and dissolution of elements. For Anaximenes, mist was the fundamental form, and other elements were formed by rarefaction or condensation.

The content of these speculations is no longer important—although we may perhaps wonder about any similarities between Anaximander's 'indeterminate stuff' and whatever it is makes up the four-dimensional space-time continuum of the Einstein universe! What is important is the spirit in which the Greek speculations were proposed and the critical debate in which they were discussed. The Greeks 'argued for the sole purpose of arriving at the truth and with argument as their chief weapon; argument used deliberately, consciously, and carefully developed into a technical method' (Richie, 1945).

The technical methods of argument were developed over several centuries, and frequently as part of, rather than separate from the content of the speculations. At this birth of what became science, questions of content and questions of methodology were inevitably mixed. Only since the Scientific Revolution of the 17th century have scientists been able to take for granted the way of carrying out scientific investigation; the history of science is also the history of the method of science.

Following the speculations of the Milesian school, with their base in everyday observation of the world, Heraclitus, of Ephesus in Asia Minor, introduced a different concept. For Heraclitus, it was not a particular *ingredient* which defines the underlying unit of nature, rather it was the *process of change*, or flux. (It was Heraclitus who said that we cannot step into the same river twice.) The ultimate unity is the *logos* or reason which orders and controls the flux. The intelligibility of the flux is that same reason which enables us to know the world as intelligible (Wartofsky, 1968). Fire is Heraclitus's example metaphor of the ever-changing process which is nevertheless governed by *logos*.

Heraclitus was a mystic, and the concept in his thinking is a difficult one. But it is also very important for the development of science because here is the articulation of a picture of the world in which unchanging laws govern the processes which underly what our senses directly perceive. Both of the concepts here: that our senses may deceive us, and that the universe is lawful rather than capricious, are fundamentally a part of the outlook we call scientific. The history of the natural sciences is the history of men attempting to get behind appearances to establish the laws governing the phenomena we observe. We could not undertake science if we believed the world to be capricious; and we would not make much progress if we accepted appearances at their face value; after all, appearances tell us that the earth is obviously at rest, and that the sun daily describes a great arc over it from East to West.

Once the clear distinction has been made between sense and reason, there can obviously be debate about which to follow. A protagonist in this debate was Parmenides of Elea, in the south of Italy. Where Heraclitus had declared that 'all is flux', Parmenides argued that 'nothing changes', that the senses are deceptive, and that observation is inferior to logical argument. Parmenides and his followers had arguments but no evidence for this view. What is important here, however, is the fact that out of the Eleatic philosophers comes a methodological conclusion, as with Heraclitus: that the criterion for reality is rational discourse, and that the condition for this is that 'what cannot be stated without contradiction cannot be' (Warkofsky, 1968).

The counter attack against Parmenides's dismissal of observational science came from Empedocles of Sicily. He was a considerable experimentalist and established air (rather than 'mist') as one of the four basic elements. He showed by experiments with water-filled vessels that the unseen air could occupy space and exert pressure, and he had answers to 'the three central problems of Greek science' (Toulmin and Goodfield, 1962): What are the stable principles behind the flux? What process is responsible for changes? What agencies control this process?

Empedocles made an important contribution to a tradition which the Greeks started and which has been reinterpreted throughout the history of science: the tradition of atomism. The Ionian Democritus had proposed a world made up of eternal atoms of different sizes and shapes, with void between them. A different kind of contribution to atomism came from a different source and from a different stream of Greek thought, the religious tradition, exemplified by the Pythagoreans, as compared with the materialism of the Ionians.

Pythagoras founded a sect, a religious brotherhood, at Croton in southern Italy. The rules of the Pythagorean order derived from a number of primitive taboos ('abstain from beans', 'do not look in a mirror beside a light', 'do not stir the fire with iron' etc.) and Russell (1946) describes Pythagoras himself as 'a combination of Einstein and Mrs Eddy'. But blended in with what to us is the nonsense, the Pythagoreans developed mathematics in the form of demonstrable deductive argument; and they were also considerable experimentalists (Jeans, 1947). Their crucial concept was that the universe could be expressed mathematically. The atom of their atomism was 'number', mathematical relations replaced physical processes, and contemplation of mathematics ensured the purity of the brothers' souls. Much of our mathematical thinking derives from the Pythagoreans' considerable achievements, and we may feel some sympathy for the members of the sect in the crisis which overtook them when they discovered that the length of the diagonal of a square ($\sqrt{2}$) cannot be expressed as a finite number of units.

Much of the experimentation of the Pythagoreans was concerned with the physics of music; they discovered, for example, the relationship between the length of a string and the pitch of the note it produces when plucked. Another area which provided something of an experimental base to Greek science was medicine.

It has often been remarked, in spite of the Pythagoreans, in spite of the likes of Empedocles, that Greek science was not experimental science: that it was science in the head rather than science in the world. The reality seems to be that the Greeks set out to explain, without recourse to divine beings as causes, the workings of the universe as they are known to anyone who has lived in the world and thereby observed it—the heavens, the changing seasons, the actions of fire and water, etc. If your problem is the way the world works, and its origins, then the idea of the special kind of controlled observation which we now call 'an experiment' is not obvious. On the other hand, if your problem is that of medicine, if you are anxious to do what you can to preserve the body's health, then the logic of the problem itself insists upon a greater reliance on observation, and in particular will force theories and conjectures to be *tested*. It may not be obvious how to test your rational explanation of the ebb and flow of the tides, but your ideas on how to preserve your patients' health will tend to be tested automatically by their mortality rate!

The transformation of Greek medicine into a scientific procedure is associated with the name of Hippocrates, who came from the island of Cos. He is said to have lived to be 100, and he had many followers. Happily, many of their writings were preserved in the Hippocratic collection in the library at Alexandria; they reveal the emergence of empiricism as a principle of science. The writings insist upon what we would recognize as the inductive method. They inveigh against *a priori* speculation, insist on patient observation, are reluctant to speculate on the unverifiable, but are anxious to generalize from actual experience. Hippocrates can sound astonishingly modern:

> One must attend in medical practice not primarily to plausible theories but to experience combined with reason. . . . I approve of theorizing if it lays its foundation in incident, and deduces its conclusions in accordance with phenomena.

Farrington (1944) recognizes in the Hippocratic doctrines the first emergence of the idea of positive science.

After the Heroic Age briefly outlined, in which so many of the components of a scientific outlook were formulated for the first time, comes a distinct change of emphasis. The great systems of thought which Plato and Aristotle developed and expounded in Athens in the period 400–300 B.C. were essentially metaphysical rather than scientific. Jeans (1947) takes an extreme view of their effect upon physics:

> While physics was still in this primitive stage of its development, it met with two major disasters in the attitudes of two great thinkers, Plato and Aristotle. . . . Plato's attitude was disaster number one for physics, but worse was to come from his pupil Aristotle. . . .

The change of emphasis begins with Plato's teacher Socrates. He was opposed to research into nature, indeed to the whole programme of the Ionian school from Thales to Democritus. His concern was man's behaviour here on earth, and what constituted behaviour which would ensure that when men died their souls would return to heaven. Socrates is, however, usually associated with the systematic use of the dialectic method, that is to say with the pursuit of knowledge by question and answer which is an important contribution to scientific method. The dialectician encourages critical discussion which will analyse arguments and premises and reveal inconsistencies. Without such critical discussion there can be no science. Popper (1963) points out that the Greek expression for dialectic may be translated 'the argumentative usage of language' and heads his paper entitled 'What is dialectic?' with a quotation from Descartes which is a sharp unspoken comment on the need for it: Descartes said 'There is nothing so absurd or incredible that it has not been asserted by one philosopher or another'.

Plato, whose Academy in Athens lasted for 900 years, did not like the world as he saw it, and was hostile to a science of nature. Because the world of appearances is deceptive, and in any case is in a constant state of flux—as Heraclitus had taught—Plato takes ultimate reality to lie in the world of intelligence, in ideas, in the concept 'redness' rather than its manifestation in any particular red object. Plato's inspiration was in geometry, which he saw as expressing through its certainties the timeless perfection of the world of ideas. Over the entrance to the Academy was the notice which read: 'No one may enter who does not know geometry', and he set his pupils to find rules by which the motion of the sun, moon, and planets could be reduced to 'perfect' spheres and circles—which became the task of astronomy up to the time of Kepler.

Plato's attitude to the material world may well have been a disaster for physics, as Jeans suggests, but it was influential, and ultimately valuable, in helping to establish the belief that to express the workings of the world mathematically is to take an important and useful step beyond mere qualitative description. And his cosmology, in terms of the basic geometrical solid figures, does show the *economy* which is prized in scientific explanation.

Plato's pupil Aristotle, born the son of a court physician in Macedonia, was probably the most influential thinker in the story of science from 600 B.C. up to the 17th century. What happened then was an intellectual revolution precisely in the sense that a complete self-consistent model of the universe, Aristotle's, had to be demolished before the crucial step to the creation of modern science could be taken. Aristotle's total world picture is included in works on physics, logic and metaphysics, ethics and politics, and biology. Ironically it is his relatively neglected writings on biology which now seem most modern scientifically; his physics, which Jeans finds so disastrous for the development of the subject, was, however, more influential.

In the period after Plato's death Aristotle worked in marine biology but, faced with the complexity of the phenomena, despaired of finding explanations in Platonic mathematical abstractions. Perfect mathematical forms are not a useful model for plants, animals, and human beings, he concluded. The *processes* going on in such things needed careful study; and Aristotle took as his basic paradigm what he saw as the directed development of living things. In opposition to his teacher's teaching, Aristotle concluded that ideas did not exist separately from their embodiment in objects in the world. Rather, the objects exhibited a striving to attain their true end *(telos)*, and the 'embodied end', or *entelechy* was the object of scientific study. (It is interesting to note that in the late 19th and early 20th centuries, in biological controversy over whether or not organisms are endowed with a mysterious vitalism, controversies very relevant to the emergence of the Systems Movement, we find the word 'entelechy' revived. Alas, in the 20th century, as in 300 B.C., creating a name does not provide an explanation. The latter, in modern thinking in science, requires the elucidation of chains of causes and effects, and testable predictions.)

Aristotle made teleological analysis the method of science as he applied it to physical as well as biological phenomena. His explanation of an eclipse of the moon becomes the statement that it is the moon's nature to be eclipsed. Thus he builds his world picture. Here on earth things were thought to be made up of the usual four elements earth, air, fire, and water, which contain pairs of the four qualities heat, cold, dryness, and moisture. Beyond the moon, matter was incorruptible and consisted of the 'quintessence', the fifth element. It was the nature of the incorruptible heavenly bodies to move around the earth in 'perfect' circular motion, being carried on crystalline spheres driven by the Prime Mover which Aristotle identified with God.

Again, the main interest is not now in the content of the speculations (conjectures which have been refuted do not hold our interest for long) but in the contributions made to the emerging tradition of scientific study of the world and its phenomena, as opposed to occult, poetic or mythological traditions.

Aristotle's emphasis on function or process, as compared with Plato's emphasis on form or structure, founds a tradition which leads to empiricism, just as Plato's thought leads to rationalism. Also, Aristotle's emphasis on the question: 'What is the fundamental nature of this object?', which leads to his view of the importance of classification by function, in addition leads to his formulation of the classic form of deductive argument, the syllogism. The familiar example: Men are mortal; Socrates is a man; therefore Socrates is mortal, may now seem trivial, but the generalizing of the *form* of an argument in this way is an important step towards conscious work on the *methodology* of science, as opposed to its content.

Finally, it should be mentioned that Aristotle himself displayed a clear understanding of the temporary nature of scientific hypotheses, and hence of the cool striving for objectivity which the logic of science calls for even though the behaviour of scientists does not always display it! In his *Metaphysics* he writes

> ... we must partly investigate for ourselves, partly learn from other investigators, and if those who study this subject form an opinion contrary to what we have now stated, we must esteem both parties, indeed, but follow the more accurate.

It is no discredit to Aristotle that his conjectures held up science, since they did so only in the sense that they were not refuted for more than twenty centuries.

Following the death in 323 B.C. of Alexander the Great (whose tutor was Aristotle) the Alexandrian Empire fragmented. The general Ptolemy created a dynasty in Egypt, and the city Alexandria became the centre of the scientific world. For 500 years it witnessed and recorded the development of science as a series of specialisms: geometry, astronomy, geography, mechanics, medicine. For the first time science emerged as a professional activity distinct from philosophy. Strato, who was one of Aristotle's successors as head of the Lyceum in Athens, was criticized for his interest in the investigation of nature rather than ethics, and with him 'we reach the point at which Greek Science fully establishes a technique of experiment' (Farrington, 1944). Euclid gathered together the knowledge of geometry which had been accumulating over hundreds of years and gave it the superbly logical presentation of his *Elements of Geometry*. No book used as a school text has ever survived so long. Aristarchus argued a heliocentric cosmology and attempted a geometrical calculation of the relative distances from the earth of sun and moon. Archimedes made his multifarious contributions to mathematics and mechanics. Hipparchus, and later Ptolemy, made accurate astronomical observations, the latter proposing a complex set of epicyclic motions to yield the observed movements of the planets. Galen, from Pergamum in Asia Minor, a rival centre to Alexandria, established by numerous dissections of animals the account of physiology which survived until the 16th century.

By the 3rd century of the Christian era Greek science was in decay. At that period in history it was possible for the scientific outlook to be abandoned, something which would not be possible after the 17th century. The declining Greek civilization could turn away from science because its science was, in the words of Singer (1941) 'a way of looking at the world rather than a way of dealing with the world'. After the 17th century science not only provided a picture of the world, it changed and transformed the world in a way that Greek science did not.

The decline of Greek science was part of the decline of the civilization which had produced it, and although Greek ideas had powerful influence upon Roman civilization, the Romans did not continue the scientific work started by the Greeks, good plumbers and engineers though they were.

If this brief summary were concerned with the content of science, with its changing theories and positive achievements, it would of course be necessary to cover periods closer to our own times in increasing detail. Because it is concerned only with the main themes in the emergence of science as a particular kind of human activity this is not necessary; and the reason it is not necessary is that the remarkable Greeks had made such astonishing progress in defining and refining this new kind of human endeavour, even though the content of much of their science had subsequently to be abandoned.

The main Greek achievement was to remove explanation of the workings of the world from the realms of religion and magic, and to create a new kind of explanation—rational explanation—which was the subject of a new kind of enquiry. The contrast has often been remarked, for example, between Greek and Babylonian astronomy. The Babylonians had many techniques for predicting the movements of heavenly bodies, but these were completely unconnected with the myths by which they explained the heavens. Crombie (1969) argues that the Greeks' bringing together of explanation and prediction was the crucial step in establishing the science movement:

> . . . . the Greeks took the decisive step in cosmology of looking for explanations deductively connected with the means of prediction, the step by which they established the European scientific tradition as distinct for example from Babylonian astronomy in which there was a total logical disjunction between the highly developed technological predictions and the myths which did service for explanations. . . .

In addition we owe much to the Greeks for their ideas of mathematical representation (even though the concept of the mathematical representation of the appearances—as opposed to a Platonic concept of geometry *as* reality—has to await later times) and we owe to them also the development of forms of rational argument. Finally there was a tradition via Empedocles and the Hippocratic physicians of the importance of *observation*.

What was lacking in Greek science, and was to be supplied by the medieval scholastics and by the 17th-century scientists, was a sharp sense of the importance of deliberately contrived observation in controlled experiments, the importance of inductive rather than deductive argument, the use of mathematics to represent the observed phenomena, and, perhaps most important, the concept of a social function of science as one of improving control over the material world and reducing the need for

Table 1a Some thinkers and experimenters important in the development of science

1 Greek Science

IONIA	Thales Anaximander Anaximenes	c.624–565 B.C. c.611–547 B.C. c.570– ? B.C.	The natural philosophers of Ionia. Creation of rational myths about the universe; critical discussion of them.
	Heraclitus	c.540–475 B.C.	'Change' as the unitary principle underlying the appearance of things; a sharp distinction between sense and reason.
	Democritus	c.470–400 B.C.	The world as atoms and void, qualities being due to our senses.

Table 1 – *continued*

MAGNA GRAECIA	Parmenides	5th century B.C.	The attack on observational science; assertion of the primacy of logical thinking. Development of conditions for rational discussion: 'what cannot be stated without contradiction cannot be' (Wartofsky, 1968).
	Zeno of Elea	5th century B.C.	
	Empedocles	c.500–430 B.C.	Defence of observational science; the world as formed by mixtures of the basic elements Earth, Air, Fire, Water.
	Pythagoras	c.582–? B.C.	The religious (anti-Ionian) tradition; the mathematical tradition. The central unity: number. The ultimate structure of the world seen as its mathematical form.
	Hippocrates (of Cos in Asia Minor)	c.460–? B.C.	Empiricism: truth as emerging from careful observation and the test of successful practice.
ATHENS	Socrates	470–399 B.C.	The dialectic method: discovery by question and answer.
	Plato	428–347 B.C.	Ultimate reality as being in ideas. The true perfection of the universe (behind outward appearances) as mathematically expressible.
	Aristotle	384–322 B.C.	Ideas not separate from their embodiment in objects. Classification by function as the fundamental principle of science: the world seen as a teleological striving of things to achieve their true nature. A comprehensive world picture which remained intact for 2000 years.
ALEXANDRIA	Strato	3rd century B.C.	Professional science, mainly in Alexandria. Elaboration and exploitation of the concepts founded in earlier schools. Work in geometry, astronomy, geography, mechanics, medicine.
	Euclid	c.330–260 B.C.	
	Aristarchus	c.310–230 B.C.	
	Archimedes	287–212 B.C.	
	Ptolemy	A.D. ?–c.168	
	Galen	A.D. 131–201	

physical labour. The failure of Greek science to develop this social function was perhaps inevitable in a slavery-based society in which science was a liberal study for the educated classes. For the concept of 'positive science' in this sense we have to wait for Francis Bacon and the 17th century. But before that stage was reached the medieval scholars had made important contributions to methodology.

Medieval Science

It used to be thought that the intellectual revolution which created modern science in the 150 years from 1600 happened suddenly, as a result of the work of a few thinkers and experimenters of genius. Johannes Kepler, himself one of those men of genius, wrote of Europe sleeping for 1000 years; and Immanuel Kant, writing in 1787, declared that 'the study of nature entered on the secure methods of science after having for many centuries done nothing but grope in the dark' (Smith, 1972). The counter view, that modern science can be seen to evolve as a result of a continuous tradition dating from the medieval times in which Greek science was recovered after the Dark Ages, has been developed as a result of researches beginning with those of the medievalist scholar Pierre Duhem (Basalla, 1968). Duhem wrote of mechanics and physics unfolding 'through an uninterrupted series of barely perceptible improvements from the doctrines taught in the medieval schools. Certainly, some of the components of what we now recognize as the scientific outlook can be seen struggling to expression in the work of medieval clerics, and it is necessary to emphasize a few of these here. Again, the chief concern is not the content of medieval science (well covered, for example, in Crombie, 1969, where there is an extensive bibliography), only the emergence of ideas we now see as helping define what we mean by 'scientific'. In their day, of course, these ideas were entertained equally with many we would now consider nonsense, but their significance is not affected by the fact that it is only with hindsight that we can separate out those ideas which in the end turned out to be significant in forming the concept 'science'.

By A.D. 750 the Arabs/ had created a Moslem Empire from Spain to China, absorbing learning as well as territory. To the Arabs we owe some optics and some alchemy; but most of all we owe to them the form in which we write out numbers, with its important invention of a sign to represent zero, which enables the position of an integer to be significant and makes mathematics as we know it possible. (The Romans simply could not multiply XVIII by XXVI!) However, in the development of modern science, the most important contribution the Arabs made lay probably in their role as preservers and translators of Greek texts. It was through translated Arabic texts that Greek learning became available in the medieval universities and in the monastic orders.

Once the 17th century had forged the weapon—experimental science—which could be used to find out how the world works, interest was directed mainly to the problems science might solve. In the 13th and 14th centuries, however, the problem was the more philosophical one: What kind of knowledge is it that science provides? Science was also philosophy of science (Crombie, 1969), and the root problem was the debate between 'realists' and 'nominalists'. Does the quality 'whiteness' have an independent existence (the realist position) or is it meaningful only in its embodiment in white objects, as nominalists argued? (Freemantle, 1954). The debate is the same one later taken up between rationalists and empiricists. Meanwhile, in 13th-century Oxford, methodological problems of induction were faced. Aristotle's biological work had led him to discuss observation and classification—a precursor to deriving generalizations by induction, but most of his thinking, like Plato's, was concerned with deducing conclusions from unquestioned premises. Now Robert Grosseteste and his followers took up as a problem the emerging methodological questions of how to investigate phenomena in order to generalize about them, and how to distinguish true from false generalizations. Grosseteste, within what was essentially an Aristotelian framework of thought, proposed rules for inductive examination of phenomena and proposed the use of deliberately arranged experiments in order to decide between rival theories (Crombie, 1953). He himself did experimental work on optics, using lenses and mirrors, choosing light as a phenomenon to investigate because of its importance in the Creation.

One of Grosseteste's followers, like him a Franciscan, was Roger Bacon, whose writings provided further encouragement for the experimental spirit. As a man of his time, his interests extended to astrology and alchemy, but nevertheless he argued that scientific knowledge should be acquired by experiment, that mathematical expression was essential; and he looked forward fancifully to practical outcomes from science, such as mechanically propelled carriages and flying machines, in a way which anticipates his namesake Francis Bacon. A quotation from Roger Bacon's *Opus Majus* illustrates his modern flavour:

> ... there are two ways of acquiring knowledge, namely by reasoning and experience. Reasoning draws a conclusion and makes us grant the conclusion, but does not make the conclusion certain, nor does it remove doubt ... unless the mind discovers it by the method of experience.... And, if we turn our attention to the experiences that are particular and complete and certified ... it is necessary to go by way of the principles of this science which is called experimental.

A third Franciscan, William of Ockham was, like Grosseteste, concerned with the logic of induction. Ockham is important in the development of the idea of science even though he is frequently unmentioned in histories of science. He is important, firstly, because his examination of induction directed attention to observation as the means necessary to discover facts about the world, secondly, because of the fundamental importance, in framing scientific explanations, of the principle we know as Ockham's Razor. This is usually stated in some such form as: 'entities are not to be multiplied without necessity' or 'when faced with competing explanations, accept the most simple'. Its history can be traced back to Aristotle (who said 'nature operates in the shortest way possible') and to Duns Scotus another Franciscan friar of Oxford, a generation older than Ockham.

Bertrand Russell (1946) says of Ockham's Razor: 'I have myself found this a most fruitful principle in logical analysis'. Its importance lies in the rigour and precision it brings into qualitative discussion. This stems from the fact that there is only *one* explanation or description of something which is *minimum*, whereas there is an infinity of explanations which bring in other entities. A proposition which is embodied in a statement which meets the principle is thus highly defined, and provides a firm base which can be amplified as and when, and only as and when, new facts emerge which demand to be incorporated. If this principle is not followed then there arise unanswerable questions concerning the criteria for including or excluding other factors. Adherence to the minimal necessary explanation or description ensures that the examination of hypotheses and evidence for or against them will remain coherent.

Ockham's boldness of expression and his incursions into papal politics got him into trouble with the Church authorities in his day. Although he made frequent protestations that he was ready to submit to the authority of the Church, his thinking followed a tradition which was gathering strength and was to become increasingly important, namely that reason and faith were separate, and that unaided reason could not arrive at a demonstration of the existence of the being the Church called 'God'.

Finally, Ockham, though writing within an Aristotelian framework, has been noted as part of the movement which eventually overthrew Aristotle's physics and the world picture embodied in it. Duhem finds in Ockham's theory of motion an early formulation of the idea of inertia (where Aristotle's theory requires, but fatally cannot find, a continuously-acting local cause of motion if a moving body is to remain in motion). This problem of motion was to be one of the major issues in the Scientific Revolution which occurred in the 17th century, after what was, relatively speaking, a hiatus in the development of science in the previous two centuries. It was during the 15th and 16th centuries, however, that social and intellectual changes took place which created the conditions for the spectacular upsurge of science from 1600: exploration of the limits of the known world, the intellectual excitement of the Renaissance, the anti-authoritarian ideas of the Reformation, and the development of technologies which led to the availability of much improved scientific instruments as well as the increased availability of printed books.

The Scientific Revolution

What has been outlined so far consists in the main of developments which with hindsight can be seen to presage the modern outlook on science. The events which provide us with *the particular kind of hindsight* we now use when we look back at Greek and medieval science are those of the 'Scientific Revolution'. It was truly a revolution mainly because it shattered a *Weltanschauung* and created a new one, the one which has created the world of the 20th century. Butterfield (1949) could hardly put it more forcefully:

> Since that revolution overturned the authority in science not only of the middle ages but of the ancient world—since it ended not only in the eclipse of scholastic philosophy but in the destruction of Aristotelian physics—it outshines everything since the rise of Christianity and reduces the Renaissance and the Reformation to the rank of mere episodes, mere internal displacements, within the system of medieval Christendom. Since it changed the character of men's habitual mental operations even in the conduct of the non-material sciences, while transforming the whole diagram of the physical universe and the very texture of human life itself, it looms so large as the real origin both of the modern world and of the modern mentality that our customary periodisation of European history has become an anachronism . . .

And what makes this period of revolution unique in the long history of the development of science is the fact that it was at this point that the new world picture was accepted, not just by exceptional individuals, but by sufficient people for there to be henceforward a steady flow of educated men with the ability and opportunity to mount a steady and co-ordinated search for the general principles which the scientist believes underpin the natural order. This was new (Whitehead, 1926).

The story of the revolution, and of the philosophical discussions which accompany it and follow from it, is well told elsewhere: see, for example, Butterfield (1949), Boas (1962), Hall (1963), Hampshire (1956), Hollis (1973), Berlin (1956). Only a few themes need be highlighted here in order to bring out the methodological issues. These are: the establishment of the heliocentric model of the solar system through Copernicus and Kepler; the development of mechanics, especially in the work of Galileo; and Newton's synthesis of terrestial and celestial dynamics. In addition it is necessary to note the development of the increasing discussion of the nature of the method of science—seen in Francis Bacon, Galileo, Descartes, and Newton.

Nicholas Copernicus, as a man of his time, was an Aristotelian, accepting that the solar system consisted of crystalline spheres upon which the planets moved in perfect circles, terminating in the sphere of the fixed stars. His work marked the start of the movement which was to destroy Aristotle's model, but this was an outcome which would have horrified him. He was a quiet canon of Frauenberg Cathedral, and was persuaded to publish his manuscript *De Revolutionibus* more than twenty years after he had written it. Koestler, in his account of his life (1959) calls him 'The Timid Canon'. His few simple instruments were even cruder than the technology of the 16th century allowed, and in any case he had little interest in star gazing! But he was dissatisfied with the complication of Ptolemy's earth-centred model, with its myriad epicyclic motions, and sought a simpler scheme. His proposed heliocentric model reduced the number of postulated circles needed to explain celestial movements from eighty to thirty-four, but even so his Aristotelian commitment to the supposedly 'natural' and inevitable uniform circular motion forced him to assume some epicyclic paths. The influence of the new model was slight. It offered a less complicated set of motions than Ptolemy's model, but in return for this required a revolutionary view of the universe, one opposed to the teachings of the most powerful institution of that time—the Church. Its lack of influence is hardly surprising. Martin Luther's comment was that Copernicus was 'the new astrologer who wants to prove that the earth moves and goes round ... the fool wants to turn the whole art of astronomy upside down' (Smith, 1972). Rome did not declare Copernicanism false until 1616, which is perhaps a measure of the general indifference.

Johannes Kepler, who spelled his name in five different ways, recorded in a horoscope he cast for himself that he was conceived at 4.37 a.m. on 16 May 1571. The contrast between the carelessness about his name and the precision about dates tells us much about the man (Koestler, 1959). For Kepler, the language of numbers was the language of the physical universe, and he was able, by prodigious efforts with the planetary observations of Tycho Brahe to produce a mathematical account of planetary motion in which no error was found for 200 years. The Danish astronomer Tycho Brahe was a great observer but a weak mathematician. Kepler—who had defective eyesight—was the reverse. For a time he was Tycho's assistant, and after Tycho's death in 1601 took possession of his astronomical papers. Kepler, who started his career by publishing astrological calendars, believed, with the Pythagoreans, that God must have created the universe according to some simple numerical pattern. This was a passionate belief which was at core mystical and aesthetic. Driven by it, Kepler sought Nature's pattern in relations between the radii of planetary orbits, using Copernicus's heliocentric model. Then he turned to geometry as the key to the pattern, and explained the planetary motions as being defined by a nest of the Pythagoreans' five regular solids. (Tycho Brahe thought this ingenious, but offered some modern-sounding advice: 'first ... lay a solid foundation ... in observation, and then ... try to reach the causes of things'.) Later he tried to make planetary orbits correspond to musical notes. Eventually, by the monumental step of abandoning the hitherto unquestioned belief that Nature must be organized according to uniform circular motion, regarded as 'perfect', Kepler was able to reduce to order the chaos of astronomical data and to summarize planetary motion in three simple mathematical laws which assumed elliptical orbits for the planets. The idea of a mechanical universe emerged, one which operated according to laws which could be expressed mathematically, laws which, for Kepler, God had created and man could discover.

If Kepler was a prophet of the scientific revolution, the first major revolutionary might be taken to be Galileo. He was a contemporary of Kepler, being eight years older, and was well aware of the importance of Kepler's astronomy. In 1610, when he was making discoveries through the use of the telescope he had made for himself (he discovered sunspots and the moons of Jupiter) he wrote 'I am anxiously waiting for what il Signor Keplero may have to say about the new marvels'. But the two men never actively collaborated.

Galileo was born into a noble but poor family in 1564. His father insisted that he study medicine but by the 1580s had agreed to his son's abandoning medical studies in favour of mathematics. So began one of the most remarkable lifetimes of intellectual effort the world has ever seen. From the age of twenty-eight Galileo held a chair at Padua, which had been one of the most important European universities since the 15th century. Padua was a very liberal university, with liberty of teaching and scholarly speculation guaranteed by the state of Venice, most anti-clerical of the Italian states (Randall, 1957). It attracted some of the best minds of its day and was ever involved in controversy. All scholarship depends upon critical discussion, but this is especially true for science; happily (for science) Galileo was temperamentally suited to argument, having an independent mind and a sharp tongue. Galileo's early education was a conventional Aristotelian one, but he was soon challenging Aristotle's physics, which means that he was challenging both the view of the world which derived from the medieval reconciliation of Aristotelianism and Christianity, and the institution whose power was linked to that view—the Church.

Galileo's work in mechanics involved a direct challenge to the Aristotelian view that motion needed a force to maintain it. In the case of an arrow flying through the air it had to be argued that the air pushed out of the way at the front somehow rushed round to the back of the arrow to provide the impetus. Galileo rolled steel balls down a groove in a sloping plank and timed their descent. He verified that the speed of fall increased uniformly with time: there was operating a law of uniform acceleration. The idea became clear that force does not *produce* motion, but *changes* it to produce acceleration. What has to be explained is not motion but changes in motion; Galileo was close to (although not quite in its modern form due to Descartes and Newton) the idea of inertia, that a state of uniform motion or a state of rest will continue until a force acts, uniform motion being just as natural as rest. Understanding the idea of inertia involves an adjustment in outlook which is fundamental, and Jeans (1947), describing the performing of Galileo's ball-rolling experiments writes: 'It was one of the great moments in the history of science'. In astronomy too, Galileo's contributions were immense. As soon as he heard of the invention of the telescope by Dutch spectacle makers, Galileo constructed one and used it to explore the skies. He was already sympathetic to Copernican views when with his telescope he observed the four satellites of Jupiter circling round the planet just as Copernicus had said the planets circle round the greater mass of the sun. (The British Ambassador, reporting home about Galileo's account of this, wrote: 'the author runneth a fortune to be either exceeding famous or exceeding ridiculous' Bronowski, 1963.) Moreover, the phases of Venus which Galileo observed were exactly those which Copernican theory required. His writings on astronomy culminated with his *Dialogue on the Two Chief Systems of the World*, in which Copernican ideas are ably defended against the doctrines of Aristotle. The book, written in vigorous Italian prose when Latin was still traditional for scholarly works, had a considerable effect in changing the picture of the universe held by educated men; and it also, of course, led to Galileo's trial by the papal authorities in 1633, after which, having been shown the instruments of torture, he was required to testify that he abjured, cursed, and detested his errors and heresies, namely his 'false opinion that the sun is the centre of the world and immovable, and that the earth is not the centre of the world and moves'. After 1633 Galileo's scientific work was on non-controversial problems and was carried out under conditions of house arrest in his villa near Florence.

Galileo's contribution was not simply an addition to knowledge; it was a change both in the conception of the universe and of the way this might be discovered. Aristotle's picture of the universe could not survive Galileo's demonstrations, and the fact that his demonstrations could be tried and tested by other people constituted a major step in the evolution of the method of science. Although it has been argued that he was more a rationalist thinker than an experimentalist, in that he did not automatically select designed experiments as a means of investigating nature (Butterfield, 1949; Santillana, 1956; Hall, 1963), Galileo wrote in 1615, in a letter to the Grand Duchess Christina of Tuscany:

> Methinks that in the discussion of natural problems, we ought not to begin at the authority of places of scripture, but at sensible experiments and necessary demonstrations.

And Galileo's own accounts of his experiments make them public knowledge which can then be tested by anyone else, in the modern manner. The accounts have a positively heroic air:

> ... having performed this operation and having assured ourselves of its reliability, we now rolled the ball only one quarter the length of the channel; and having measured the time of its descent, we found it precisely one half of the former. Next we tried other distances, comparing the time for the whole length with that for two thirds, or

three fourths, or indeed for any fraction; in such experiments, repeated a full hundred times, we always found that the spaces traversed were to each other as the squares of the times . . . .

This particular description illustrates the other great contribution which Galileo made. Kepler had reduced the great mass of astronomical observations to three simply expressed mathematical laws, and Galileo was now doing the same in a demonstration or experiment. He was in fact doing what comes 'naturally' to modern physical scientists—conceptualizing a problem in an abstract world of nevertheless measurable 'primary' qualities (as opposed to 'secondary' qualities such as taste, smell, etc.). This was the new method which, in the hands of Newton, was to complete the scientific revolution.

Galileo, in spite of performing some crucial experiments, may not have made a principle of the need for experimental investigation of the natural world, but his contributions to the methodology of science were a part of the steady growth of the concept of deliberate experimentation. We may note two important instances of this: William Gilbert's work on magnetism and William Harvey's work on the workings of the heart. Gilbert, Queen Elizabeth's personal physician, was intellectually a man of his time. He believed in astrology, and was ready to accept animistic interpretations of nature. But his fascination in the phenomenon of magnetism, and his practice of talking with the technologists of his day, the miners, navigators, and instrument makers, led to an approach to practical experimentation which now seems modern. In writing *de Magnete* in 1600, six years before Galileo's first publication and five years before Francis Bacon advocated an experimental approach in *The Advancement of Learning*, Gilbert produced the first book on natural science based on observation and experiment (Zilsel, 1957). He is sceptical of authority, tests anything reported by others, and derides superstitious ideas, such as the belief that the lodestone's power could be destroyed by garlic. Cardanus had written that 'the magnet feeds on iron'. Gilbert refuted this hypothesis by magnetizing a piece of iron, storing it with iron filings, and weighing the filings periodically. Zilsel sees as a decisive event in the history of science 'the social rise of the experimental method from the class of manual labourers to the ranks of university scholars'; certainly there is a technological, rather than a philosophical air about Gilbert when he is writing on magnetism. For example:

We had a twenty pounds' heavy lodestone dug and hauled out after having first observed and marked its ends in its vein. Then we put the stone in a wooden tub on water, so that it could turn freely. Immediately the surface which had looked to the North in the mine turned itself to the North on water.

What Gilbert misses is only mathematical expression of the phenomena investigated.

We have noted earlier that the first insistence upon observation came from the Hippocratic physicians, and it is perhaps not surprising to find medical teaching making a further major contribution to the Scientific Revolution in the 17th century. William Harvery wrote *On the Motion of the Blood* in 1628, twenty-five years after his medical studies at Padua. It was at that remarkable university that Harvey absorbed the experimental tradition; there, medicine rather than theology was 'queen of the sciences', and a chair had been held by Vesalius, founder of the modern subject of anatomy. It had become the practice in the 16th century for the professors to read from the ancient texts while an assistant simply pointed appropriately to a cadaver. Vesalius wielded the knife himself, and this tradition persisted at Padua. When Harvey came to write his masterpiece he declared that he both learned and taught anatomy 'not from books but from dissection' (Butterfield, 1949). He treated the heart as a machine, and, applying Galileo's principle of measurement, calculated from its dimensions and pumping rate that in an hour it threw out 540 pounds of blood! The body's blood could only be like a stage army, constantly circulating, and with this discovery Harvey made possible a new start in the biology of living creatures.

Significant contributions to the methods of experimental science were also made by the greatest of the scientists of the 17th century, Isaac Newton, but his contributions were so great that it is necessary to consider them as a whole. The new picture of the universe which the development of science produced was essentially Newton's picture, just as the picture it replaced was essentially Aristotle's. The events which Butterfield describes, in the quotation given earlier, as 'the real origin both of the modern world and of the modern mentality' may be summarized as the acceptance by educated men of the new *Weltanschauung* which Newton created out of the two great strands of 17th century physical science, Kepler's astronomy and Galilean mechanics. Sir Isaac's achievement did not go unnoted. The doyen of empiricist philosophers, John Locke, wrote that in an age which produced such masters as 'the incomparable Mr Newton' he was happy to be employed as an under-labourer in 'removing some of the rubbish that lies in the way of knowledge'. And Alexander Pope, in one of his heroic couplets declaimed:

Nature and Nature's Laws lay hid in night:
God said, Let Newton be! and all was light.

Newton was born in Lincolnshire, posthumous son of a yeoman farmer, in the year in which Galileo died, 1642. He was what is now known as a 'late developer', showing no particular distinction at school or as an undergraduate. In 1665 when Cambridge University was closed because of fear of the plague, Newton returned home to Woolsthorpe and there started his remarkable career, doing original work on mathematics and optics as well as working out an early formulation of the law of gravitation. None of this work was published at the time but his teacher, Barrow, recognized a pupil more able than himself. When he resigned his Chair in 1669 he secured it for Newton, then aged twenty-seven and this position gave Newton the freedom to follow his own intellectual inclinations. (He was required to give only one lecture a week, and frequently had no audience for that!) He tackled major problems of science, making major mathematical inventions in so doing, but seemed content to do this for his own personal satisfaction. As a major biographer says: 'while he made inventions of first rate importance, he made them for his own use; he rarely developed them systematically and had no desire to publish them' (More, 1934). Later on, his epoch-making published work led him into bitter public controversies and he seemed to live suspended between a hunger for honours and an autocratic fear of criticism (Karp, 1972). The honours came; he became Master of the Mint, President of the Royal Society, was knighted by Queen Anne, and when he died in 1727 was buried in Westminster Abbey. Many of his last years he devoted to theological and alchemical studies, leaving behind long unpublished manuscripts on these subjects.

When Newton was at Woolsthorpe during the plague year he took up an idea with which Kepler had toyed, namely that there was some force acting between the sun and orbiting planets. Kepler thought of it as analogous to magnetism. Newton applied the idea to the motions of earth and moon, invented calculus as a mathematical tool for dealing with the problem, and calculated fairly accurately the period of the moon using an inverse square law of gravitation. He published none of this at the time but nearly twenty years later, urged on by the astronomer Halley, began to write out the mathematical proofs. The resulting book *Mathematical Principles of Natural Philosophy*, the *Principia*, is undoubtedly the most celebrated scientific work ever written. In it Newton sets out the definitions and concepts to be used (and states the three laws which underpin classical dynamics); discusses the motions of bodies in a vacuum (providing the basis of celestial mechanics); discusses the modifications introduced by motion in fluids; and in Book Three demonstrates 'the frame of the system of the world'. Book Three shows that the ideas of Book One, applied to the solar system, accurately predict all the known facts about the motion of the planets. Adoption of a universal law of gravitational attraction had led to a single description of the world under a single set of simple laws. Terrestial and celestial mechanics were united, and here was a testable mathematical model of the workings of the universe conceived as an elegant, ingenious, and majestic clockwork. In Buchdahl's summary

(1961):

> This synthesis of empirical data and abstract mathematical relations which here united to lead to accurately verifiable observations, impressed Newton's contemporaries by seemingly bestowing the certainty of mathematics upon man's knowledge of physical phenomena, and gave them a new sense of power over nature.

Newton's 'system of the world' was purposeful only in the sense that it fulfilled the purposes of its architect, God: its features were not explained by deducing them from any purpose (Koyré, 1965). Animistic and teleological explanation were demonstrably no longer necessary, it seemed. (It is an argument of this book that in the last thirty years systems thinking has rehabilitated teleology as a respectable concept.)

The science of the next two centuries may be seen as an application of Galilean–Newtonian methodology to the study of phenomena other than motion: heat, light, magnetism, and electricity, for example, with the aim of formulating mathematically-expressed general laws (Smith, 1972).

Newton was as brilliant an experimentalist as he was theoretician, and he became a methodologist also, as a result of the fierce controversies in which he became involved. In defending his results and methods, and scoring points off opponents, he set down many principles of what became accepted scientific practice, as well as exemplifying them in his approach to experimental work. Thus we find him expounding Rules of Reasoning in Philosophy which are an elaboration of the principle of Ockham's Razor; and writing in his *Opticks*:

> As in Mathematics, so in Natural Philosophy, the investigation of difficult Things by the Method of Analysis ought ever to precede the Method of Composition. This Analysis consists in making Experiments and Observations, and in drawing general conclusions from them by Induction, and admitting no Objections against the Conclusions but such as are taken from Experiments or other certain Truths . . .

Here is a thorough-going commitment to the principle of experimentation beyond that of Galileo. But Newton was as firm as Galileo in his belief in the importance of expressing Nature's behaviour in the language of mathematics. It is the combination of both attitudes, in fact, which makes Newton supreme.

The importance of the designed experiment in science is best illustrated by an example from Newton's work on optics, subject of a bitter dispute with Robert Hooke. It has been known for many hundreds of years that a prism produces coloured light; the general theory was that the prism changed the light passing through it. Newton's careful experimentation led to his view that the action of the prism was physically to *separate* the light into its components. He then designed a *crucial experiment* in which a second prism was introduced into the refracted beam. It caused no further dispersion, but simply refracted the incident light by the same amount as the first prism without change of colour. This experiment was a test which refuted Hooke's theory, a test which Newton's survived. Newton 'had shown himself to be a new kind of experimenter, who understood how to form a theory and to test it decisively against alternatives' (Bronowski, 1973).

These practical contributions to methodology came at a time when the human activity of science was sufficiently well appreciated for its principles of method to be a subject for scholarly debate. The methodology of the activity had already been the subject of serious exposition by, in particular, Francis Bacon and René Descartes at the time when Newton spoke from his experience. Bacon (1561–1626) was not himself a practising scientist, but he had a keen sense of the power of science to transform the material conditions of life if positively directed (Farrington, 1949). Men must consult Nature not books, he argued, and in his manifesto-like works he laid down the programme to be followed. The basic need was for designed and directed experiments which were recorded and reported, so that technical progress might be cumulative. History has shown that Francis Bacon was right about the power of science to transform the physical world. But he was wrong in his estimation of the time needed to do it—he thought it might take a decade or two; and intellectually he failed to grasp the crucial contribution which mathematics was to make to experimentation, his experiments being information-collecting exercises followed by the generalization of principles by induction. Nevertheless he was a prophet of the exploitation of science, and he influenced his contemporaries' readiness to experiment.

René Descartes, whose insistence upon deductive chains of reasoning—on the pattern of geometry—complements Bacon's advocacy of induction from collected facts, is the most important figure in the development of modern science apart from Newton. In the 18th century they were taken to be rival physicists offering accounts of the mechanism of the universe (Koyré, 1965, reviews the dispute) but we no longer see them in that light. Newton demolished the content of Descartes's physics, with its vortices of particles filling the entire universe, and today the latter's scientific importance lies in his work on co-ordinate geometry, and in his being the lucid exponent of scientific rationalism, the *methodologist* whose principle of reductionism has deeply permeated science for 350 years. (The Systems Movement, in fact, may be seen as a reaction against just this principle.) Newton was taught at Cambridge by teachers who accepted Descartes's general thesis about the material world (Ronan, 1969) and was obviously much influenced by the idea of a mechanical model of the workings of nature, and by Descartes's insistence on rational argument expounded in terms of clear, well-defined concepts.

Descartes, a member of the minor French nobility, was educated at a Jesuit college which taught literature and sciences as well as theology. He quickly revealed his independence of mind and his determination to work things out for himself. After a few years as a soldier he decided to devote himself to the problems of mathematics and physics, and wrote in 1619 that he wished to establish 'an absolutely new science enabling one to resolve all questions proposed . . .' But he was a cautious as well as an intellectually self-confident man, and when his most distinguished contemporary, Galileo, met the wrath of the church, Descartes, living quietly in liberal Holland, decided not to publish his major scientific work *De Mundo* which included the Copernican theory of the rotation of the earth. A few months later, however, he decided to publish some samples of the work, and wrote, in the vernacular, a short preface to them, the *Discourse on Method*; Butterfield calls it 'one of the really important books in our intellectual history', which it undoubtedly is, although at Descartes's death in 1650 the 300 copies for general distribution were still not exhausted! (Sutcliffe, 1968). Descartes emphasized not the facts of science but the scientific way of thinking. He rejected the untested assumptions of scholastic philosophy and sought the truth by careful deductive reasoning from basic irreducible ideas. It is exactly the opposite approach to that of Francis Bacon. Descartes starts from a position of extreme scepticism, of absolute doubt. The world we perceive, for example, might be a dream. He decides that the one certainty is that *he doubts*; and this remains true even if he doubts his doubt! Hence he arrives at the most celebrated proposition in Western philosophy: *Cogito, ergo sum*, 'I think, therefore I am'. That this is the only certainty does not deter him. No, he thinks that by analysing the process by which he has become certain of his own existence he can discover the general nature of the process of becoming certain of anything (Pritchard, 1968). In the Second Discourse, Descartes gives four rules for 'properly conducting one's reason'; the first covers the avoidance of 'precipitancy and prejudice', the acceptance only of clear and distinct ideas; the third requires an orderly progression from the simple to the complex; the fourth calls for complete analysis, with nothing omitted. It is the second rule which is most significant, however, since it encapsulates a prime characteristic of the scientific way of thinking as it has been practised for three centuries:

> The second (was) to divide each of the difficulties that I was examining into as many parts as might be possible and necessary in order best to solve it.

Here is the principle of analytical reduction which characterizes the Western intellectual tradition.

Most of the literature on Descartes concentrates on his position as the founder of modern philosophy, the rationalist complement to the empiricist tradition stemming from John Locke. But a recent study by Rée (1974) usefully pays more attention to Descartes's physics and the thinking underlying it. The core of Descartes's approach to science, he points out, was 'reductionist', in the sense that 'science should describe the world in terms of "simple natures" and "composite natures", and show how the latter could be reduced to the former". The process of identifying the simple natures in complex phenomena was what Descartes meant by 'analysis', and he excluded from physical science any explanation in terms of purpose. Rée argues that 'although [Descartes] did not take much interest in organised scientific observation or experiment, his reductionist "mechanical philosophy" was an integral part of the seventeenth-century revolution in physical science', and concludes

> The influence of his reductive ideal on the development of science has been so enormous that it justifies even his most boastful claims about the importance of his work. Newton took over Descartes' belief that the purpose of science was to reduce everything to the 'universal qualities of all bodies whatsoever'. And it was in the spirit of Descartes that he wrote dismissively of the things to which 'the Aristotelians gave the Name of occult Qualities'.... Similarly, in advocating the 'investigation of things by the Method of Analysis' he echoes Descartes in speaking of it as moving 'from particular Causes to more general ones, till the Argument end in the most general'.

This seems a fair judgement: the reductive ideal can be seen in virtually all science of the 18th and 19th centuries. Not until the 20th century have significant challenges to reductionism been made; it will be argued in this book that the Systems Movement is the most serious of these challenges.

The revolution brought about by men like Copernicus, Kepler, Galileo, Newton, Bacon, and Descartes gave men a new picture of the universe which replaced the medieval world-view, and a method of investigating Nature which demonstrably worked. It is not necessary to review the exploitation of that method in the centuries which followed. In fact, in completing this sketch of the development of science, it is necessary to include only a single development from more recent times; but that is a very important one for what it teaches about the nature of scientific knowledge. Newton's results came to be seen not as a plausible model of the workings of the physical world but as an unshakeable account of how the physical world *actually did work*. It is thus especially significant that the main development in science in the 20th century has been the downfall of Newton's model and its replacement by that of Einstein, the latter being preferred because it can yield all of Newton's results and more besides.

Newton had assumed that space provided an absolute framework, stationary and immovable, within which physical bodies moved. With the development of wave theories of electromagnetic radiation a postulated 'ether' was the carrier of the waves; it provided the fixed frame of reference which Newton's cosmology required. When the crucial Michelson–Morley experiment in 1887 failed to detect any difference in the velocity of light such as would be caused by the earth's passage through the ether, classical physics was thrown into confusion. It was suggested that the result would be explained if moving objects contracted in the direction of their motion but this seemed a far-fetched notion. Although at the time he proposed his first theory of relativity (1905), Einstein did not know of the Michelson–Morley experiment (Bernstein, 1973), his theory required just this, for in that theory the relativity of time and distance makes the velocity of light a fixed universal constant and, to accommodate that, moving objects must shrink. Descriptions of mechanics require mass as well as distance and time, and it too is predicted to vary with its motion; and since motion is a form of energy, energy and mass are also interchangeable, the relationship being the famous equation $E = mc^2$ (c being the velocity of light), an equation which is unhappily confirmed in every nuclear explosion. Later extension of the relativity ideas to gravitation and inertia, in the general theory of relativity, provided a formulation in which the universe is not independent matter located in independent space and time, but is a variable four-dimensional space-time continuum. No one can form a clear *picture* of this model in his mind, and relatively few can follow Einstein's physics, but everyone can understand that from it Einstein made predictions which were publicly tested. The planet Mercury is close to the sun and travels very fast. There are anomalies in its motion which Newtonian physics cannot explain. Einstein's laws give calculations in perfect agreement with observations of the planet. Even more dramatically, Einstein's theory predicts that light rays will be

Table 1b   Some thinkers and experimenters important in the development of science

2   Medieval Science

Avicenna	980–1037	Greatest of the early Muslim philosophers. Systematic writings on medicine.
Averroes	1126–1198	A long series of commentaries on the works of Aristotle. The recovery of Aristotelian science.
Grosseteste	c.1169–1253	Discussion of the methodology of inductive science. Experimental work on optics.
Roger Bacon	c.1214–c.1294	Proposed a theory of experimental science as a way of establishing truth.
William of Ockham	c.1300–1349	An extreme nominalist. Principle of Ockham's Razor: Do not multiply entities unnecessarily.

3 The Scientific Revolution and Beyond

Copernicus	1473–1543	Suggested a heliocentric model of the universe which reduced the complications of earlier earth-centred models.
Gilbert	1540–1603	An important emphasis on an experimental approach to practical problems (e.g. navigation).
Bacon	1561–1626	Declaimed the power of experimental science to give power over the material conditions of life and urged the method of experimentation.
Galileo	1564–1642	A major challenge to the Aristotelian world picture through work on mechanics. Experimental demonstrations expressed in the language of mathematics.
Kepler	1571–1630	Brahe's atronomical observations reduced to three mathematical laws. Number seen as nature's language.
Harvey	1578–1657	Experimentation and practical observation in medical research yields the theory of the circulation of the blood; the conclusion backed by calculation.
Descartes	1596–1650	The methodology of rationalism established; reductionism as the aim of scientific explanation.
Newton	1642–1727	The culmination of the move to experiment and mathematical expression started by Galileo. The method unites terrestial and celestial mechanics and finally demolishes the Aristotelian picture of the world. Principles of experimentation established.
Einstein	1879–1955	Establishes a new world picture which explains observations more completely than Newton's, and which survives stringent tests.

bent when passing close to a massive body. In 1919 an eclipse of the sun allowed the apparent position of stars which we see as situated close to the sun to be photographed. Einstein had suggested the experiment, and of course had staked his theory on the result. The observed deflections were in very close agreement with Einstein's predictions. It is public tests of this kind which confirm Einstein's model as better than Newton's, even though for terrestrial calculations and, indeed, for moon flights, Newtonian calculations are accurate enough.

The lesson for science from this 20th century experience is thus that the results from scientific work are never absolute, but may be replaced in time by better models having greater descriptive and predictive power. When the results from the 1919 eclipse experiment became known

Professor Littlewood sent an excited note to Bertrand Russell: 'Dear Russell, Einstein's theory is completely confirmed'. It would have been better if he had written 'Einstein's theory has survived this severe test'. Scientifically acquired and tested knowledge is not knowledge of reality, it is knowledge of the best *description* of reality that we have *at that moment in time.*

The Method of Science

If systems thinking and 'a systems approach' are serious, if they are more than a temporarily fashionable piece of claptrap—and I believe they are —then it is necessary to establish what exactly systems thinking is, and what it means to adopt a systems approach to a problem. It is in order to do that that I have found it necessary to outline the development of science, the human activity which is 'the origin of the modern world and the modern mentality' (Butterfield's words) and within which the Systems Movement has emerged during the last thirty to forty years. In order to see systems thinking as a complement to the main mode of thinking in science, it is necessary to summarize the essential elements in the activity of science and then to see where systems thinking includes or excludes them: hence we arrive at a view of the Systems Movement in relation to the Science Movement which spawned it.

In the previous sections of this chapter I have traced the emergence of science as a consciously recognized, organized human activity. As such it is itself 'a system'. It is an institutionalized set of activities which embody a particular purpose, namely the acquiring of a particular kind of knowledge (Bernal, 1939; Ravetz, 1971; Skair, 1973). Science is an *enquiring* or *learning* system, a system to find things out about the mysterious world we find ourselves inhabiting. The crucial characteristics of this learning system derive from its history. The Greeks contributed the invention of rational thought, breaking with the idea of the irrational authority who is not to be questioned; the medieval clerics started the conscious development of methodology and provided the beginnings of the experimental approach; and the age of Newton united empiricism and theoretical explanation in a way which 'dealt with necessity and contingency at the same time', which 'made the real world comprehensible through ideas' (Hall, 1963). The 20th century has reminded us that the knowledge gained is always provisional. These are the strands which unite in science as we now perceive it, and our account of that activity as a whole would be in some such terms of these: science is a way of acquiring publicly testable knowledge of the world; it is characterized by the application of rational thinking to experience, such as is derived from observation and from deliberately designed experiments, the aim being the concise expression of the laws which govern the regularities of the universe, these laws being expressed mathematically if possible.

In making this outlook a reality, a particular pattern of human activity has evolved and been institutionalized. The professional role 'scientist' has been created, and within the institutional framework scientific work can be initiated, carried out, and appraised. This particular pattern of human activities can be summarized, I suggest, in three crucial characteristics, each of which can be traced back to the history of the development of science. The three characteristics which define the pattern of activity are reductionism, repeatability, and refutation. We may *reduce* the complexity of the variety of the real world in experiments whose results are validated by their *repeatability,* and we may build knowledge by the *refutation* of hypotheses.

There are three senses in which science is 'reductionist'. Firstly, the real world is so rich in variety, so messy, that in order to make coherent investigations of it, it is necessary to simplify it, to select some items to examine out of all those which could be looked at. To define an experiment is to define a reduction of the world, one made for a particular purpose. Secondly, as William of Ockham emphasized, there is much to be gained in logical coherence by being reductionist in explanation, accepting the minimum explanation required by the facts to be explained. Thirdly, more broadly, the scientific outlook has absorbed deeply Descartes's advice to break down problems and to analyse piecemeal, component by component. 'Scientific thinking' is almost synonymous with 'analytical thinking' in this sense.

Experiments are a special kind of observation. Initially in science, as we have seen, the problems were those of cosmogony, and the observations relevant to the reasoning were the common facts of everyday life. Later the idea of the contrived experiment emerged, being relevant to a more detailed examination of the workings of nature. In such experiments the experimenter, by means of his reduction, aims at a complete control over the investigation, so that the changes which occur are the result of his actions, rather than the result of complex interactions of which he is unaware. Given this control, questions can be asked of nature. In a section on 'Observation and Experiment' in his discussion of inductive thinking, Mill (1884) explains

> the first and most obvious distinction between Observation and Experiment is, that the latter is an immense extension of the former. It . . . enables us . . . to produce the precise *sort of* variation which we are in want of for discovering the law of the phenomenon. . . . When we can produce a phenomenon artificially, we can take it, as it were, home with us, and observe it in the midst of circumstances with which in all other respects we are accurately acquainted. . . .

When Newton made a small hole in the blind in his room in Trinity College and passed the resulting beam of light through his prism, when

Galileo rolled polished wooden balls down the groove in his inclined plank, they were 'taking the phenomena home with them' in order to investigate these narrow selections from the world's variety. That is the reductionism of experimentation. It applies equally to 'Baconian' experiments of a fact-gathering kind and to 'Galilean' experiments (the distinction is Medawar's, 1967, 1969) which test hypotheses or distinguish between possibilities.

The reductionism of explanation is clear enough in the principle of Ockham's Razor, but it is an extension of this which leads to the concept most usually thought of as 'scientific reduction', namely the explanation of complex phenomena in terms of simpler ones. Obviously if we can do this—if biological phenomena, for example, can be wholly explained in terms of physics and chemistry—the principle of the Razor has been met: we have not multiplied entities unnecessarily. The reductionist ideal would be an explanation of social science in terms of psychology, of psychology in terms of biology, of biology in terms of chemistry, and of chemistry in terms of physics, most basic of the sciences. (This is the ideal which underlies Lord Rutherford's famous remark: 'There is physics and there is stamp collecting'.) Longuet-Higgins (1972), seeking to establish that the phenomena of the mind are in fact irreducible to neurophysiology or behaviourist psychology, gives a satirical account of the thinking of the reductionist, who 'notes with satisfaction that, after many centuries of illusory independence, chemistry has been fitted into the framework of physics'. What is more, the recent successes of molecular biology show that 'biology is really chemistry', and surely neurophysiology is really molecular biology, and psychology is really neurophysiology? And so on. 'Sociology is really psychology, economics is really sociology, history is really economics, and there the trail becomes indistinct.' Obviously reductionism in this sense is a perfectly proper aspiration for science, and when, for example, the work of Maxwell and Boltzmann allows the laws describing the thermodynamics of gases to be derived from the assumptions of mechanics, there is a proper feeling that progress has been made. What is most interesting, however, is the very fact that the reductionist ideal is expressed in terms of a hierarchy of the sciences—physics, chemistry, biology, psychology, social science—which is intuitively convincing to everyone. No one would ever argue that the place for psychology lies between chemistry and biology. It seems convincing to everyone to describe the knowledge we have of the world in terms of different *levels of complexity*. Laws which operate at one level seem to be higher order with respect to those of lower levels. This is the kernel of the concept of 'emergence', the idea that at a given level of complexity there are properties characteristic of that level (*emergent* at that level) which are irreducible. The doctrine of emergence is closely related to the idea of reduction and, as we shall see in the next chapter, the debate of reductionism–versus–emergence is a prime source of the thinking which became generalized as 'systems thinking'.

The second major characteristic of the learning system which is science is the repeatability of the experiments. This is a crucial characteristic which puts any knowledge which may properly be called 'scientific' in a different world from, say, the literary knowledge which is embodied in books of literary criticism. A critic who wishes to convince us that D. H. Lawrence is a great novelist will explain *why* he thinks this. He will propound his criteria for making value judgements, analyse Lawrence's novels according to these criteria and so try to influence our opinion of the books. We may find his arguments convincing and agree with him, or we may not. Whether we do or not will be dependent upon our tastes and feelings, both of which may change with time, as may the critic's also. Even if there were to be a general consensus among educated people and literary critics that Lawrence *is* a great novelist, this would tell us something about literary tastes in a particular society at a particular time rather than something about the novels themselves. In this example we could of course replace 'literary knowledge' with 'appreciation of music', 'hortatory statements about religion or politics', or many other kinds of knowledge. Knowledge of this kind remains 'private knowledge' in the sense that the choice is ours to accept it or not. On the other hand, scientific knowledge is 'public knowledge' (Ziman's definition, 1968); we have no option but to accept what can be repeatably demonstrated in experiments. The inverse square law of magnetism, discovered, say in Boston, is found still to be an inverse square law when the experiments are checked in Basingstoke. This is one of the great and crucial strengths of science, although it is often misunderstood. It is important to realize what it is that 'has to be accepted': it is the *happenings in the experiment,* and it is only that. Every single opinion about the experiment, or the theory which makes it meaningful, may be disputed. No one has to agree with any interpretation of the results, such interpretations being the same kind of statement as those made in literary criticism. But if the experimental happenings can be checked and found to be repeatable by other disinterested people, then they count as 'scientific'. Although you may dispute the interpretations I put upon, say, some experiments I say I have carried out with magnets and iron filings, if I hypothesize that magnets attract iron filings because they are made of iron rather than because of their shape, you may repeat my experiments with iron and with plastic filings of the same shape and see for yourself. We may continue to argue about interpretation but we shall be forced to agree about the happenings in the experiment, which are repeatable. The example is trivial but the principle is not. It is the repeatability of the experimental facts which places this knowledge in a different category to opinion, preference, and speculation. It gives the activity of science a solid core which is unaffected by the irrationality, the emotionalism, and the foolishness of human beings—including scientists, who are not less human than any other group. We must not expect human beings to behave entirely rationally; they will not. But science has, through the public knowledge embodied in repeatable experiments, a means of insulating itself from the consequences of human folly which is not available to any other kind of knowledge.

Consider as an example a speculative area of biology which fascinates the layman: the possibility of creating living material in the laboratory. There is probably now a consensus among biologists that the present knowledge of molecular biology shows that this is in principle possible (Monod, 1972; Jacob, 1974). This consensus is not, in the strict sense, scientific. The creation of living material from non-living starting materials in the laboratory will become a scientific fact when it has been done in experiments which are shown to be repeatable in other laboratories. At present the general view is that life could have evolved on this planet by known chemical mechanisms and that the way to create life in the laboratory is to follow the same sequence. What is needed is, firstly, the formation of organic molecules of moderate complexity from chemically simple starting materials, secondly, the formation of long chain molecules (polymers) from the organic molecules, and thirdly, the collection together of appropriate polymers in a physical situation which allows development of a particular internal (cell-like) organization. At present the first two stages are scientific fact (Bernal, 1967; Orgel, 1973; Miller and Orgel, 1974). Miller, for example, exposed mixtures of simple molecules to electrical discharges and obtained mixtures of the kind of organic molecules involved in life: amino acids, sugars, etc. Butlerov and Katchalsky among others, have shown that high yields of organic polymers, required for the second stage, can be obtained from organic molecules with surprising ease, in Katchalsky's case by bringing certain amino acids into contact with a common clay mineral in an aqueous environment. Concerning the third stage, not so far accomplished in the laboratory, Miller and Orgel write

> we are convinced that natural selection, acting on a system of polymers (some of which are able to replicate) was responsible for the emergence of organised biological structures.

But the distinction is clear: the first two stages are scientific fact, the third is speculation which will become accepted science when it has been carried out and shown to be repeatable in other laboratories.

Also connected with the repeatability criterion for science is the importance of measurement. Measured values can be recorded, and repeated, more easily than qualitative findings. Whether a repetition has been achieved or not is much clearer if the experimenter is dealing with measured quantities, quite apart from the fact that measurements, standing for properties or relationships, enable mathematical theory to be used to deduce all the logical consequences of the information embodied in the experimental results (Richie, 1945). Hence the potentially most powerful scientific facts are those expressed as the quantitative results of experiment.

The third major defining characteristic of science is that cumulative progress may be achieved as a result of sequences of experiments of the 'Galilean' type in which hypotheses are subjected to testing against experience. What is 'public' about the public knowledge of science is that the results of the (reductionist) experiments are described in a way which enables other people to test them. But the criterion applied to experiments is not only 'Is this repeatable?' After all we are not likely to be very interested in scientific facts which are true but trivial. The logical value system which operates in science has to be one in which one experiment is more valuable than another if it is less trivial, more significant. And the significance comes from the degree to which the experiment provides through its design a difficult test for some hypothesis of wide scope. A good experiment is one in which some significant conjecture is *at risk*. This means that every experiment which is not of a random fact-gathering kind implicitly or explicitly embodies some theory, and that this theory stands to be corroborated or refuted by the happenings in the experiment. We have seen a sensational example of this in Einstein's pointing out that according to his theory light passing near the sun should be bent by it, and that this might be tested during a total eclipse. He was putting his theory of relativity up for public testing in this instance—a kind of testing which literary critics, historians, and scholars of politics do not have to endure! This is a major revolutionary example, but the logic of the progress of science requires that scientific experiments should if possible be of this kind. More modest fact-collecting experiments have a role, of course, but the progress of science will be determined by significant experiments in which significant conjectures stand to be refuted.

Even in a modest fact-finding investigation it will still be true that the experimenter's selection of what to observe—his definition of 'a fact'—implies that his investigation is based upon the implicit acceptance of some body of previous knowledge which makes meaningful his particular investigation. At the present time, for example, no one would carefully test the inverse square law of magnetism when the moon is full and again at the new moon to see if the law is still of that form; this would not be sensible because there is no body of theory which suggests any plausible connection between the two, and much tested theory which suggests that the two phenomena are totally unconnected. Kuhn (1962) refers to the body of currently accepted knowledge which makes particular experiments meaningful as 'a paradigm', and in his very influential book describes the history of science as the history of periods of 'normal' science carried out under the influence of a particular paradigm interspersed by revolutionary shifts in the paradigm. Kuhn's account of a paradigm has it as an achievement or set of achievements which a scientific community 'acknowledges as supplying the foundation for its further practice', achievements which 'attract an enduring group of adherents away from competing modes of scientific activity' and are 'sufficiently open-ended to leave all sorts of problems for the redefined group of practitioners to resolve'. At the highest level, Newton and Einstein were responsible for revolutionary paradigm shifts. In between such shifts what goes on is 'normal science' in keeping with the prevailing paradigm.

Thus we may describe in the following (slightly idealized) terms what happens when a piece of scientific work is planned and carried out—or rather this is the logic of what happens or should happen: a real instance may not follow the logic!† The scientist, as a result of his choice of problem, decides what section of the world's variety to examine. He makes his reduction, designing an artificial situation within which he can examine the workings of a few variables while others are held constant. His experimental design will 'make sense' in terms of some particular view of, or theory about, that part of the world's variety he is investigating, and his particular experiment will constitute the testing of a hypothesis within that theory. The question the experiment poses is: will it pass the test? And the artificiality of the experimental situation is such that, when the results are carefully described, analysed, and interpreted, well-defined critical discussion among interested scientists can take place. Human thought is polemic; new thoughts are necessarily the foes of those they replace, and the experimental approach of science generates, contains, and orchestrates critical debate.

Finally, when a hypothesis is subjected to test in the way described, it must in logic be the case that we are more interested in a refutation than a corroboration. This stems from the impossibility of *proving* anything by induction. With deductive argument there is no problem. We can conclusively deduce that Socrates is mortal, because 'all men are mortal', and 'Socrates is a man'. But there is no equivalent to the syllogism in inductive argument. The fact that every night of our lives has been followed by dawn may cause us to expect that the same will happen tonight—we may be prepared to bet on it—but there is no way to *prove* this, as David Hume pointed out in the *Treatise on Human Nature*; and multiplying confirmatory observations does not, in logic, get us nearer to proof. For this reason a hypothesis refuted is a more valuable experimental result than one in which the hypothesis survives the test; hence the experimentalist ought to be trying to test his hypothesis to destruction, setting it the most severe test he can think of.

---

†Feyerabend (1975), arguing against the accuracy or usefulness of any structured account of science, maintains that the only principle which does not inhibit progress is *anything goes*. This is certainly true with regard to *the creation of hypotheses worth testing* (which a scientist might get from a dream, or a chance remark by his mother-in-law) but beyond that the activity recognizable as 'science' must have some structural characteristics if it is to be the recognizable activity it is. The core of its structure, I am arguing, is the generation of repeatable experimental happenings which represent tests of hypotheses.

This view of the nature of science derives to a major extent from the philosopher C. S. Pierce, who wrote: 'The conclusions of science make no pretence to being more than probable', from the methodologist William Whewell, whose *History of the Inductive Sciences* (1837) saw scientific progress as a continuous refining of necessary truths about the physical universe based on observation and experimentation, and, more recently from the trenchant writings of Sir Karl Popper (1959, 1963, 1972). Popper, who believes that he has solved Hume's problem of induction, describes the method of science as 'the method of bold conjectures and ingenious and severe attempts to refute them', and takes the demarcation between science and other activities which might be mistaken for it (for example, astrology or psychoanalysis) to be the criterion that science must produce conjectures which can be (publicly) falsified. It is sometimes assumed that there is a conflict between Popper's account of science and Kuhn's description of the history of science as being a sequence of periods of 'normal science' according to a certain paradigm and revolutionary periods in which the paradigm is overthrown (Lakatos and Musgrave, 1970; Schilpp, 1974). However, the conflict is more apparent than real. Kuhn's account stems from an historical study of how actual scientists have behaved in the past; Popper's account concerns the logic of the activity. We ought not to be surprised that actual scientists, being human, may often be observed seeking evidence which supports rather than refutes a hypothesis with which they personally identify! Popper's message to them would be: 'Do not be satisfied with normal science; try to find ways of challenging the paradigm'. In fact a number of very distinguished scientists including Medawar, Monod, Eccles, and Bondi have attested to the importance of Popper's ideas in their thinking (Magee, 1973).†

We have a picture of science, then, as a method of enquiring, or learning, which offers us, at any moment of time a picture of our understanding of the world's reality which consists of certain conjectures, established in reductionist repeatable experiments, which have not yet been demolished. The most pleasing image of this activity comes not from a scientist or a philosopher of science but from a novelist, Vladimir Nabokov. In his story 'Ultima Thule', he writes:

> When a hypothesis enters a scientist's mind, he checks it by calculation and experiment, that is by the mimicry and pantomime of truth. Its plausibility affects others, and the hypothesis is accepted as the true explanation for the given phenomenon, until someone finds its faults. I believe the whole of science consists of such exiled or retired ideas;

† We may doubt whether Popper has solved the philosophical problem of induction, since, for example, it is an inductive step to assume, as Popper does, that a theory which has passed a severe test is a better guide to the future than one which has passed only a modest test (Ayer, 1973). But at a practical level, as Eccles *et al.* proclaim, the refutability of hypotheses does provide a cogent demarcation between science and non-science.

and yet at one time each of them boasted high rank; now only a name or a pension is left.

The outline account of science above provides a background which serves two purposes. First, in the next chapter, it will enable the Systems Movement to be seen as a response to certain problems within science. Second, it will be relevant to later discussion of whether or not systems thinking can contribute to the solution of the difficult problems which face a social science which aspires to be scientific in the full sense of that word.

Chapter 3

Science and the Systems Movement

The method of science has turned out to be powerful enough to create the modern outlook; and the exploitation of science in Western technology has largely created the modern world in a physical sense. This ascendancy is due to the fact that the potent combination of rational thinking and experimentation demonstrably works. It provides knowledge different in kind from that derived from consulting the oracle at Delphi, reading the entrails of slaughtered goats, casting horoscopes or turning up the Tarot cards. The fact that in Western societies in the 1970s there is a reaction against science, evidenced, for example, in the polemics of Roszak (1970, 1973) and in the lively business in Tarot cards done in university bookshops, is itself a testament to the dominance of science. The present cult of unreason is a not surprising reaction to the astonishing success of the cult of reason as embodied in modern science, especially as the certain fruits of science and technology are seen to be at a material level only. But the method of science is not all-powerful even at a material level, and there is much to be learnt by examining the limitations of the scientific method not least because an examination of the bounds of science, carried out in the rational enquiring spirit of science itself, might provide a more civilized base for cultural change than that likely to be provided by a mindless cult of unreason.

The crucial problem which science faces is its ability to cope with complexity. Descartes's second rule for 'properly conducting one's reason', i.e. divide up the problems being examined into separate parts—the principle most central to scientific practice—assumes that this division will not distort the phenomenon being studied. It assumes that the components of the whole are the same when examined singly as when they are playing their part in the whole, or that the principles governing the assembling of the components into the whole are themselves straightforward. We now know, on the basis of the success of physical science so far, that these seem to be reasonable assumptions, at least for many of the physical regularities of the universe. Suppose, for example, that we separate out the phenomena of heat from the total complexity of the physical world, and suppose that we take steps to discover in the laboratory the laws governing the transfer of heat. Subsequent evidence suggests that for everyday purposes this separation, or reduction, is legitimate. This is shown by the fact that in the real word, in which the transmission of heat is mixed up with other physical phenomena such as light, sound, and the action of gravitational forces, as far as heat is concerned, the same laws of heat transmission established in the laboratory are obeyed. This finding justifies the initial reduction, and this separability of many physical phenomena is now an unquestioned part of our picture of the world. But if we move beyond the physical regularities of the universe to apparently more complex phenomena—such as those of human society—how to make the separation, and how to know whether in the end it is legitimate to do so, are much harder questions to answer. If the investigation is to concern 'heat transfer', then it is not difficult, we now know, to draw boundaries round the area of investigation and to design experiments; if, however, the investigation is to be into, say, the phenomenon 'voting' these things are much less clear; how to separate the phenomenon, what to leave in, what to leave out, are much harder to define.

The interesting question is: To what extent can the method of science cope with complexity? Where does it fall down, and why? This chapter examines some of the limitations of the method of science as the complexity of the subject matter examined increases. After examining some of the problems of complexity within the physical sciences, attention is drawn to the problems of two other areas rich in complexity: the social sciences and 'problems in the real world' the latter being viewed as problems of 'management', defining that term very broadly. It is as a response to these problems of complexity that systems thinking develops.

Problems for Science: Complexity

Cursory inspection of the world suggests that it is a giant complex with dense connections between its parts. We cannot cope with it in that form and are forced to reduce it to some separate areas which we can examine separately. Our knowledge of the world is thus necessarily divided into different 'subjects' or 'disciplines', and in the course of history these change as our knowledge changes. Because our education is from the start conducted in terms of this division into distinct subjects, it is not easy to remember that the divisions are man-made and are arbitrary. It is not nature which divides itself up into physics, biology, psychology, sociology, etc., it is we who impose these divisions on nature; and they become so ingrained in our thinking that we find it hard to see the unity which underlies the divisions. But given that our knowledge has to be arranged in this way—and it is inevitable given our limited ability to take in the whole—then it is useful in the interests of coherence to arrange the classification of knowledge according to some rational principle. Many possible classifications have been proposed based on a number of different principles, and it would be foolish to expect any one version to be generally agreed, given the different purposes for which the classification may be carried out.

Kotarbinski (1966) points out that a main motivation for devising classifications of knowledge has been the professional concern of librarians for order in libraries! Aristotle is known to have advised on library organization, and the catalogue of the great library at Alexandria ran to 120 chapters, apparently being based on the nature of the authors' roles: poets, lawyers, historians, orators. Francis Bacon proposed a classification as part of his scheme for the organized scientific research which was to transform man's lot. Ampère, having undertaken to teach physics, decided it was necessary to draw the demarcation lines of that discipline and hence proposed a classification of the sciences in order to place physics within it.

For my purposes it is useful to recall the classification of the sciences proposed by Comte in the 19th century. Comte's didactic purpose was to establish a then non-existent science of society which he christened 'sociology'. (See Kolakowski, 1972, for an account of Comte's 'positivist' philosophy and its subsequent history; Abraham, 1973, reprints long extracts from Comte's major works.) Comte's grandiose aim was to establish a uniform organization of the totality of human knowledge in order to provide a base for the new science of sociology, through which it would be possible to transform social life. The classification would of course also form the basis of teaching, and through it students would absorb a coherent picture of science as a whole. Comte's doctrine was that human thought in any subject area passed through three phases: a *theological* phase dominated by fetishist beliefs and totemic religions; a *metaphysical* phase in which supernatural causes are replaced by 'forces', 'qualities', 'properties'; and finally a *positive* phase in which the concern is to discover the universal laws governing phenomena, leading to as near certainty as man can hope to attain. Comte argued that the sciences had all passed through, or were passing through, this sequence. For example, astronomy emerged from the metaphysical fog with the work of Copernicus, Kepler, and Galileo; chemistry passed from its theological phase (when it was alchemy) and achieved the status of positive science in the 18th century; biology in the 19th century turned its back on teleological explanation and explanation through 'vital forces', and began the positive investigation of laws relating to organisms in an environment. This account of the history of the sciences led Comte to place the sciences in a natural order which went as follows: mathematics, astronomy, physics, chemistry, the biological sciences, and finally sociology. This sequence, which was that in which Comte would have them taught, represented the view that each science, more complex than those before it, presupposes the less complex sciences which precede it, but shows its own irreducible laws. The unity of the sciences in this scheme stems from their interdependence as historically differentiated aspects of a single social reality which is to be the concern of his new science of society.

The principles behind the classification are: the historical order of the emergence of the sciences; the fact that each rests upon the one which precedes it and prepares the way for the one which follows; the increasing degree of complexity of subject matter; and the increasing ease with which the facts studied by a particular science may change. (With regard to this latter principle, the 'facts' of social science are constantly changing; some of the facts of biology change with evolution in a relatively short time; but the experimentally determined properties of sulphuric acid and similar chemical facts remain relatively unchanged, as do the facts of physics.) The classification remains, with a few modifications, a useful framework for the examination of some of the difficulties and limitations of the scientific method. The modifications concern the inclusion of psychology (which Comte omitted) as a science linking biology and the social sciences, and the exclusion of mathematics, regarding it not as a science in its own right but as a language which any science may use, and which is more used the lower down the hierarchy we travel. Mathematics is very much the language of modern physics, and chemistry is increasingly mathematical; but in the social sciences mathematics will usually appear in those forms, such as statistics, which have been developed to cope with rather fuzzy measurements and quantities. This gives us a classification of the experimental sciences into, in sequence, physics, chemistry, biology, psychology, and the social sciences. In this sequence physics is the most basic science, being concerned with the most general concepts, such as mass, motion, force, and energy. Chemical reactions obviously entail these phenomena, and are increasingly explained in terms of them, but when hydrochloric acid gas and ammonia mix, and a white solid not previously present settles out, we have a phenomenon which intuitively is more complex than those of physics. And though a biological phenomenon such as the growth of a plant from a seed entails much chemistry (as well as physics), the reproductive ability of the plant again brings in a new level of complexity. Psychology, and the concept of consciousness, bring in a higher level still and social life exhibits yet higher levels.

In the discussion of the previous chapter it is physics which is most clearly successful as a science. It exemplifies most clearly the characteristics of the scientific method, and examination of the main difficulties for this particular method of gaining knowledge becomes an examination of the degree to which reductionism, repeatability, and refutation can characterize the study of increasingly complex phenomena as we ascend the hierarchy. Can the scientific method cope with phenomena more complex than those of physics? Where does it break down?

An interesting example of a science faced with a new complexity occurred in chemistry during the last century. Chemists at that time isolated, purified, and determined the constituents of the complex carbon compounds which where known to characterize living material, the so-called 'organic' chemicals. Such molecules as brucine, $C_{23}H_{26}O_4N_2$, analysed in 1831, were so complex that it was supposed that they could be made only by the intervention of a mysterious 'vital force' which, it was supposed, all living organisms possessed. The vitalist controversy was settled when Wöhler was able to show that a known organic compound, urea (NH_2CONH_2) could be made simply by heating the substance ammonium cyanate (NH_4CNO) which had been made from wholly inorganic starting materials. This transformation was repeatable, and other similar reactions followed. It was soon accepted that the crucial characteristic of organic chemicals was their *structure*, and this episode is a satisfactory piece of reductionism, the experimental findings leading to elimination of the need to 'explain' the complexity by invoking vitalism.

Interestingly, the same vitalist controversy broke out in biology rather later. In the 1880s and 1890s Roux had suggested that hereditary 'particles' in the egg of an organism were divided unevenly during the cell divisions which form the multi-celled embryo from the single-celled egg. Thus different parts of an embryo would carry different hereditary potentialities. Driesch, working at the Zoological Station at Naples, split up the eggs of a sea urchin when they were at the two-cell stage. He allowed each cell to develop separately and expected to obtain deformed embryos. To his astonishment each cell produced a normal, though somewhat small, sea urchin larva, a finding which refutes Roux's hypothesis. Driesch had to conclude that cells had an inherent ability to adapt themselves to varying circumstances; an embryo at an early stage of its development seemed to be a self-adjusting whole. He suggested that a cell had to be regarded as a 'harmonious equipotential system' (Allen, 1975). A few years later Driesch carried out some even more remarkable experiments with other embryos. He showed, for example, that if the future tail of a newt embryo were removed and grafted into a position where a leg would normally grow, the would-be tail in its new position grew into a leg! In the 1920s similar experiments showed that at a later stage of development the parts of an embryo were apparently committed to their development path. Young tissue from the stump of an amputated newt-tail became a leg if grafted into a leg position, but older tissue grew into a tail wherever it was grafted on (Koestler, 1945). These results caused consternation among biologists. If the development of organisms was deterministically controlled by physico-chemical laws, as was firmly believed to be the case, how could these results be explained? It was almost impossible to avoid describing the problem in anthropomorphic terms: how did the tail tissue 'know' that in its new position it should become a leg? How did it 'know' that after a certain point it was committed to becoming a tail, and could 'persuade' its host tissue to behave accordingly? The controversy between vitalists and mechanists which broke out around the turn of the century continued until the 1930s. Hans Driesch and the vitalists believed that results from experimental embryology established that life involved something beyond the physical. Driesch himself despaired of finding a causal-mechanical explanation and argued in *The History and Theory of Vitalism* for an Aristotelian notion that the egg cell contained a vitalistic agency, a mysterious *idea* of the complete organism, which he called an *entelechy* which no amount of experimentation could analyse or describe.

> Driesch's vitalism was a real revival of the old animistic theory of Aristotle, including the concept and the term *entelecheia,* meaning the perfect and complete idea of the organism, which exists before its actual material realisation (Montalenti, 1974).

Driesch actually gave up experimental biology and became a somewhat mystical philosopher. Meanwhile the vitalism controversy reached deadlock, with most biologists regarding the invoking of *entelechy* as a sham solution. Biologists came to accept that the key to explaining living material lay in its degree of organization, and they believed and hoped that the eventual experimental unravelling of the molecular mechanisms in the cells of living materials would provide the explanation for *vital* behaviour. The recent triumphs of molecular biology have to some extent done this (Monod, 1972; Watson 1968). Roux's 'heredity particles' and Driesch's *entelechies* can now in part be equated with 'programmes' which direct development and which are specifically encoded in the sequence of organic bases in the long chain molecules called nucleic acids. There is still much to be learnt about the macro-organization of the development of individual organs in an organism (Pattee, 1970) but biologists accept that in principle the problem is solved, and the solution has been obtained without invoking anything beyond known principles of physical chemistry. So vitalism, which will be finally dead when living material is created in the laboratory, as discussed in the previous chapter, is already a doctrine without adherents.

In the two examples described above we see the sciences chemistry and biology facing new complexities in their subject matter. The reaction is typically scientific: to try to provide explanations for the new unknown in terms of the known. This is a sensible reductionist approach, a wise wielding of Ockham's Razor which should be applauded. In the present instances the approach was successful in the case of chemistry, and is generally believed to be not far from final success in the case of biology and vitalism. Nevertheless these pieces of history draw attention to some interesting problems which remain even after reductionist explanations for the mechanisms of the new phenomena have been found.

The main puzzle which remains is that a new problem, such as that of the increased complexity of organic compared with inorganic molecules, is seen to be *a problem of chemistry* and of the particular level of phenomena with which that science deals. Explaining a phenomenon of chemistry in terms of the physics (the masses, energies, and force fields) of molecules provides an explanation of the observed phenomenon without introducing new concepts, and we should be glad of this reduction; but it does nothing to explain *away* the fact that the phenomenon of chemistry exists—and is capable of being investigated experimentally—at a higher level of complexity than that encountered in physics. Physics can provide an account of the mechanism of some chemical phenomena, but cannot explain the existence of problems of chemistry as such. The puzzle which remains is that of the apparent existence of a hierarchy of levels of complexity which we find it convenient to tackle through a hierarchy of separate sciences. Again, at a level above chemistry, the problems of the development of embryos and of hereditary are problems of biology. Explanations in terms of chemistry, though welcome, do not explain away biology: we are left with a level of complexity which is characterized by its own autonomous problems. Popper (1974) in a section called 'Darwinism as Metaphysics' in his 'intellectual autobiography' seems to be making much the same point when he writes:

> I conjecture that there is no biological process which cannot be regarded as correlated in detail with a physical process or cannot be progressively analysed in physicochemical terms. But no physicochemical theory can explain the emergence of a new problem ... the problems of organisms are not physical: they are neither physical things, nor physical laws, nor physical facts. They are specific biological realities; they are 'real' in the sense that their existence may be the cause of biological effects.

Given the messy richness of 'biological effects' it is perhaps not surprising that biologists, much more than chemists, have been conscious of the fact that an unsolved problem is presented to science by the very existence of a set of phenomena which are higher order with respect to those of chemistry and physics. The existence of the problem of the emergence of new phenomena at higher levels of complexity is itself a major problem for the method of science, and one which reductionist thinking has not been able to solve.

Another aspect of the problem of complexity for science has been discussed by Pantin (1968) who makes a useful distinction between 'restricted' and 'unrestricted' sciences. In a restricted science such as physics or chemistry a limited range of phenomena are studied, well-designed reductionist experiments in the laboratory are possible, and it is probable that far-reaching hypotheses, expressed mathematically, can be tested by quantitative measurements. The wider and more precise the quantitative predictions, the greater the possibility of failure and hence the greater confidence possible in a hypothesis which actually passes a severe test. In an unrestricted science such as biology or geology, the effects under study are so complex that designed experiments with controls are often not possible. Quantitative models are more vulnerable and the chance of unknown factors dominating the observations is much greater. Pantin writes:

> The physical sciences are restricted because they discard from the outset a very great deal of the rich variety of natural phenomena. By

selecting simple systems for examination they introduce a systematic bias into any picture of the natural world which is based on them. On the other hand, that selection enables them to make rapid and spectacular progress with the help of mathematical models of high intellectual quality. In contrast, advance in the unrestricted sciences, as in geology and biology is necessarily slower. There are so many variables that their hypotheses need continual revision.

There is obviously the possibility that the scientific approach based upon reductionism, repeatability, and refutation will founder when faced with extremely complex phenomena which entail more interacting variables than the scientist can cope with in his experiments. The social sciences are all 'unrestricted' in Pantin's sense, and present considerable problems for the method of science. And they introduce a new kind of difficulty beyond that of mere complexity.

Problems for Science: Social Science

Even when the natural sciences face the problem of extreme complexity, the professional scientists concerned are convinced that there is no fundamental issue of principle at stake. Climatologists and meteorologists, for example, deal with large and complex systems involving more variables than can be handled analytically, and it will be a long time, it appears, before meteorologists can tell us whether or not it will rain in Bolton-le-Sands, Lancashire, England, on Thursday next. On the other hand, forecasts of the general weather pattern for England as a whole are now made one month ahead, and are improving in quality. The problems still to be tackled are of a recognized kind, concerning instrumentation, data collection, and data analysis, rather than fundamental issues which might cast doubt on the possibility of accurate weather forecasting. The future of meteorology as a science is certainly not an issue.

This is a very different situation from that in those unrestricted sciences we call 'social'. The status of disciplines like anthropology, economics, sociology, political science, etc. *as sciences* is a question which is still problematic. To an outsider, for example, initial inspection of the literature of sociology gives the impression of its being strong on social findings of the kind which can be collected by asking people to fill in questionnaires, relatively weak on substantive accounts which are the result of sociologists *doing* sociology in the way that chemists *do* chemistry, and relatively strong on discussions of the nature of sociology, what constitutes an 'explanation' in sociology, whether or not a value-free sociology is possible, and similar questions (for some recent British examples see Winch, 1958; Rex, 1961; Cohen, 1968; Ryan 1970, 1973; Emmet and MacIntyre, 1970; Gellner 1973; Giddens 1974, 1976, 1979; Keat and Urry, 1976). It was a sociologist, Cohen, not an outsider, who wrote '. . . there is a depressing tendency for social theorists to discuss the nature of social theory rather than the nature of social reality'. By contrast, the literature of a natural science like chemistry is heavily dominated by substantive contributions within the discipline; simple data collection and discussion of the intellectual status of chemistry are noticeably absent.

This concern with questions *about* rather than *in* the disciplines of social science is an indication not that social science attracts dilettantes, but rather that exceptionally difficult problems arise when the methods developed for investigating the natural world which exists outside ourselves are applied to the social phenomena of which we are a part. The founding fathers of social science were not in doubt that the new science was to be science in the image of the natural sciences. Comte saw the subject he named 'sociology' as the apex of the pyramid of all the sciences. Emile Durkheim, while referring to the new word as a 'barbarous neologism' nevertheless sought to make sociology 'a distinct and autonomous science'. It was assumed that empiricism applied to the facts of social life would yield generalizations and predictions on the pattern of natural science. This has not yet happened, and though a century is not a long time in the history of an idea, it does appear that the hiatus in the substantive development of the social sciences is due to the special kind of difficulties they face. Near the end of his long examination of the logic of scientific explanation, Ernest Nagel (1961) turns his attention to the social sciences and finds the position to be as follows:

> ... in no area of social inquiry has a body of general laws been established, comparable with outstanding theories in the natural sciences in scope of explanatory power or in capacity to yield precise and reliable predictions ... many social scientists are of the opinion, moreover, that the time is not yet ripe even for theories designed to explain systematically only quite limited ranges of social phenomena ... social scientists continue to be divided on central issues in the logic of social inquiry. ... The important task, surely, is to achieve some clarity in fundamental methodological issues and the structure of explanations in the social sciences ...

After careful examination of the formidable difficulties facing social science Nagel concludes that the problems of determining systematic explanations of social phenomena are not *in logic* insuperable. He admits that 'problems are not resolved merely by showing that they are not necessarily insoluble', and the fact is that the problems of a social science patterned on natural science have hardly been solved.

It is obvious that one aspect of these problems is that already met with in biology and other unrestricted sciences—complexity of subject matter. The phenomena involved are ones with dense connections between many different aspects, making it difficult to achieve the reduction required for a meaningful controlled experiment. As in some of the unrestricted physical sciences, such as geology, the opportunities for experimentation designed to give unequivocal results are very limited. Social institutions are not available for experimentation even if experiments complete with controls could be designed. But in social science it is not only complexity and this non-availability of experimental objects which bring problems. There is also the greater problem of the special nature of the phenomena to be studied.

Our knowledge of the world acquired by science (and perhaps the world itself?) does seem to show symmetries, patterns, regularities. It would seem surprising if social phenomena were not similarly patterned. Indeed on a common-sense inspection they obviously are, since social life would not be possible if the behaviour of our fellows did not in general meet our expectations. But given the 'messy' nature of social phenomena as they appear to us, we can expect the findings of a scientific approach to the investigation of social reality to have certain characteristics which distinguish them from the findings acquired by the natural sciences' investigation of the physical world.

Firstly, we must expect any generalizations to be imprecise compared, say, with Ohm's Law. This refers to a well-defined relationship between well-defined variables which is exhibited in the very special environment of an electrical circuit, which may be precisely defined. Compare this with the finding of a European sociologist that in passing sentence on criminals *some* judges tend to have more regard to the social class of the criminal than they do to the nature of the offence. As MacIntyre (in Magee, 1971) describes this situation:

> When we ask what follows, what generalisations can be framed, we are not at all clear, because we are not at all clear how we should understand the phenomena of a legal system in such a way as to be able to generalise about them, and this is because we are not at all clear at what level we should be looking for a theory. A theory of evaluative behaviour in general? Or a theory of the behaviour of officials such as judges? How ought we to group together the phenomena of social life? How ought we to categorise them in such a way that we can begin to frame explanations? We do not know.

The point is that in the case of the relationship between voltage applied, circuit resistance, and current flowing in an electrical circuit we *do* know how to 'group together' and 'categorize' the phenomena. We can reduce the context of these phenomena to a single explicit form, generalize Ohm's Law, and invite anyone who is interested to check the experimental findings for himself. The variety of possible viewpoints which is always confusingly available in the case of social phenomena is drastically reduced in the case of natural science.

Closely related to the availability of many possible interpretations of social phenomena is the second feature which will distinguish the findings of social science from those of natural science: the special nature of the component of the system studied by the social scientist. The component is the individual human being, and even if we depersonalize him as an 'actor' in a 'role' he will be an active participant in the phenomena investigated, attributing meanings and modifying the situation in a potentially unique way. The chemist studying the properties of ammonia has no way of telling one ammonia molecule from another; he can assume total similarity between the properties of one molecule and the next, and he can be totally confident that he can conduct the experiment he wants: the individual molecules will not manipulate him! Even the biologist studying animal behaviour is in a simple position compared with the social scientist. Animal behaviour is apparently programmed to a far higher degree than dare be assumed (happily) for the human being. The social scientist is in the position in which the biologist studying the life cycle of the cuckoo would find himself if, one year, for a change, some cuckoos suddenly built themselves nests and sampled the experience of hatching eggs and rearing young cuckoos. The cuckoo does not in fact behave in this erratic manner, and ethologists have not found it necessary to evoke the idea that the phenomena studied are affected by the attribution of *meaning* by the participants. Social scientists do have to evoke this idea, and it has been a source of much controversy since Max Weber (a vigorous proponent of a 'value-free' social science, in the sense that the scientist would expound what the actor *could* rather than *should* do (Weber, 1904)) argued that the social actions of men could not be observed and explained in the same way that physical phenomena (or animal behaviour) could. The scientist's observation must necessarily include interpretation in terms of its meaning for the actors, and the social scientist needed a sympathetic appreciation of the situation from the viewpoint of the actors themselves. Weber himself, in *The Protestant Ethic and the Spirit of Capitalism*, gave an account of the rise of capitalism in terms of the changes in values and motivation which Calvinist Protestantism engendered. Such accounts may be thought insightful or nonsensical, but they are not testable—they are not 'public knowledge'. Weber himself argued that the values implicit in the actor's attribution of meaning are 'empirically discoverable and analysable as elements of meaningful human conduct'. His claim is that only the *validity* of such values 'can *not* be deduced from empirical data as such' (Weber, 1904). But there is clearly an important distinction between, on the one hand, a social scientist's acceptance of the participants' own account of the meaning of a social action for them (or his mounting of a plausible argument that a particular meaning is implicit in the happenings) and, on the other hand, a natural scientist's description of happenings which any other observer may check by repeating them.

A third difficult feature of social science, which is implicit in those already discussed and also adds to the distinction between it and natural science, is the problem of making predictions of social happenings. Partly this must be a matter of sheer complexity, the fact that what happens in social systems is always a mix of intended and unintended effects. But additionally there are other problems. For one thing, as is clear from the everyday observation of events, predictions of the outcome of observed happenings in social systems may change the outcome. Physical systems cannot react to predictions made about them; social systems can. A prediction that the building of improved roads giving access to and within a national park will destroy the beauty which visitors seek, might well lead a government body to introduce a traffic management scheme in order to make sure that the beauty is preserved (and, incidentally, that the prediction is falsified). Also, more fundamentally, there is the argument that laws (which are a way of embodying predictions) in social systems are not possible. That such laws, which would be laws of history, are non-existent, is the main concern of Popper's *Poverty of Historicism* (1957) and is sharply summarized in the Preface to that book. The argument is roughly as follows: the happenings in social systems are strongly influenced by the growth of human knowledge; the *future* growth of knowledge is in principle unpredictable since we cannot know the not-yet-known; therefore the future of social systems cannot be predicted. 'This means that we must reject the possibility of a *theoretical history*; that is to say, of a historical social science that would correspond to *theoretical physics*. There can be no scientific theory of historical development serving as a basis for historical prediction.' Aside from this stark argument, Popper elsewhere in the book actually takes the same line as Nagel, namely that *in logic* there is no distinction between natural and social science, both needing to proceed by the severe testing of hypotheses. The fact remains that the social sciences have not made the same kind of practical progress which is so apparent in the other sciences.

All of these crucial distinctions between the established sciences and the would-be sciences can be summarized in the fact that at the core of the phenomena studied by social science is the self-consciousness of human beings and the freedom of choice which that consciousness entails. This irreducible freedom stems ultimately from the experimental fact, upon which neurophysiologists agree, that all our mental activity—choosing, believing, etc.—is associated with specific electrical activity in the brain. The arguments are not easy to summarize (see Thorpe, 1974, for a more detailed discussion, mainly of arguments due to MacKay) but hinge on the fact that an observer can *never* obtain an up-to-date account of the state of mind of an agent he is observing which it would be correct for the agent to accept.

As soon as the agent agreed the correctness of the observer's account, his act of belief would render that account out of date by changing the state of his brain. Now suppose the observer were able to make a detailed prediction of the agent's future action based upon some miraculously perfect detailed knowledge of the state of the agent's mind. The agent himself would actually be wrong to believe that prediction *before* he makes up his mind what action to take (!) since his belief would render the perfect account of his state of mind, upon which the prediction is based, obsolete. Nothing can remove from the agent his freedom to select his action, there is no one outcome which he would be correct to regard as the only possible one. This kind of argument suggests that at best social systems will reveal 'trends' rather than 'laws' and that the social scientist will be reduced to studying not exactly social reality but only the logic of situations, producing findings of the kind 'In situation A, a likely outcome is B', without any guarantee that this will hold in any particular situation. And over the years, with the growth of human knowledge, the 'logic of situations'—which will involve actors' attribution of meaning—will gradually change.

This discussion hopefully indicates that the method of science, so powerful in the natural sciences, has not yet, and will not easily be applied to the investigation of social phenomena. An outline of a few of the difficulties has been sketched in. They are further discussed in Chapter 8. Perhaps it is finally worth reminding ourselves here of the splendid richness of the phenomena with which a would-be social science must deal by recalling a bravura passage from Nabokov's novel *The Eye*, which in fact refers to many of the arguments expressed here:

> It is silly to seek a basic law, even sillier to find it. Some mean-spirited little man decides that the whole course of humanity can be explained in terms of insidiously revolving signs of the zodiac, or as a struggle between an empty and a stuffed belly; he hires a punctilious Philistine to act as Clio's clerk, and begins a wholesale trade in epochs and masses; and then woe to the private individuum, with his two poor u's, hallooing hopelessly amid the dense growth of economic causes. Luckily no such laws exist: a toothache will cost a battle, a drizzle cancel an insurrection. Everything is fluid, everything depends on chance, and all in vain were the efforts of that crabbed bourgeois in Victorian checkered trousers, author of *Das Kapital*, the fruit of insomnia and migraine.

Problems for Science: 'Management'

If we had available a social science on the pattern of the natural sciences, with hypotheses and laws which have been well tested and have survived, and a body of theory which tells a story in which the hypotheses and the laws are meaningful and coherently linked, then that social science would surely help us in the solution of 'real-world problems', just as natural science is available to help technologists and engineers solve their problems. It would not *have* to be called on, of course, just as technology and engineering do not have to call on science: there are some brilliant and successful technologists and engineers who act intuitively. Nevertheless it is not unreasonable to anticipate that, were it available, an established social science would help us when, as decision takers, we face problems in social systems. By 'real-world problems' I mean problems of decision, in social systems, which *arise*, which we find ourselves facing, in contrast to the scientist's problem in a laboratory which he can define and limit. The scientist selects the most difficult problem which in his judgement offers a reasonable chance of solution; real life pushes its problems upon us. Real-world problems are of the kind: What should the British Government do about the supersonic passenger aircraft project, Concorde? How should we design our schools? Should I marry this particular girl? Shall I change my career? Such problems are in fact problems of *management*, broadly defined. The management process, not interpreted in a class sense, is concerned with deciding to do or not to do something, with planning, with considering alternatives, with monitoring performance, with collaborating with other people or achieving ends through others; it is the process of taking decisions in social systems in the face of problems which may not be self-generated.

This being so, we might expect that so-called 'management science' is in fact a body of scientific knowledge and principles relevant to the management process. Management science certainly exists in the sense that there are professionals who would think of themselves as 'management scientists', and institutions to which they belong; there are books, journals, and conferences devoted to management science as well as courses and qualifications. What is noticeably lacking is any great feeling on the part of the people who would apparently get most help from such a science that it really can help them. Managers in industry and in the public sector, leaders of trade unions and shop stewards, politicians and administrators, have not been convinced by its fruits that management science is more than marginally relevant to their task. Most of them would agree with Drucker (1974) that 'Management is a practice rather than a science. It is not knowledge but performance.' Management scientists accept that this view prevails. When the Operational Research Society held a public celebration to commemorate thirty years of OR the *Financial Times* said 'the keynote was one of disillusionment', and the President of the Society commented: 'perceptive FT' (Beer, 1970).

The failure of the scientific approach to make much progress so far in its application to the process of management can usefully be examined by looking briefly at the example of operational research, which is the closest management science comes to having a hard scientific core. All definitions of OR emphasize its scientific nature, and of course its origins lie largely in the attachment of professional scientists to wartime groups responsible for military operations (see McCloskey and Trefethen, 1954, for an account of the origins of OR, with contributions from a number of the pioneers; also

Morse, 1970). Blackett in a paper written in 1941 to inform the Admiralty of developments which had occurred in the RAF, described the objective in having scientists in close touch with operations: the operational staffs provided the scientists with the operational outlook and data, the scientists applied scientific methods of analysis to these data and thus were able to offer useful advice (Blackett, 1962). These elements have survived the transfer of OR into non-military activities, and the OR Society's official definition reads (my italics):

> Operational Research is the application of the methods of *science* to complex problems arising in the direction and management of large systems of men, machines, materials and money in industry, business, government, and defence. *The distinctive approach is to develop a scientific model of the system*, incorporating measurements of factors such as chance and risk, with which to predict and compare the outcomes of alternative decisions, strategies or controls. The purpose is to help management determine its policy and actions *scientifically*.

This statement well expresses the brilliant solution which OR has adopted to the difficult problem of applying the methods of science to parts of the real world, as opposed to the artificial situations created in the laboratory. Engineers use the same solution: to carry out 'experiments' not on the real-world object of study—which is usually not available—but on a model of it, if possible a quantitative model. The strategy of OR is to build a model of the process concerned, one in which the overall performance is expressed in some explicit measure of performance (often economic), then to improve or optimize the model in terms of the chosen performance criterion, finally to transfer the solution derived from the model to the real-world situation. This is an heroic attempt to be scientific in the real world (as opposed to the laboratory) and the difficulties are great. The strategy obviously ought not to be pressed unless the model can be shown to be valid. In the case of a well-defined production process this may not be too difficult—if the model, when fed with last year's demand, can generate last year's output then we may feel reasonably confident that it reflects reality. But instances as clear cut as this are extremely rare. And equally obviously no single performance criterion can possibly unite within itself the myriad considerations which actually affect decisions in social systems. Thus what OR can provide is one crucial contribution to a management decision, a rational story of the form: 'if you adopt x as the measure of performance then you may optimise with respect to x by the following actions . . .', but it can hardly generate the kind of irrational decision which, in a management situation, often turns out to be a good one.

What has happened historically is that OR has concentrated most of its efforts on refining its quantitative tools and developing them for specific situations. The argument implicit in the development of OR in the last thirty years is that *problem situations recur*. Once the *form* of the problem has been idealized then the algorithms for its solution may be worked out and refined. Hence we find most OR texts devoting successive chapters to the problem types which are believed to recur (for example, Churchman *et al*, 1957) This outlook is expressed by Wild (1972) as follows, in a book about the management of production facilities:

> ... although problems tend to differ in practice this difference often derives from their content details rather than from their form. ... it is generally recognised that the following problem forms exist:
> Allocation Problems
> Inventory Problems
> Replacement Problems
> Queueing (or Waiting Line) Problems
> Sequencing and Routing Problems
> Search Problems (i.e. concerned with location)
> Competitive or Bidding Problems

But it is the fact in real life that 'the problem' is usually perceived as such because of the content details which make it unique, rather than because of the form which makes it general. This is probably the explanation of the intellectual gap which exists between practising managers and management scientists; and it indicates the formidable problems which still face both OR and management science as a whole as they try to extend the application of the method of science in areas of extreme complexity.

The three problems for science considered above—complexity in general, the extension of science to cover social phenomena, and the application of scientific methodology in real-world situations—have not yet been satisfactorily solved, although some progress has been made. Had they been solved, it is unlikely that systems thinking and the systems movement which unites systems thinkers in many different fields, would exist in their present form. This is not to say that systems thinking has made spectacular progress—indeed, on the whole the substantive results from the systems movement are still meagre—but the existence of the movement at all is a response to the inability of reductionist science to cope with various forms of complexity. Systems thinking is an attempt, within the broad sweep of science, to retain much of that tradition but to supplement it by tackling the problem of irreducible complexity via a form of thinking based on wholes and their properties which complements scientific reductionism.

Systems Thinking: Emergence and Hierarchy

'Systems thinking' is not yet a phrase in general use. Eventually, I believe, systems thinking and analytical thinking will come to be thought of as the twin components of scientific thinking, but this stage of our intellectual history has not yet been reached. It is at present necessary to establish if we can the phrase's credentials. This and the following section will argue that systems thinking is founded upon two pairs of ideas, those of *emergence and hierarchy*, and *communication and control*.

In discussing the problem which complexity presents to the method of science, the controversy concerning vitalism, first in organic chemistry and later in biology, was noted. Biology is an 'unrestricted' science, in Pantin's sense, and its phenomena are of a complexity which has severely tested scientific method. Biologists, in fact, have been among the pioneers in establishing ways of thinking in terms of wholes, and it was a biologist, Ludwig von Bertalanffy, who suggested generalizing this thinking to refer to any kind of whole, not simply to biological systems. The second strand in systems thinking comes from a very different source, from electrical, communication, and control engineers. A number of other groups have also made contributions, and we can trace the development of systems ideas in, for example, psychology, anthropology, and linguistics. But the four main ideas have emerged most clearly in biology and in communication and control engineering, and I shall present them in the context of work in those areas.

Aristotle argued that a whole was more than the sum of its parts, but when Aristotle's picture of the world was overthrown by the Scientific Revolution of the 17th century, this seemed an unnecessary doctrine. Newton's physics provided a mechanical picture of the universe which survived severe tests, and Aristotle's teleological outlook, in which objects in the world fulfilled their inner nature or purpose, seemed tainted by quite unnecessary metaphysical speculation. Nevertheless the history of modern biology is the history of the reinstatement of purpose as a respectable intellectual concept. The reinstatement is not, however, exactly a reinstatement of the concept *teleology*, the doctrine that structures and behaviour are determined by the purposes they fulfil. Biologists are extremely wary of offering fulfilment of purpose as a *causal explanation* of biological performance. Rather the rehabilitation is of what Medawar and Medawar (1977) describe as 'a genteelism' derived from teleology, namely *teleonomy*. This latter word has none of the metaphysical connotations of teleology. Behaviour described *as if* it fulfilled a purpose is 'teleonomic'. Now, making this distinction between the two words is not mere academic pedantry. In biology during this century there has been an often truculent debate about the nature of an organism. This debate has been one version of a wider debate between reductionism and holism. The terms of that debate have changed, and the change signals the emergence of systems thinking.

On the reductionist side in the argument has been the mechanist position, for which living systems are simply complex machines. Initially, the holist side of the argument was taken by the vitalists, such as Hans

Driesch. For Driesch and the other vitalists the development of a whole organism from a single egg must mean that in each developing organism resides a mysterious spirit-like *entelechy* which somehow directs and controls the growth of the whole. The mechanists were justifiably scornful of the unscientific stance adopted by the vitalists—hence the modern preference for teleonomy rather than teleology—but gradually the holist side of the argument changed. Gradually the argument was built that opposing reductionism did not mean embracing vitalism. The old argument of mechanism—versus—vitalism is now dead, even though some writers still appear to believe that to reject reductionism is to embrace vitalism—see for example Beck (1957) and Bunge (1973). But no professional experimental scientist would now deny that the evidence is that living systems in all their mechanisms obey the established laws of physics and chemistry. No professional scientist seriously invokes *entelechy*. But this does not mean that reductionism has carried the day; it does not mean that biological phenomena are *nothing but* physics and chemistry. Biology is now established as an autonomous science which is not reducible to chemistry and physics. Establishing this has established systems thinking.

The modern science of biology emerged from the Aristotelian view that living things (and, indeed, for Aristotle, inanimate objects as well) functioned to fulfil their innate purpose. A decisive act which made biology potentially a modern science was Harvey's calculation that the heart must be a pump which continuously circulates a limited quantity of blood, as described in the previous chapter. Harvey's work thus started the debate as to whether living organisms were complex machines or entities imbued with a special vital force. He himself appears on both sides of the argument, believing both that the heart was a pump and that there dwelt in the blood 'the vital principle itself' (Beck, 1957). The debate has continued ever since the 17th century, its present-day form being discussion of the question: is biology an autonomous science or is it in principle reducible to physics and chemistry?

Practical advances in the science of living things accelerated with the invention of the microscope. Microscopic examination of plants and living tissue gave the new science its central discovery—that of the cell—a discovery which leads on to the modern view that there is in living things a hierarchy of structures in the sequence: molecules, organelles (specific entities making up the organization of the cell), cells, organs, the organism. In this hierarchy the organism itself seems intuitively to mark a boundary, organisms having an obvious identity as whole entities, in that they have a boundary which separates them from the rest of the physical world, even though there may be transports across the boundary. The question is whether or not the organism as the ultimate object of concern in biology makes biology autonomous. Or will physical and chemical explanations demolish this autonomy? One of the crucial arguments which counters this reductionism was discussed above with regard to organic chemistry; this time the argument is that a demonstration that all the workings of cells and organs can be explained in terms of known facts of physics and chemistry does not explain *away* the existence of organisms as entities-to-be-explained. But the strength of this argument (in essence epistemological) was not easily established and the issue has remained a live one.

Modern discussion of it started in the late 19th and early 20th centuries with the emergence of the school of biology known as 'organismic'. The organismic biologists, for whom the organism was the irreducible object which a purely analytical approach could not explain away, are sometimes referred to as 'organicists', that word pairing with the doctrine of 'organicism'. But there is confusion to beware of here, organicism being also the use of the living organism as an *analogy* or metaphor for other larger entities such as families, societies, and civilizations (Phillips, 1970). This is a metaphor which—as I shall argue later—has been very bad for social science; here, however, concern is only for the debate initiated by organismic biologists and for the parallel philosophical writings. The debate began in the second half of the 19th century, and some of the first tentative expositions of what later became 'systems thinking' were written in the 1920s: for example, C. D. Broad's *The Mind and its Place in Nature* in 1923, J. C. Smuts's *Holism and Evolution* in 1926, and J. H. Woodger's *Biological Principles* in 1929. Woodger later translated, and helped the author to reformulate, L. von Bertalanffy's *Modern Theories of Development: An Introduction to Theoretical Biology* (Floyd and Harris, 1964). It was Bertalanffy who in the mid 1940s generalized organismic thinking ('the system theory of the organism', as he called it) into thinking concerned with systems in general, and in 1954 helped to found the Society for General Systems Research, initially the Society for the Advancement of General System Theory (Bertalanffy, 1972).

Because some organismic biologists, such as Driesch, were also vitalists, the argument between reductionism and holism was for many years mixed up with that between mechanistic explanation of living properties and vitalism. But the biologists who can now be seen as the early systems thinkers were anxious to dissociate themselves from vitalism because it is not a scientific concept; in the terms of the argument developed in Chapter 2, it is not scientific because it does not ask any question of nature which can be examined experimentally, it is an untestable assertion rather than a testable conjecture. In the book which Woodger translated in the early 1930s, Bertalanffy was making the point that the distinguishing feature of living things appeared to be their degree of organization: (my italics)

> The essential feature of the vital phenomenon which has still not received sufficient attention is that the processes of metabolism ... etc. occur exclusively in relation to well-individualised material objects *with a definite organisation*. This organisation of the processes

is the clearest and indeed the only decisive distinguishing feature between the vital happenings and the ordinary physico-chemical processes (Gray and Rizzo, 1973).

It is the concept of organized complexity which became the subject matter of the new discipline 'systems'; and the general model of organized complexity is that there exists a hierarchy of levels of organization, each more complex than the one below, a level being characterized by emergent properties which do not exist at the lower level. Indeed, more than the fact that they 'do not exist' at the lower level, emergent properties are *meaningless* in the language appropriate to the lower level. 'The shape of an apple', although the result of processes which operate at the level of the cells, organelles, and organic molecules which comprise apple trees, and although, we hope, eventually explicable in terms of those processes, *has no meaning* at the lower levels of description. The processes at those levels result in an outcome which signals the existence of a new stable level of complexity—that of the whole apple itself—which has emergent properties, one of them being the apple's shape.

The classic philosophical expression of the theory of emergence is that of Broad (1923). (for a cogent modern account, see Wartofsky, 1968, Chapter 13.) Broad discusses three outlooks: what he calls the 'Substantial Vitalism' of Driesch *et al*; 'Biological Mechanism', the view that living things are merely machines; and what he terms 'Emergent Vitalism'—what is now usually referred to as 'emergence' or 'emergentism'. Some of the arguments now have a quaint ring, such as the argument that one reason emergence is superior to mechanism is that where the latter requires the existence of God as nature's designer, the former does not *require* this but is compatible with it should God be proved to exist! However, the book is extremely important in that it disentangles the emergence concept from crude vitalism and helps to establish the modern position that while a (reductionist) elaboration of mechanisms at the physico-chemical level is extremely useful, the fact is that the existence of organisms having properties as wholes calls for different levels of description which correspond to different levels of reality. This is so even if the properties of parts, together with the laws of combination, can explain the mechanism of the whole. The point is, the whole was there to be explained in the first place. In a sentence, we can express Broad's position (which is central to all subsequent discussion of the issue) briefly as follows: 'Neither a one-level epistemology nor a one-level ontology is possible'. Smuts in 1926 covered much the same ground, being anxious to eliminate metaphysical concepts such as *entelechy* or *élan vital* by the concept of organized complexity, which is what he means by 'holism':

> Every organism, every plant or animal, is a whole with a certain internal organisation and a measure of self-direction. Not only are

plants and animals wholes but in a certain limited sense the natural collocations of matter in the universe are wholes; atoms, molecules and chemical compounds are limited wholes. . . . A whole is a synthesis or unity of parts, so close that it affects the activities and interactions of those parts. . . . The parts are not lost or destroyed in the new structure. . . their independent functions and activities are grouped, related, correlated and unified in the structural whole.

Given the state of biological knowledge in the 1920s, Smuts was not able to give an account of the mechanisms whereby the organism becomes an entity, and when he says that 'Holism is the inner driving force behind (evolutionary) progress' a whiff of the metaphysical enters the argument. Nevertheless the book is a remarkable one considering that it was written by a professional politician. Smuts tells us that it was a lost election which provided him with the time to write it! More professional is Woodger's *Biological Principles* (1929) which sought to re-think rigorously the intellectual basis of biological science. Part II of the book examines what constitutes an explanation in biology. This leads Woodger to consider the antithesis between vitalism and mechanism, and he too points to the importance of the concept 'organization', marshalling arguments from cell studies which show that 'organization *above* the chemical level is of great importance in biology'. This leads to a recognition that the architecture of complexity is one of hierarchical organization and that the levels of complexity are fundamental to any account of the organism. Moreover: (my italics)

 . . . from what has been said about organization it seems perfectly plain that an entity having the hierarchical type of organization such as we find in the organism requires investigation at all levels, and investigation of one level cannot replace the necessity for investigations of levels higher up in the hierarchy . . . a physiologist who wishes to study the physiology of the nervous system must have a level of organization above the cell level *to begin with*. He must have at least the elements necessary to constitute a reflex arc, and in actual practice he uses concepts appropriate to that level which are not concepts of physics and chemistry.

He concludes that vitalism is not the only alternative to mechanical explanations, and that truly biological explanations are possible, being dependent upon the existence of hierarchies of levels of organization in living organisms. Hierarchical organization, and the emergent properties of a given level of organization are consonant with a process of evolution which is *creative*.

The writings of Broad, Smuts, and Woodger illustrate (rather than themselves constitute) the emergence of a new mode of thought which we now call *systems thinking*. Other examples could have been selected. Philosophically, the work of Bradley (1893), or Whitehead's difficult 'process philosophy', which argues against the validity of common-sense predication in describing reality, represent the same movement. (*Process and Reality*, 1929; Sherburne, 1966, provides an accessible exegesis.) Among other biologists writing in this vein in the early decades of the 20th century were J. S. Haldane and C. Lloyd Morgan, the biochemist L. J. Henderson, who later turned to teaching a sociology based on physiological analogies (Lilienfeld, 1978), and W. Cannon, whose oft-quoted *Wisdom of the Body*, (1932) describes the 'homeostatic' or self-regulatory mechanism whereby the animal organism is able to maintain constant such properties as levels of blood sugar, oxygen, body temperature, etc.

This strand of holistic thinking in biology, which began in the second half of the 19th century, has continued throughout the 20th century. Its modern form is still the discussion of the autonomy of biology, and there is a continuing current literature on this (see, for example, Smith, 1966; Elsasser, 1966; Polanyi, 1968; Koestler and Smythies, 1969; Breck and Yourgrau, 1972; Bunge, 1973; Grene, 1974; Ayala and Dobzhansky, 1974; Colodny, 1977; Thorpe, 1978).

A sample of the modern debate, which neatly summarizes the anti-reductionist position is provided by Grene (1974). She points out that in principle a one-level ontology—the belief, for example, that with increasing knowledge all science will become an account of the world in the language of, say, atomic events—contradicts itself. This is so because such a belief, to be meaningful, requires an ontology which admits both atomic events *and* cognition. Here at once a second level is smuggled in! It is logically possible, of course, that there might be no levels in between those of atomic events and cognition (that in essence is Descartes's position) but the sciences of chemistry and biology consist of some well-tested conjectures that there are such intermediate levels as are represented by molecules, cells, organelles, organs, and organisms. That there is a live issue is illustrated by contemporary debate in molecular biology. The major success of molecular biology is its elucidation of the process by which DNA is replicated in cell division, thus providing a detailed molecular account of the process by which offspring acquire inherited characteristics from their parents. We now know that the transmission to a new cell of a particular characteristic, such as blue eyes or black hair, is the result of the replication of a particular sequence of organic bases along the molecular chain of deoxyribonucleic acid (DNA). This is the now famous 'genetic code'. One of the discoverers of the structure of DNA, and hence of a possible mechanism for the operation of the code, Francis Crick (1966), adopts in *Molecules and Men* a crudely reductionist position. 'The ultimate aim of the modern movement in biology' he claims, 'is in fact to explain *all* biology in terms of physics and chemistry'. What he fails to notice, as Grene (and also Polanyi, 1968) points out, is that *any* arrangements of the organic bases is compatible with the laws of physics and chemistry. But in biological terms the actual arrangement is *crucial*: certain sequences constitute 'a code'. The investigator, even if he is Crick himself, in recognizing a code, is accepting that 'all biology' cannot, in principle, be explained as physics and chemistry. The code is an emergent property at the level of complexity represented by biological phenomena.

The idea that the architecture of complexity is hierarchical and that different languages of description are required at different levels has in recent years led to a developing interest in hierarchy theory as such, although much of the interest still centres on the biological hierarchy from cell to species (Simon, 1962; Pattee, 1968–72, 1973; Whyte, *et al*, 1969; Mesarovic *et al.*, 1970; Anderson, 1972; Milsum, 1972). Simon points out that the time required for the evolution of a complex form from single elements depends critically upon the numbers and distribution of intermediate forms which are themselves stable. Simple mathematical analysis of probabilities shows that the time required for a complex system to evolve is much reduced if the system is itself comprised of one or more layers of stable component sub-systems. Turning the argument round: the age of planet earth is such that only hierarchically organized entities have had time to evolve! The structure of the world we inhabit *could* only be hierarchical.

Hierarchy theory is concerned with the fundamental differences between one level of complexity and another. Its ultimate aim must be to provide both an account of the relationships between different levels and an account of how observed hierarchies come to be formed: what generates the levels, what separates them, what links them? Such theory is still in its infancy (see Pattee, 1973, for the best discussion of what is required of it) but whatever form it eventually takes, it seems likely to be built upon the fact that emergent properties associated with a set of elements at one level in a hierarchy are associated with what we may look upon as *constraints* upon the degree of freedom of the elements. The emergent properties resulting from application of the constraints will entail a descriptive language at a meta-level to that describing the elements themselves. Thus, in the language of chemistry, any arrangement of the bases in DNA obeys the laws of physical chemistry. But it is constraints upon the ordinary chemistry of the base-sequences which produce the specifically biological property of genetic coding, an emergent property which marks a transition from the level we call 'chemistry' to that called 'biology'. This imposition of constraints upon activity at one level which harnesses the laws at that level to yield activity meaningful at a higher level, is an example of regulatory or *control* action. Hierarchies are characterized by processes of control operating at the interfaces between levels.

This brings us to the second pair of root ideas in systems thinking. To emergence and hierarchy we must add *communication* and *control*. These will be discussed in the next section. I end this one with a quotation from a distinguished contemporary biologist, Francois Jacob, which neatly summarizes the picture of emergence and hierarchy which biology has built up. Writers on systems have frequently felt the need to invent a new word for the entities which are wholes at one level of a hierarchy while simultaneously being parts of higher level entities. Gerard (1964) speaks of *orgs*, Koestler (1967, 1978) of *holons*. Jacob (1974) uses the concept of the *integron*.

> ... biology has demonstrated that there is no metaphysical entity hidden behind the word 'life' ... From particles to man, there is a whole series of integrations, of levels, of discontinuities ... investigation of molecules and cellular organelles has now become the concern of physicists ... This does not at all mean that biology has become an annex of physics, that it represents, as it were, a junior branch concerned with complex systems. At each level of organization, novelties appear in both properties and logic. To reproduce is not within the power of any single molecule by itself. This faculty appears only with the simplest integron deserving to be called a living organism, that is, the cell. But thereafter the rules of the game change. At the higher-level integron, the cell population, natural selection imposes new constraints and offers new possibilities. In this way, and without ceasing to obey the principles that govern inanimate systems, living systems become subject to phenomena that have no meaning at the lower level. Biology can neither be reduced to physics, nor do without it. Every object that biology studies is a system of systems. Being part of a higher-order system itself, it sometimes obeys rules that cannot be deduced simply by analysing it. This means that each level of organization must be considered with reference to adjacent levels ... At every level of integration, some new characteristics come to light ... Very often, concepts and techniques that apply at one level do not function either above or below it. The various levels of biological organization are united by the logic proper to reproduction. They are distinguished by the means of communication, the regulatory circuits and the internal logic proper to each system.

Systems Thinking: Communication and Control

In treating the living organism as a whole, as a system, rather than simply as a set of components together with relationships between components, von Bertalanffy drew attention to the important distinction between systems which are *open* to their environment and those which are *closed*. He defined an open system (1940) as one having import and export of material. More generally, between an open system and its environment there may be exchange of materials, energy, and information. Organisms, he pointed out, are unlike closed systems in which unchanging components settle in a state of equilibrium; organisms can achieve a steady state which depends upon continuous exchanges with an environment. What is more, the steady state may be thermodynamically unlikely, creating and/or maintaining a high degree of order, where closed systems have no path to travel but that towards increasing disorder (high entropy). In a hierarchy of systems such as that represented by the sequence from cell organelle to organism, or, in general, in any hierarchy of open systems, maintenance of the hierarchy will entail a set of processes in which there is *communication* of information for purposes of regulation or *control*.

This is reflected in the language used in discussions of modern molecular biology. DNA is regarded as 'storing' and 'coding' information (and a very great deal of it—in the forty-six chromosomes of each somatic cell of a man, according to Pratt, 1962, are instructions which are the informational equivalent of forty-six volumes each containing millions of words). The genetic processes entail chemical 'messages' which carry instructions to 'activate' or 'repress' further reactions, and constitute control processes which guide the development of the organism. Indeed it is intuitively obvious that a hierarchy of systems which are open must entail processes of communication and control if the systems are to survive the knocks administered by the systems' environment. This is even more obvious if we consider, not the natural hierarchy of living systems, but man-made hierarchical systems such as machines and industrial plant. The designer of an industrial plant to make a particular chemical product, for example, is forced to be 'a systems thinker'. He has to consider not only the individual reactor vessels, heat exchangers, pumps, etc., which make up the plant but also, at a different level of consideration, the plant as a whole whose overall performance has to be controlled to produce the desired product at the required rate, cost, and purity. He will have to ensure that there exist means by which information about the state of the process can be obtained and used to initiate action to control the reaction within predefined limits; hopefully, by knowing the variability of the starting materials and the possible environmental disturbances to which the process will be subject, it will be possible to manipulate automatically a few so-called *control variables* according to a control strategy for the plant as a whole.

Devices incorporating automatic controllers based upon the communication of information about the state of the system were known to the ancient Greeks (Mayr, 1970) but the working out of the theory of control is relatively recent. The mathematics of Watt's 1788 centrifugal regulator for governing the speed of a steam engine, most famous of all control devices, was worked out towards the end of the 19th century but it is only in the last forty years that control engineering and control theory have become established as, respectively, a professional activity and a related academic discipline. Nevertheless the ideas from control theory and from information and communication engineering have made contributions to systems thinking no less important than those from organismic biology.

A link between control mechanisms studied in natural systems and those engineered in man-made systems is provided by the part of systems theory known as *cybernetics*, the word coming from the Greek word meaning *steersman*. Plato used the word in making an analogy between a helmsman steering a ship and a statesman steering the 'ship of state', and Ampère in the 19th century used it in discussing the place of political science in a classification of the sciences. Unaware of this, Norbert Wiener suggested the word in 1947 to cover the subject matter of a group of studies which had then been underway in America since the early 1940s. These studies were concerned with the theory of messages, and of message transmission for purposes of control in many different contexts (Wiener, 1948, 1950). Recognizing the generality of the notions of communication and control (whether, for example, in a specific context they refer to the regulation of blood sugar concentration, the motions involved in an owl catching a mouse, or the manipulation of temperatures and pressures in a chemical reactor to control the rate of product formation), Wiener defined cybernetics as 'the entire field of control and communication theory, whether in the machine or in the animal' (1948). Under this dictum there is a range of interpretations (Pask, 1961). For Ashby, the leading theoretician in the 1950s and 1960s, cybernetics deals with 'all forms of behaviours in so far as they are regular, or determinate or reproducible' (Ashby, 1956). Ashby's interest is not in the real-world manifestations of controllable wholes but in the abstract mathematically-expressible image of mechanism which underlies the real embodiments. He works out the synthetic operations which can be performed upon the abstract image and so establishes the formal rules for controllability. For Ashby, the relation between cybernetics and actual machines parallels that between geometry as an abstract study and real objects in three-dimensional space. Geometry, no longer dominated by terrestrial space, provides a general mathematical framework within which terrestrial objects may be treated as a special case.

> Cybernetics is similar in its relation to the actual machine. It takes as its subject matter the domain of 'all possible machines' ... What cybernetics offers is the framework on which all individual machines may be ordered, related and understood (Ashby, 1956).

And the 'machines', of course, may be either natural or man-made.

Wiener's work covers well the core of the ideas contributed to systems thinking by study of 'communication and control'. Much of the decade of work which led to the publication of Wiener's book in 1948 was carried out in collaboration with J. H. Bigelow and A. Rosenblueth, the latter a medical scientist at the Harvard Medical School who had himself been a colleague and collaborator of W. B. Cannon, well known for his work on the body's internal control processes. Wiener the mathematician and Rosenblueth the medical scientist both felt that useful developments were likely in the intellectual no-man's land between the various established disciplines. In the early 1940s Wiener was trying to develop computing machines for the solution of partial differential equations. At the beginning of the war Wiener worked with Bigelow on the problem of improving the accuracy of anti-aircraft guns. The difficulty there is to compute automatically the future likely position of the aircraft, to keep the computation up-to-date as the aircraft moves, and to arrange for a shell to detonate at the appropriate spot. For the second time Wiener found himself working on mechanical/electrical systems designed to usurp what was previously thought of as a human function, in this case the huntsman's intuitive assessment of where to aim relative to a moving quarry. Wiener and Bigelow realized the importance and ubiquity of what control engineers call the process of *feedback*, namely the transmission of information about the actual performance of any machine (in the general sense) to an earlier stage in order to modify its operation. (Usually, in *negative* feedback the modification is such as to reduce the difference between actual and desired performance, as when the increasing speed of a steam engine causes the flying pendulum of the governor to reduce the steam supply and hence lower the speed; positive feedback induces instability by reinforcing a modification in performance, as when a conversation between two people in a crowded room is conducted in louder and louder tones as their output, increasing the general noise level, makes it increasingly difficult for them to hear each other.) Examination of situations in which excessive feedback causes oscillatory hunting about the desired performance led Wiener and Rosenblueth to recognize the essential similarity between hunting in mechanical or electrical control systems and the pathological condition ('purpose tremor') in which the patient, trying to perform some simple physical act, such as picking up a pencil, overshoots and goes into uncontrollable oscillation. This focused attention in neurophysiology on 'not solely the elementary processes of nerves and synapses but the performance of the nervous system as an integrated whole', on the kind of description, in fact, which had begun to enter medical science with the work of Claude Bernard in the 1860s (Olmsted and Olmsted, 1952). Thus was initiated Wiener's interest in the general problem of control. The work with Bigelow, which extended to the construction of machines which tried to predict the extrapolation of curves, also convinced Wiener that 'the problems of control engineering and of communication engineering were inseparable'. Both centred on a core concern not for mechanical equipment or electrical circuitry but the more fundamental abstract notion: that of a message and its transmission. Now, because a message may be represented as a discrete or continuous sequence of events in time—what statisticians call a time series—Wiener was led to consider the statistical nature of message transmission. Work on the separation of the intended message (the 'signal') from background distorting 'noise' (both of which may be represented by time series) then led to a view of information transmission as a transmission of alternatives, with the unit *amount of information* defined as that transmitted as a single decision (a 'yes' or 'no', a one or a nought, a dot or a dash) between equally probable alternatives. The idea of a quantitative statistically-based theory of information transmission was occurring simultaneously to the statistician R. A. Fisher, to Wiener, and to Claude Shannon, a communications engineer of Bell Telephone Laboratories. For Wiener, the essential unity of the set of problems centring on communication, control, and statistics called for a new name for the field: he was 'forced to coin at least one artificial neo-Greek expression'. Cybernetics was born.

In spite of the perfervid enthusiasm with which some devotees seize upon the general notions of cybernetics, usually in order to apply them promiscuously in other fields, it has not been easy to use the logico-mathematical formulations of cybernetics to derive propositions testable in real systems. Nevertheless, it does provide a potentially novel view of many problem areas in which reductionism has previously held sway. For example, the mind–body problem which has dogged philosophy for hundreds of years, viewed cybernetically, seems to be wrongly posed (Bateson, 1971; Sayre, 1976). In a cybernetic analysis of the process in which a person thinks, acts, and modifies subsequent behaviour in the light of preceding acts, all these items (including the acts themselves) may be seen as information processing. Bateson argues that the total self-corrective unit which does this processing is not, however, the human being; it is a system whose boundaries extend beyond the human body. The system is a network of information—transmitting pathways including some external to the actor; on this view, mind is not simply associated with the human body but is immanent in brain, plus body, plus environment.

The practicalities usually associated with cybernetics are those to do with the design of controllers for man-made systems. This is, of course, the province of control engineers, and much of the experimental work in cybernetics seeks to construct machines which exhibit 'intelligent' behaviour, as a means of exploring possible mechanisms for models of the brain. Ashby himself, although responsible for many of the formal theorems of the subject, undertook this work because of a prime interest in biological mechanisms, especially those by which living systems regulate their own behaviour. It is certainly the case that the nature of the ideas *control* and *communication*, and their link to *emergence* and *hierarchy* are most richly apparent in biological systems.

Following arguments advanced by Pattee (1973), a physicist interested in the problem of the origin of life, we may picture the concept *control* through the following picture of biological systems.

The physical universe is a hierarchy; at the lowest levels are nuclear particles, then atoms, then molecules, three levels differentiated by the forces involved: between nuclear particles are forces of about 140 million electronvolts, between atoms in a molecule are 'valency' forces of about 5 electronvolts, between molecules weaker forces of about one-half of an electronvolt. The hierarchy at this stage is characterized by these energy levels, and it is these differences which suggest that it is not foolish to treat systems of nuclei, atoms, and molecules as different kinds of whole. Above the molecular level, some of the examples of systems consisting of groups of interacting molecules exhibit special properties which cause us to name them 'living' systems. Such systems are characterized, among other things, by their exhibiting differentiated processes which may be brought into play by appropriate messages and which serve to create, and re-create, the systems themselves; in the longer term such systems may adapt and evolve more varied forms. In order to understand these properties it is necessary to invoke the idea of hierarchical control. Control is always associated with the imposition of constraints, and an account of a control process necessarily requires our taking into account at least two hierarchical levels. At a given level it is often possible to describe the level by writing dynamical equations, on the assumption that one particle is representative of the collection, and that the forces at other levels do not interfere. But any description of a control process entails an upper level imposing constraints upon the lower, as, for example, when the cell as a whole constrains the physico-chemical possibilities open to DNA and makes it 'the bearer of a code'. The upper level is a source of an alternative (simpler) description of the lower level in terms of specific functions which are emergent as a result of the imposition of constraints. Thus some otherwise ordinary molecules in a cell are constrained by the cell to bear the function 'repressor' or 'activator': these functions are not inherent chemical properties, they are the result of a process of hierarchical control. In general, hierarchical control requires three conditions to be met. Firstly, imposition of a constraint must impose new functional relations; this will involve an 'alternative description' of the lower level which ignores some of its detailed dynamics. Secondly, the imposition of the constraint—if the system is to be 'living'—must be optimal in the sense that it is neither so tight that it leads to rigidity (as do the constraints which govern crystal formation, for example) nor so loose that specific functions are not generated at the lower level (in which case we simply have what physicists recognize as a 'boundary condition'). Thirdly, the constraint must act upon the detailed dynamics of the lower level. Finally, in living systems, the variety of alternative descriptions or specific functions which may be developed does not appear to be fixed—over a long time period new forms and functions may evolve. A complete theory of hierarchical control would explain exactly how it is that collections of elements may spontaneously separate out persistent special functions under the constraints of the collection as a whole. As Pattee points out, no such theory yet exists. These are still early days in the systems movement.

This brief examination of the subject matter of cybernetics and of the basic nature of a control mechanism, indicates how close is the link between control and communication. All control processes depend upon communication, upon a flow of *information* in the form of instructions or constraints, a flow which may be automatic or manual. In the case of the steam engine governor the whirling pendulums which automatically open or shut the steam valve replaced a human operator who previously controlled the valve by hand. Both he and the automatic governor may be thought of as a receptor of information about engine speed which 'takes a decision'—completely preprogrammed in the case of the automatic governor—and feeds back 'an instruction' to the valve. Note that the action of the governor has virtually nothing to do with energy considerations, the energy involved in the control process being negligible compared with that of the steam engine itself. And yet the action of the regulator is a crucial one in the system as a whole, its power residing simply in its ability to receive and transmit information. Similar considerations apply in living systems at all levels. Monod and his co-workers, for example, have elucidated the control mechanisms governing protein synthesis in the bacterium *Escherichia coli* (Monod, 1972). The feedback mechanisms involve 'recognition' of 'chemical signals': a repressor molecule inactivates protein production but is inactivated in its turn by an 'inducer'. Monod calls the phenomenon, which enables the laws of chemistry to be transcended, if not escaped, 'wonderfully and almost miraculously teleonomic'. His explanation could not be written in the language of energetics, only in the language of information transmission. In general, the idea of information is prior to that of feedback, any feedback mechanism in a viable system consisting of a sensor capable of detecting potentially disruptive environmental changes and an effector capable of initiating remedial action. Some of Ashby's most important work has been his demonstration that continuing effective control in a changing environment requires a controller with a variety of response which can match the variety of the environmental information—the so-called Law of Requisite Variety. A final example of the importance of the communication of information for system behaviour comes from the work of James Miller. He has devoted a scientific career to defining the characteristics of living systems. According to his model, any member of the class 'living system' has nineteen critical sub-systems (Miller, 1978). No fewer than eleven of these process information (some also process matter–energy): 'reproducer, boundary, input transducer, internal transducer, channel and net, decoder, associator, memory, decider, encoder and output transducer'. Even if this model is not accepted, it is difficult to see how any alternative account of the process of living could exclude the concept of information processing.

Many would argue, in fact, that the concept of information is the most powerful idea so far contributed by the Systems Movement, comparable in importance with the idea of energy. Both are abstractions; both have considerable explanatory power; both generate conjectures which can be put to the test experimentally. Physics would be a chaotic subject without the idea of energy, defined as the capacity to do work. Systems thinking, similarly, could not do without the idea of information, although its precise definition raises some problems which are not yet solved. It is important in reaching an understanding of the nature of systems thinking to acknowledge both the importance of information as a concept and the severe limitations of information theory as it exists at present.

Although the rise of civilizations has to a large extent depended upon man's ability to establish long-term continuity of communication beyond mere moment-to-moment discourse, it is only in the last sixty years that communication has been the subject of scientific study, and information has emerged as an important scientific concept. (See Cherry, 1957, for a rich general essay on human communication.) The work done independently by Wiener and by Shannon in the late 1940s, already noted, which established information theory as a scientific subject, built upon the earlier work of communication engineers, notably the work of Nyquist (1924) and Hartley (1928).

The basic conceptualization in communication system engineering is that an information source produces a 'message' which is coded to produce a 'signal'; this signal is transmitted along a 'channel', which will inevitably introduce some unwanted disturbances called 'noise'; signal plus noise then pass to a 'decoder' which regenerates the original message, hopefully little distorted, for the receiver. Now it is obvious that engineers concerned with such systems will be interested in comparing one with another, and that this will require some way of measuring what is transmitted. In particular the measurement of the rate at which the coded message passes through the channel—'channel capacity' as it is called—will require a quantitative measure of what is being transmitted, a quantitative measure of 'information'. It is the triumph of information theory that it provides this measure, Shannon's 1945 paper being the classic exposition (reprinted in Shannon and Weaver, 1949). The accompanying misfortune, however, is that what is measured under the name of 'information' in information theory actually bears very little relation to what we understand by the word in everyday language. In normal usage information is obtained by our attributing meaning to data: I convert the data provided by the positions of the hands of a clock into the information that I am late for an appointment. The quantitative measure of 'information' in information theory, however, has nothing at all to do with meaning, the reason for this lying in its origin among engineers. Their concern is for the efficiency of the process whereby a message—*any* message—becomes a signal which is transmitted and received. They are not concerned, *as engineers*, with the content of the message, hence the limitations on the technical definition of 'information' in statistical information theory, which applies indifferently to both significant and trivial messages.

Given the nature of the general model of a communication system, and ignoring any consideration of the value of the message to a human being receiving it, it is obvious that measures of communication system performance could derive from several different aspects of it. Firstly, the information capacity of a communication channel, if measurable as the quantity of information transmitted per second, would tell us something highly significant about the system. Alternatively, concentrating on the code as a signal source, the information content of the signal transmitted will presumably bear some definite relation to the set of all possible signals which the code symbols could provide. Finally, there must in principle be a measure proportional to the degree of confidence that the message received is that transmitted and not some distortion of it. In fact all three measures are used, and all involve an expression for 'information content' of the same mathematical form. Since my concern here is only to illustrate the nature of information theory, I need only briefly refer to one of these measures. (For fuller treatments, see Shannon and Weaver, 1948, and many other texts, for example, Cherry, 1957; Rosie, 1966; Bell, 1968; both Sayre, 1976, and Chapman, 1977, give short treatments developed for a 'systems' rather than an 'engineering' audience.)

Consider a source code containing sixteen equally used symbols. It is easily checked that four binary (yes/no) questions are required to specify (i.e. to select) one symbol from sixteen. (The first tells you whether the symbol is in the first eight or the last eight, the second whether it is in the first (or the last) four, etc.) This arises from the relationship $2^4 = 16$, an equation which may be rewritten $\log_2 16 = 4$. Thus it seems that the informational nature of this source, namely that a symbol requires for its selection four units of information (or 'bits' as they are called) can be expressed as $\log_2 16$. In general for a code of N equally-likely symbols the expression will be $\log_2 N$. Since the probability of selection p for any one of these symbols is $1/N$ this may be written as $-\log_2 p$. The theory has shown that in general, for codes having any number of symbols which are not equally likely to be selected, the expression is still of this form, being $-\Sigma_i p_i \log p_i$. This is the expression for information content which is derived for each of the three potential measures referred to above. The suggestion that the logarithmic expression be used to measure information was made by Hartley. Shannon's great contribution was to extend the theory to include such things as the effects of noisy channels, and to deduce a number of theorems which are basic to the design of communication systems. For example, one of Shannon's theorems demonstrates to design engineers that any channel has a maximum capacity related to channel constraints which the engineer cannot ever exceed: his job is thus to define and approach that limit as closely as possible.

Statistical information theory is a considerable and spectacular achievement, bringing into the domain of science a new area of human activity. Simply because of this we should note rather carefully its limitations and discount some of the grander claims made for it. The theory covers more than adequately those aspects of information which are crucial to the communication engineer. In doing this it necessarily assumes that any 1000 words of prose has the same information content as any other 1000 words. The measures of information content have to do only with the frequency of occurrence of symbols, nothing at all to do with what they symbolize. This latter is usually called 'semantic information', but semantic information theory is in its infancy. Although Wiener treated 'amount of information' and 'amount of meaning' as synonyms (Bar-Hillel, 1955, discusses this), the distinction was recognized by Shannon, who insists that 'these semantic aspects of communication are irrelevant to the engineering problem' and by Weaver, whose early paper discussing this was reprinted with Shannon's classic paper in 1949. Weaver suggests that

> The concept of information developed in [Shannon's] theory at first seems disappointing and bizarre—disappointing because it has nothing to do with meaning, and bizarre because it deals not with a single message but rather with the statistical nature of a whole ensemble of messages . . .

He defines three levels of problem: level A, that of 'the technical problem' of signal transmission (with which the theory deals); level B, 'the semantic problem' of how precisely the symbols convey the desired meaning; and level C, 'the effectiveness problem' of how the meaning affects the recipient's conduct. He points out the hierarchical nature of the levels, in that levels B and C 'can make use only of those signal occurrences which turn out to be possible when analyzed at' level A, and hopes that his discussion has 'cleared the air . . . ready for a real theory of meaning'. This has been very slow to emerge, and represents an extremely difficult problem. Bar-Hillel and Carnap (1953) have made a modest step towards it based upon Carnap's work on inductive probability; they try to set up a measure of the *a priori* probability of very simple declarative statements, and hence are concerned with Weaver's level B in a particularly simple instance. Ackoff (1957) and Ackoff and Emery (1972) bring in level C; they set up a series of tentative definitions and measures related to the objectives of an individual who receives information, his valuation of his objectives, his possible courses of action and the probabilities that he will choose a particular course of action to pursue a particular objective. For example in the 1972 work they define information as a communication which produces a change in any of the receiver's probabilities of selecting a particular course of action, and motivation as a communication which 'produces a change in any of the relative values the receiver places on possible outcomes of his choice' of a course of action. The unsolved problem is to make such definitions operational. The authors comment:

> Such applications are not easy. They are time-consuming and costly, and may require a degree of control over subjects that is difficult, if not impossible to obtain.

Nevertheless the hope is that

> ... the existence of such measures, even when not practical or easy to apply ... provides an *objective standard* for which indexes can be sought.

It is appropriate to say again: these are early days in the systems movement.

The Shape of the Systems Movement

The two previous sections have discussed the emergence in different areas of study of two pairs of ideas which are the core of systems thinking: emergence and hierarchy, communication and control. They provide the basis for a notation or language which can be used to describe the world outside ourselves. Together these ideas provide an outline both for a systems account of the universe and for 'a systems approach' to tackling its problems, this latter being complementary to the reductionist approach embodied in the method of the natural sciences. A systems account of the observed world and a systems approach to its problems are found in many different disciplines; together all these efforts constitute what I mean by 'the systems movement'. It is the set of attempts in all areas of study to explore the consequences of holistic rather than reductionist thinking. The programme of the systems movement might be described as the testing of the conjecture that these ideas will enable us to tackle the problem which the method of science finds so difficult, namely the problem of organized complexity.

It might have happened that the exploration of holistic thinking developed in various disciplines using the language appropriate to each different subject. What in fact has happened is that wholes in many different areas of study, from physical geography to sociology, have been studied using the ideas and the language appropriate to systems of any kind. That the systems movement is, even on a jaundiced view, at least a loose federation of similar concerns—linked by the concept 'system'—is the main achievement of Ludwig von Bertalanffy. The greatest individual contribution to this minor intellectual revolution of the 1940s (it is too early to know whether it might eventually be seen as a major one) is probably that of Norbert Wiener. But it was Bertalanffy who insisted that the emerging ideas in the various fields could be generalized in systems thinking; hence it is he who is recognized as the movement's founder. (Gray and Rizzo's *Festschrift*, 1973. gives a panoramic view of his work and interests.) His own thinking as reflected in his writings shows little development from the 1940s until his death in 1972, and Bertalanffy's work is not unfairly described by Lilienfeld (1978) as 'rather repetitious and even static in character'. Bertalanffy's unchanging view, which might fairly be described as his *vision*, was that there would arise as a result of work in different fields a high-level meta-theory of systems, mathematically expressed. This aspiration is clear in the founding documents of what is now the Society for General Systems Research. In the Journal *Philosophy of Science*, 1955, for example, is an announcement (page 331) that ('A Society for the Advancement of General Systems Theory is in the process of organization'. Those concerned were the biologist Bertalanffy, together with an economist (K. E. Boulding), a physiologist (R. W. Gerard), and a mathematician (A. Rapoport). The purpose was to encourage the development of 'theoretical systems which are applicable to more than one of the traditional departments of knowledge'. The aims of General Systems Theory (GST) were to be:

1. To investigate the isomorphy of concepts, laws, and models in various fields, and to help in useful transfers from one field to another;
2. To encourage the development of adequate theoretical models in areas which lack them;
3. To eliminate the duplication of theoretical efforts in different fields;
4. To promote the unity of science through improving the communication between specialists.

The main achievement of the organization resulting from this initiative has been the annual publication of the wide-ranging papers in the *General Systems Yearbook*. The general theory envisaged by the founders has certainly not emerged, and GST itself has recently been subject to sharp attacks, notably by Berlinski (1976) and Lilienfeld (1978). Chiding Lilienfeld for bothering to attack 'a mirage', Naughton (1979a) suggests that there is nothing approaching a coherent body of tested knowledge to attack, rather 'a mélange of insights, theorems, tautologies and hunches . . .'.

The difficulties of GST, and the kind of temptation it faces, are illustrated by a teasing problem in information theory. The mathematical form of the expression for information content is the same as that for entropy in statistical mechanics but with the sign reversed. Now, since entropy measures degree of disorder, and information may plausibly be regarded as that which reduces uncertainty and hence increases order, it is extremely tempting to equate information with negative entropy ('negentropy'). In an influential little book, Schrödinger (1944) declared that the living organism 'feeds on negative entropy', and Wiener (1948) had no doubt that the order which an organism extracts from its environment is synonymous with information:

> Just as the amount of information in a system is a measure of its degree of organisation, so the entropy of a system is a measure of its degree of disorganisation; and the one is simply the negative of the other.

In fact this apparent link between information and entropy, which is simply assumed in much of the literature, is exceptionally difficult to establish rigorously (Bell, 1968, and Chapman, 1977, discuss the problem, which hinges on the different dimensions of the constants in the two expressions). The similarity of mathematical form between information and negative entropy does not establish any physically meaningful connection between the two concepts. Mathematical analogies never can establish such connections, but GST has little content beyond such analogies.

The problem with GST is that it pays for its generality with lack of content. Progress in the systems movement seems more likely to come from the use of systems ideas within specific problem areas than from the development of overarching theory.

Although GST does not itself provide a means of picturing the totality of work going on in the systems movement, the distinction just made—between the development of systems thinking as such and the application of systems thinking within other areas, or disciplines—can be extended to provide a reasonable map of all the activity of the movement (Checkland, 1979a). This I find useful in 'placing' any particular piece of work and in keeping a grasp on the growing literature of this meta-discipline.

To construct the map a number of distinctions have to be made. Firstly, make a distinction between the development of systems ideas as such (as in, say, cybernetics) and in the application of systems ideas within an existing discipline (as in the Cambridge geographers' re-writing of geography from a systems point of view—Chorley and Kennedy, 1971; Chapman, 1977, Bennett and Chorley, 1978). This gives two broad areas of systems work. Secondly, within the work on systems thinking as such, distinguish between purely theoretical development of systems ideas and their interrelationships, and work based on the notion of developing the ideas by seeking to 'engineer' systems in the real world, using that word in its broad sense. GST is an example of the former, the development of systems engineering methodology an example of the latter. But 'hard' systems engineering (to be discussed in Chapter 5) is only one example of the development of systems thinking by attempts at problem-solving. There are others, and this leads to a third distinction: that between (a) engineering 'hard' systems as such; (*b*) using systems ideas as an aid to decision-making (as in operational research); and (*c*) using systems thinking to tackle 'soft' ill-structured problems (as will be discussed in Chapters 6 and 7). These distinctions are set out below.

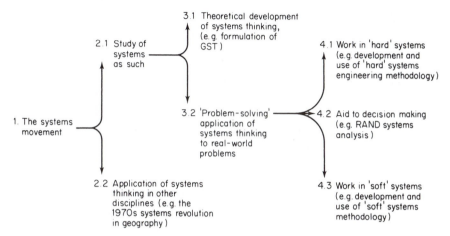

These seven activities within the systems movement make the picture in Figure 2. Also indicated are the major influences of one body of systems work upon another. Thus the engineering of hard systems has been powerfully affected by the development of control theory and by information theory as developed by communication engineers; and systems engineering itself gave a powerful impetus to the work on a 'soft' systems methodology for ill-structured problems described in Chapters 6 and 7. Significantly missing is an arrow from 3.1 to 4.2; on the whole the RAND/OR/management science world has been unaffected by the theoretical development of systems thinking, it has been *systematic* rather than *systemic* in outlook, in spite of the insistence of the first textbook of operational research that 'the comprehensiveness of OR's aim is an example of a "systems" aproach' (Churchman *et al.*, 1957) and its admirable sentiment that 'OR is concerned with as much of the whole system as it can encompass' within the constraints of time and resources.

The influences shown are internal to the systems movement. Figure 2 can also be extended to include the major influences of external bodies of knowledge. This is done in Figure 3 which indicates, for example, that the major influences on RAND systems analysis, OR, and management science generally have been the methods of the natural sciences (understandable given the historical origins of OR) and economics.

It is important to be clear about the status of this 'map of the systems movement'. It is not, as such, a picture of the real-world systems movement: any real-world systems project may well cut across several of the map's categories. In itself, the map is no more than a set of logical distinctions; it is not even a set of distinctions which *have* to be made; it is

96

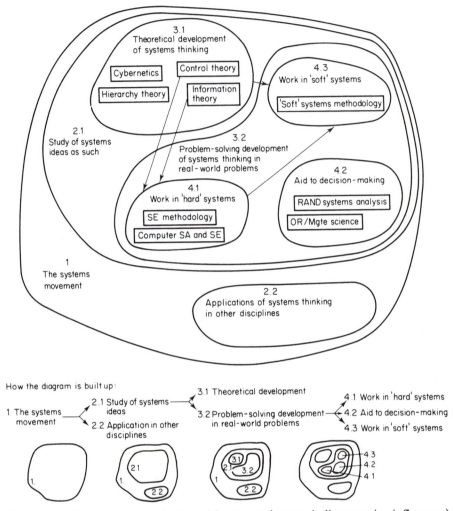

Figure 2. The shape of the Systems Movement (arrows indicate major influences)

a set I have chosen to make. What this particular set provides, I argue, is a picture of the systems movement which does map well on to current real-world systems activity, its intellectual efforts and its literature, and which also enables any particular piece of work or literature to be placed in the context of the movement as a whole.

Looking back over the whole history of the science movement and the emergence within it of the systems activity displayed in Figures 2 and 3, the nature of this development can be understood in terms of changing concepts of *a machine*. The concept of a machine developed by Newtonian physics is that of a clock-like mechanism, deterministic and

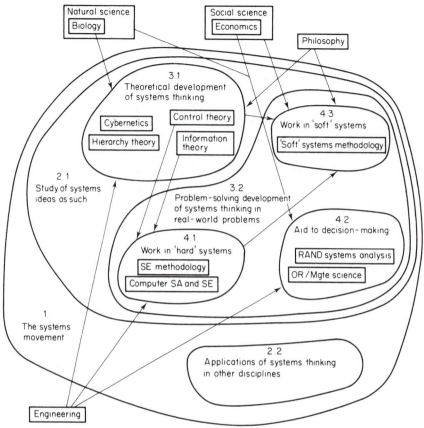

Figure 3. The shape of the Systems Movement indicating major external influences

preprogrammed. Modern science, with the creation of quantum mechanics contributes the idea of the statistical machine, the fine detail of whose behaviour is not deterministic. The systems movement then adds the concept of the self-regulated machine which has autonomous control over its own behaviour. Systems thinking has now been exploring the implications of this concept for about thirty years.

The results so far might be fairly described as 'significant but not spectacular'. Why has progress not been more rapid? One reason is undoubtedly the strong grip which reductionist thinking has on anyone educated in Western civilization. It is not simply that reductive analysis is, in the phrase of Medawar and Medawar (1977) 'the most successful explanatory technique that has ever been used in science', it is also the case that the counter position to reductionism, based on holism and emergence, does not provide as clear a philosophy. (The rather woolly Messianic air of some of the GST writings is counter-productive in this respect.) The point is strongly made by Marjorie Grene (1974):

> ... the anti-reducibility position is, for many people, impossible to accept. Why? Because it breaks through the defences of a simple, one-level physicalism without providing an alternative metaphysic to take its place. To think anti-reductively demands thinking in terms of hierarchical systems, of levels of reality and the like; but we don't know any longer how to think in that way—and to be told, even to *know*, that the contrary position is absurd does not in itself allow us to embrace wholeheartedly what ought to be the more reasonable alternative. For anti-reductivism is 'reasonable' only in the perverse sense that its negation is self-contradictory, not in the more substantive sense of fitting smoothly into a *Weltanschauung* in which, as people educated in the ideal of a 'scientific' world view, we can feel at home.

It must be the current task of the systems movement to work towards a version of systems thinking within which we can feel 'at home' in a positive sense.

Chapter 4

Some Systems Thinking

The systems movement comprises any and every effort to work out the implications of using the concept of an irreducible whole, 'a system', in any area of endeavour. The value and limitations of the concept can be examined in virtually all of the arbitrary divisions of human knowledge which we presently know as the separate disciplines, so that we find that there are systems thinking scientists, technologists, engineers, economists, managers, management scientists, psychologists, sociologists, anthropologists, geographers, political scientists, historians, philosophers, artists . . . and many more. Because systems ideas provide *a way of thinking* about any kind of problem, systems thinking is not itself a discipline, except to the extent that there will be a few people whose professional concern is with systems concepts as such. The number of this latter group, the 'general systems theorists', is likely to remain small, given the sheer generality, and hence the lack of specific content in systems ideas as such. Their activity has been fairly described by Boulding (1956) as:

> a level of theoretical model-building which lies somewhere between the highly generalised constructions of pure mathematics and the specific theories of the specialised disciplines.

This activity, if it is to be useful in any practical sense, will have to interact constantly with the work of the systems thinkers within particular disciplines, and similarly the results of systems thinking in, say, management science need to be accessible to interested professionals in disciplines such as sociology or political science. There is therefore a need for a basic *language* of systems ideas which is meta-disciplinary, and also, perhaps, for an agreed broad-level account of the world in systems terms.

It would be naive to imagine that any such language or any such general systems model will be consciously adopted by systems thinkers in widely different disciplines. Rather, over a period of time, there is likely gradually to emerge a consensus on the ideas which have been found useful and on the language in which they are expressed. This is a process which probably cannot be artificially accelerated, and I shall not attempt to do so here. What follows in this chapter is an account of some very basic systems thinking; in particular it includes the particular systems thinking which underlies the systems practice to be described in Part 2.

Some Basic Systems Thinking

First let us assume, with the realists, the proposition which subjectivists would deny, namely that there is outside ourselves a reality which does actually exist. Billie Holiday sings: 'for all we know, this may be a dream . . . ' and it is hard to prove her wrong. But when I observe an ant walking along the ground climbing over a stone in his path, and when independent witnesses (with whom I am not in collusion) or, for that matter, automatic cine-cameras confirm that the ant climbs over the stone, I am prepared to accept that something we perceive as ant and stone do exist.†

Now let us imagine an observer, a would-be describer of that world which exists outside ourselves. Suppose that the observer is 'a systems thinker'. What does that imply about his response to the problem of providing a description of the external world?

He may well appreciate the potency of Descartes's dictum that every problem should be broken down into as many separate simple parts as possible, and he may well agree with Medawar and Medawar (1977) that 'reductive analysis is the most successful explanatory technique that has ever been used in science'. But being a systems thinker he will also be aware of the problems with which the reductionist method of science cannot cope, especially the problems of the real world, as opposed to those defined in the laboratory; and he will take seriously the arguments outlined in the previous chapter, that nature will be hierarchically organized with emergent properties at various levels of complexity. Above all he will be ready to try out the usefulness of thinking in terms of coherently organized entities which cannot properly be reduced merely to an aggregate of their components. He will seek an account of the structure of reality and of the processes observed going on within it in terms of whole entities ('systems') *which he will define*.

The structure of this basic picture of an external reality and an observer/describer of it who will couch his description in particular terms is important. It emphasizes the status of systems ideas as a *language* by means of which reality may be described. These are early days in the systems movement, and the stage reached is that the movement is still testing the proposition that systems concepts can be the basis of a fruitful epistemology. Eventually we may reach the stage at which there are examples of well-tested 'public knowledge' which is *systems* knowledge, knowledge which could have been gained only by means of systems thinking. When this happens we shall have objective systems knowledge in what Popper (1972) calls 'World 3', the world containing the products of

---

†Popper (1972) gives convincing arguments for accepting realism rather than subjectivism.

the human mind (as opposed to 'World 1', the physical world, and 'World 2', the world of experience or thought in the subjective sense). When this stage is achieved we shall have a recognized systems epistemology *without a knowing subject*, and achieving it ought to be a prime objective of the systems movement. It is a legitimate criticism of the systems movement, in fact, that too much of the work within it so far has been a self-indulgent development of systems language for its own sake, or an elaboration of untestable assertions. But the ultimate objective is clear: the attainment of public knowledge of the kind which science accumulates, by means of a modified scientific method in which a form of holism replaces reductionism. Until such knowledge is accumulated our basic model has to include both an external reality and an observer/describer who will, *for his own purposes*, use systems thinking as a means of arriving at his description. Hence the reason for the italicized phrase at the end of the previous paragraph. We are at present sufficiently far from achieving an agreed systems account of reality that in appraising systems accounts of the world we need to know the observer's purpose in making the description.

Our observer may have various motives for making his systems description, and the latter will reflect the nature of his motivation. He may be motivated by *curiosity*, aiming only to observe and describe in order to ascertain whether clear and intelligible descriptions are possible in systems terms. Or he may want to *make use* of a systems description in some kind of problem-solving. Or he may want to bring about changes in a part of reality, his motive may be one of *design*. These motivations reflect three different roles for the observer. He may be a 'natural historian', describing and classifying, a 'manager', or 'an engineer'.

We cannot say very much about the observer and his systems description which will be true regardless of his role and purpose. All that we can say at this general level is that he will identify (or define) some entities which are coherent wholes. He will perceive (or invent) some principles of coherence which makes it meaningful to draw a boundary round an entity, distinguishing it from its environment; and he will identify (or envisage) some mechanism of control by means of which the system-entity retains its identity at least in the short term. The existence of the system boundary defines as 'inputs' or 'outputs' anything which crosses it, and these flows may be physical, e.g. materials, people, machines, money, or abstract, e.g. information, energy, influences. Similarly the components of the system itself may be physical entities or abstractions; in either case the components will show some degree of organization beyond that of a random aggregate of components. And finally, any whole conceived as 'a system' is, in general, at least potentially a part of a hierarchy of such things—it may contain 'sub-systems' and itself be a part of 'wider systems'.

We can also say, in general, that the observer/describer will be able to describe the behaviour of his systems in two ways. He may concentrate exclusively upon the inputs and outputs, in which case the system is treated as a so-called 'black box' embodying a transformation process which converts the one into the other. Or he may describe the internal *state* of the system in terms of suitable variables, and the trajectory of it from one state to another under the influence of external conditions (Ashby 1956; Klir, 1969, contains a useful short 'summary of attempts to formulate general systems theory' which covers approaches to systems behaviour deriving from both cybernetics and generalized electrical circuit theory).

This is about the limit of what we can say about every example of systems thinking. In summary, there will be: an observer who gives an account of the world, or part of it, in systems terms; his purpose in so doing; his definition of his system or systems; the principle which makes them coherent entities; the means and mechanism by which they tend to maintain their integrity; their boundaries, inputs, outputs, and components; their structure. Finally their behaviour may be described in terms of inputs and outputs or via state descriptions.

All this is sufficiently general to cover examples ranging from, say, physical systems like alarm clocks or fire engines, through systems involving human agents, such as a football match (where, for example, the coherency principle might be taken to be the players' readiness to keep to the arbitrary rules of the game as administered by a sometimes fallible referee) to entirely abstract examples like Mary Hesse's (1976) account of science as a learning process. In cases like this latter one (the social science literature brims with them) in which the system description 'is intended only to indicate what relations must subsist between scientific learning and the external world if the empirical character and objectivity of science is to be maintained', the components are logically connected abstractions and the coherency principle is really no more than the observer/describer's allegiance to his or her model. Others will accept it as useful if they applaud the observer's purpose and find the working out of the description defensible, or if some problem can be solved by the use of it.

Given the systems hypothesis that it will be insightful to take the apparently chaotic universe to be not a set of phenomena (whose laws can be established by the reductionist experimental approach) but rather a complex of interacting wholes called 'systems', it is not surprising that a number of general attempts to describe and classify the possible types of system have been made: living and non-living systems, concrete and abstract systems, closed and open systems, etc. There is as yet no generally accepted classification, and many suggestions reflect a particular outlook, interest or purpose; hence lists of basic system types frequently mix logical categories, something which might not matter in a particular area of application but which will invalidate any general systems description of the world. Jones (1967), for example, gives a systems classification claimed to be useful from an ergonomist's point of view:

Manual, Mechanised, Automatic, Collaborative man-machine systems,

Mechanical sub-systems, Administrative, Voluntary, Environmental, Biological, Physical and Symbol systems.

Burton (1968) on the other hand, writing on a systems approach to the subject of international relations, offers a very different framework:

Basic, Operational, Behavioural, Purposeful and Controlling systems, together with linked systems and administrative systems.

Both of these classifications might be useful within their particular areas of application, but both are too confused conceptually to offer a basic account of the world in systems terms. More useful from this latter point of view are the general accounts of system types given by Boulding (1956) and Jordan (1968). These were also the starting point for the development of the fundamental systems thinking (Checkland, 1971) which underlies the systems research to be described in later chapters.

A Hierarchy of Systems Complexity

Boulding's paper of 1956, 'General Systems Theory—the Skeleton of Science' is very well known, and is a frequent starting point for discussion of systems ideas. It is perhaps surprising that it has not led to further work starting from the base it furnishes — that it has not is a measure of the immaturity of systems thinking as an area of endeavour. And it is instructive that the problem it poses has not yet been solved, as we shall see.

In the paper Boulding introduces the idea of General Systems Theory (GST) and argues that its concern is not with a single self-contained 'general theory of practically everything' since we pay for generality by sacrificing content, and 'all we can say about practically everything is almost nothing'. Nevertheless there ought to be a level at which a general theory of systems can achieve a compromise between 'the specific that has no meaning and the general that has no content'. Such a theory could point out similarities between the theoretical constructions of different disciplines, reveal gaps in empirical knowledge, and provide a language by means of which experts in different disciplines could communicate with each other. As evidence for the need for such general meta-disciplinary theories, Boulding cites the emergence of an increasing number of hybrid disciplines of mixed parentage such as cybernetics, information theory, organization theory, and management science, hybrids which have connections with many different fields of study. He suggests that the approach adopted to the development of GST might be either to establish theories of very general phenomena, such as birth-growth-death or the interaction of an individual entity with an environment, or to

> arrange the empirical fields in a hierarchy of complexity of organisation of their basic 'individual' unit of behaviour, and to try to develop a level of abstraction appropriate to each.

He adopts this latter course and presents a preliminary hierarchy of the individual 'units' found in empirical studies of the real world, the position of an item in the hierarchy being determined by its degree of complexity as judged intuitively. The hierarchy is summarized in Table 2. Boulding points out the absence of adequate systems models above level 4, and suggests that the use of the hierarchy is in pointing to gaps in knowledge and in serving as a warning that we should never accept as final 'a level of theoretical analysis which is below the level of the empirical world which we are investigating'. Finally GST is 'the skeleton of science' in that it

> aims to provide a framework or structure of systems on which to hang the flesh and blood of particular disciplines and particular subject matters in an orderly and coherent corpus of knowledge.

Let us now examine this piece of systems thinking in terms of the rudimentary analysis of the previous section. Boulding is here an observer/describer whose concern is 'the general relationships of the empirical world'. His purpose is to make connections between the different areas of empirical investigation, and he hopes to do this by describing some aspects of a system—an abstract one—whose boundary is that of the empirical investigation of the world. Within his overall system are components which are the levels of complexity of empirical studies. He hopes to pursue this purpose of connecting the different empirical disciplines by identifying the entities which are investigated at the different levels. He takes it that the entities studied are themselves systems and he describes the overall system by giving mainly concrete examples at the various levels. The entities exemplifying his overall system are a coherent system in his eyes because he sees them as things of the same kind (in fact, 'systems') ranked in order on a scale which is a measure of complexity; this gives his overall system its structure. His method of approach is to start, not from the real-world disciplines, but from an intuitive account of the levels of complexity which he subsequently relates to the different empirical sciences.

The whole systems picture which Boulding paints is not itself an empirical finding; we can ask of it only: Is it convincing and does it help resolve any problems?

Taking the latter question first, the hierarchy can certainly be a source of insight. For example it provides a means of appreciating the history of the hybrid discipline management science. The pioneer of 'scientific

Table 2   An informal intuitive hierarchy of real-world complexity (after Boulding, 1956)

Level	Characteristics	Examples (concrete or abstract)	Relevant disciplines
1. Structures, Frameworks	Static	Crystal structures, bridges	Description, verbal or pictorial, in any discipline
2. Clock-works	Predetermined motion (may exhibit equilibrium)	Clocks, machines, the solar system	Physics, classical natural science
3. Control mechanisms	Closed-loop control	Thermostats, homeostasis mechanisms in organisms	Control theory, cybernetics
4. Open systems	Structurally self-maintaining	Flames, biological cells	Theory of metabolism (information theory)
5. Lower organisms	Organized whole with functional parts, 'blueprinted' growth, reproduction	Plants	Botany
6. Animals	A brain to guide total behaviour, ability to learn	Birds and beasts	Zoology
7. Man	Self-consciousness, knowledge of knowledge, symbolic language	Human beings	Biology, psychology
8. Socio-cultural systems	Roles, communication, transmission of values	Families, the Boy Scouts, drinking clubs, nations	History, sociology, anthropology, behavioural science
9. Transcendental systems	'Inescapable unknowables'	The idea of God	?

Notes: (1) Emergent properties are assumed to arise at each defined level.
     (2) From level 1 to level 9: complexity increases; it is more difficult for an outside observer to predict behaviour; there is increasing dependence on unprogrammed decisions.
     (3) Lower level systems are found in higher level systems—e.g. man exhibits all the distinguishing properties of levels 1–6, and emergent properties at the new level.

management', F.W. Taylor (1947 reprints), treated engineering workshops as 'level 2' systems. He believed, mechanistically, that a workman 'deliberately plans to do as little as he safely can' but is 'glad to work at maximum speed for 30–100% more than the average wage of his trade'.

Taylor purports to show how to obtain the extra productivity which will more than pay for the extra wages. During and after the Second World War, and especially following the publication of Wiener's (1948) book the development of cybernetics put emphasis on 'level 3' feedback control systems. Most recently there has been a considerable attempt to bring in behavioural science in order to treat management problems at levels 7 and 8. Thus the historical development of management science can be seen as an attempt to treat its problems as being those of ever more complex systems. This can be a useful perspective. It serves as a reminder, for example, that a typical management science model constructed in terms of multiple interacting feedback loops, even if complicated, is only a level 3 model and hence can cover only certain aspects of a management problem at level 8. Management scientists have been known to claim more.

But the most interesting question is the first one: is Boulding's hierarchy convincing? Actually the most appropriate question is: Why is it so convincing? There is an interesting puzzle here which seems barely to have been noticed. Everyone agrees that the hierarchy is convincing; everyone can recognize intuitively the entities whose emergent properties signal a new level; and no one ever argues that the ranking order is wrong. Yet this unanimity is itself problematic, in that, more than twenty years after Boulding's paper was published, we still have no definition of the nature of the scale of 'system complexity' which everyone finds so convincing. Hence we still cannot even argue intelligently about the relative sizes of the gaps between levels.

Initially an approach based upon identification of the new factor needed to specify a system which characterizes a new level of the hierarchy seems promising. To specify a structure, length and perhaps mass are needed; clock-works require lengths, masses, and *time*; a thermostat specification requires lengths, masses, time, and a measure of *information*. But thereafter, not surprisingly, this essentially reductionist approach breaks down, and the problem remains that we have no adequate account of systemic complexity. The revelation of this ignorance is itself a useful outcome of Boulding's piece of systems thinking.

A Systems Taxonomy

A second example of very general systems thinking is Jordan's (1968) attempt to construct a systems taxonomy. Descartes would have approved of this, since Jordan's strategy is that embodied in his rules for thinking. Descartes was most impressed by the reasoning methods of Euclidean geometry; he advocated not only breaking down problems into separate parts (the reductionist second rule) but following this by building up gradually from the parts 'as far as the knowledge of the most complex', on the pattern of 'these ... chains of reasonings which geometers are accustomed to using' (Sutcliffe translation, 1968). Jordan starts from intuitive guesses at three organizing principles which might enable us to perceive a group of entities as 'a system'. The principles are *rate of change*, *purpose*, and *connectivity*. Each principle defines a pair of systems properties which are polar opposites, thus: rate of change leads to the properties 'structural' (static) and 'functional' (dynamic); purpose leads to 'purposive' and 'non-purposive'; and the connectivity principle leads to the properties of groupings which are densely connected ('organismic') or not densely connected ('mechanistic' or 'mechanical'). There are eight ways of selecting one from each of the three pairs of properties, giving eight cells which are potential descriptions of groupings worthy of the name 'system', e.g. structural/purposive/mechanical, functional/non-purposive/organismic, etc. Having built up this framework Jordan now looks around the real world for examples of systems meeting the requirements of each cell. Having found them to his satisfaction (see Table 3) he argues that in talking about systems we should use only 'dimensional' descriptions of this kind, and should avoid especially phrases such as 'self-organizing system' which are no more than 'verbal magic'. He cites the transfer of the control engineering concept of feedback to the description of living organisms—with the brain as a kind of servomechanism — as an analogy which has been a cause of conceptual confusion because it has been taken to be more than that, to be in fact a scientific description:

> The conceptual confusion accompanying 'self-organizing system' results from the fact that explicitly and/or implicitly the scientists in defining this term wish to lump together both a certain set of physical systems and living organisms. This cannot be done, since the set of physical systems they have in mind is only superficially similar to living organisms . . .

Jordan concludes that descriptions of systems should be sparse, if verbal magic is to be avoided:

> The only things that need be common to all systems are identifiable entities and identifiable connections between them. In all other ways systems can vary unlimitedly.

This is a needed warning, one too little heeded in the systems movement. It parallels Berlinski's (1976) justified complaint about the purely 'ceremonial' use of mathematics by general system theorists.

The overall argument of Jordan's paper is that there is a core meaning to the word 'system' which makes it proper to attach it to many different things perceived in the real world outside ourselves. (He begins by quoting fifteen different dictionary definitions of the word.) The core meaning, he argues, includes a set of entities and connections between them; his three 'bipolar dimensions' describe the information needed to specify a given

108

Table 3   Dimension-based taxonomy of systems (after Jordan, 1968)

Three principles lead to three pairs of properties:

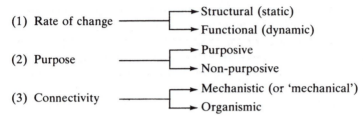

(1) Rate of change ⟶ Structural (static) / Functional (dynamic)

(2) Purpose ⟶ Purposive / Non-purposive

(3) Connectivity ⟶ Mechanistic (or 'mechanical') / Organismic

In (3), in a mechanistic system the remaining elements are unchanged when some elements (or the connections between them) are changed, removed or destroyed. In an organismic system a change in one affects all.

The three 'bipolar "dimensions"' generate eight cells:

Cell	Example
1. Structural Purposive Mechanical	A road network
2. Structural Purposive Organismic	A suspension bridge
3. Structural Non-purposive Mechanical	A mountain range
4. Structural Non-purposive Organismic	A bubble (or any physical system in equilibrium)
5. Functional Purposive Mechanical	A production line (a breakdown in one machine does not affect other machines)
6. Functional Purposive Organismic	Living organisms ('The most parsimonious way to understand life at all its levels ... is by means of purpose')
7. Functional Non-purposive Mechanical	The changing flow of water as a result of a change in the river bed (or, perhaps, the pattern of shadows thrown on a lawn by a tree)
8. Functional Non-purposive Organismic	The space–time continuum

example of a system. The preceding three sections of the paper discuss cognition. In them it is argued that we have the ability to distinguish a figure and a background, and to segregate 'various different figures and backgrounds depending upon, among other things (our) *interests'*. Given this recognition of the need for an observer with a particular interest, it is surprising that the weakest part of Jordan's argument is that in which systems are described as 'purposive' or 'non-purposive'. Surprisingly, Jordan leaves out the observer/describer of a system and ascribes the purpose, or lack of it, to the system itself. For example it is said of a road network, a specimen of a cell 1 (structural, purposive, mechanical) system, that 'it has an obvious purpose'. What is on the contrary obvious is that the *designers, builders*, and *potential users* of the road network had a purpose. The road network itself, as far as intrinsic purpose is concerned, is similar to a mountain range, the cell 3 (structural, non-purposive, mechanical) example. The important difference here is that one system serves a designer's purpose, the other is a system without purpose, a natural system which, as such, could not be other than it is.†

Consideration of Jordan's taxonomy thus re-emphasizes the need, in systems accounts of the world, to start from an observer/describer, and also draws attention to the distinction between natural and designed systems. This was found to be an important distinction in the systems thinking which underlies the systems research described later.

A Systems Typology

The research intention was to examine the applicability of systems ideas within all of the areas which were described in the previous chapter as problems beyond the scope of the method of science, namely complexity, the extension of science to cover social phenomena, and the problem of scientific methodology in real-world problems rather than in those which the scientist can himself define in a laboratory. The intention was to examine 'management' problems, broadly defined.

According to Boulding's hierarchy such problems will inevitably be of immense complexity, being set in level 8 'socio-cultural systems'. In terms of Jordan's system dimensions, any problem perceived to arise in the real world, will do so in situations which can only be described as functional, (multi-) purposive, and organismic. Again the complexity is emphasized; Jordan's example for this cell is a living organism but it could equally well have been a culture. Neither example of systems thinking provides any direct guidelines for this research, though Boulding's warning that we should never accept analysis at a level below that of the problem faced is useful in a negative sense. Similarly Jordan's taxonomy, though not directly helpful provides a useful lesson in the realization that his omission of the

---

†It would be more 'verbal magic', which I join Jordan in condemning, to claim that a mountain range 'serves the purpose' of some supernatural being!

human observer leads him to imply—falsely in my view—that 'purposive', meaning 'serves a purpose' is the same as 'purposeful', meaning 'according to an act of will'. The importance of this distinction was in fact found to be critical.

If we accept that as a result of the process of evolution, both inorganic and organic, the universe contains some entities which show emergent properties, and hence are integral wholes rather than mere aggregates of components, then it is reasonable to look for different classes of entity depending upon their origin. This was done initially by developing an idea of Blair and Whitston (1971) who, in a book on industrial engineering, chart some physical systems seen to exist in the world. The approach, like those of Boulding and Jordan, is at core intuitive, starting from an account of 'how the world is' which seems convincing to a human observer or experimenter.

Systems Classes

Let us start with the physical systems which apparently make up the universe. These range from the subatomic systems of atomic nuclei as described by physics, through the physical framework of this and other planets and the living systems observed on earth, to galactic systems at the other extreme. All these are *natural systems*, systems whose origin is in the origin of the universe and which are as they are as a result of the forces and processes which characterize this universe. They are systems which could not be other than they are, given a universe whose patterns and laws are not erratic.

There are also many other observed entities which are similar to natural systems in respects other than this last one: they could be other than they are. These are the systems which are the result of conscious design. They are the *designed physical systems* which man has made, the class stretching from hammers via tram cars to space rockets. They are designed as a result of some human purpose, which is their origin, and they exist to serve a purpose, even though, as in the case of an artist's painting, for example, the purpose may be hard to define explicitly.

But man's design capability is not restricted to the construction of physical artefacts. We also see in the world a large number of what may be described as *designed abstract systems* such as mathematics or poems, or philosophies. They represent the ordered conscious product of the human mind. They are in themselves abstract systems, though thanks to previous successful design activity they can now be captured in designed physical systems such as books, films, records, blue prints. Again they will exist as a result of a positive act related to some objective—elucidation, maybe, or the enlargement of knowledge, or an inner urge to express the inexpressible.

The human act of design is itself an example of a fourth possible system class: the *human activity system*. These are less tangible systems than natural and designed systems. Nevertheless, there are clearly observable in the world innumerable sets of human activities more or less consciously ordered in wholes as a result of some underlying purpose or mission. At one extreme is a system consisting of one man wielding a hammer, at the other the international political systems needed if life is to remain tolerable for the human race on this small planet. The range covered by this class of system is very large indeed. What every member of the class has in common is that it consists of a number of activities linked together as a result of some principle of coherency. This will as a minimum consist of the observer's interest in viewing the set as a whole. For example, a dietitian might study the human activity system which consists of 'the eating habits of the octogenarians of Basingstoke', in which case the people whose habits are studied will probably be unaware of their involvement with this system. Or an observer might take as a system a football team seeking to win a championship; here the team members will themselves know of their involvement as crucial to the system's purpose, and will in fact have their own definitions of the purpose or mission which links the system's activities and marks its boundary. The components of all such systems I take to be human activities. In the initial version of the typology (Checkland, 1971) these were combined with the natural and designed systems which will inevitably be closely linked to the human activity described—for example a 'taking-leisure system' will consist of human activities involving various natural and designed physical and/or abstract systems such as playing fields, cricket bats, rules of games, etc. The research work has shown, however, that it is better to restrict the definition of the human activity system to the activities themselves, naming and describing other associated systems if appropriate at the time.

Beyond natural, designed physical, designed abstract, and human activity systems there has to be a category to include the systems beyond knowledge. Following Boulding we may term these *transcendental systems*.

This completes a simple systems map of the universe which, as far as system classes is concerned, is itself complete. It is summarized in Figure 4. Any whole entity which an observer sees as a figure against the background of the rest of reality, may be described either as a system of one of these five classes or as a combination of systems selected from the five. Pursuing systems thinking becomes a matter of ascertaining the properties of systems of each class, and the way in which they combine and interact to form wider systems showing emergent properties. The long-term programme of the systems movement may be taken to be the search for conditions governing the existence of emergent properties and a spelling out of the relation between such properties and the wholes which exhibit them.

Natural Systems

Our experience of the natural world gives us a profound belief in its orderliness. The sun *always* rises in the East, the colours of the rainbow

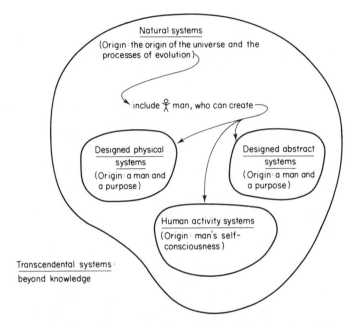

Figure 4. Five classes of system which make up a systems
map of the universe. We may — investigate, describe, learn
from, natural systems — create and use designed systems —
seek to 'engineer' human activity systems

always appear in the same pattern. Without this belief in a universe which
is not capricious, scientific investigation would be a meaningless enterprise.
With it, we may sensibly attempt to classify the systems of the natural
world, knowing that all members of a group will have similar properties,
and will remain members of that group.

Starting with atomic nuclei there is an obvious hierarchy through atoms
to molecules. Molecules in combination then give rise to a branched
hierarchy. In one branch we have inorganic crystals, rocks, minerals, the
non-living world; in the other branch are the special molecular wholes
which we call 'living things'. Enough is now known about their molecular
organization to make it possible to dispute precisely where this branch
starts, but certainly it will encompass a hierarchy which includes
single-celled creatures, the whole of the plant and animal kingdoms, and
the systems of ecology. An observer could also include here systems of
animal behaviour, such as courtship displays and animal communication
systems; and because this branch will also include human beings, we might
argue from this that 'human behaviour systems' should be included also.
This is reasonable to the extent that man is 'a naked ape'; but it will be
argued below that it is the fact that man is more than a naked ape which
makes it necessary to create different kinds of system beyond those which
are natural. These latter systems are systems which could be other than

they are, whereas the distinguishing characteristic of natural systems is that they could only be as they are, given their origins in a rationally intelligible universe.

In this systems typology I am claiming only that natural systems are the evolution-made, irreducible wholes which an observer can observe and describe as such, being made up of other entities having mutual relationships. They are 'irreducible' in the sense that meaningful statements can be made about them as wholes, and this remains true even if we can describe their components and the relationships between the components with some precision. Carbon dioxide is not reducible in this sense to carbon and oxygen, in that however much we know about interatomic distances and bond angles, carbon dioxide remains a higher level whole having properties of its own.

(These are limited claims, deliberately avoiding the kind of hard-to-defend broad claims which are sometimes made. Laszlo (1972a, see also 1972b), for example, describes natural systems not only as wholes with irreducible properties but goes on to add other characteristics: that they maintain themselves in a changing environment, 'create themselves in response to the challenge of the environment' and are 'coordinating interfaces in nature's hierarchy'. Such statements are best regarded as hypotheses which need to be expanded into statements which can be tested empirically. My concern here is only with the number and name of the classes of system needed to describe the universe.)

Many of the entities which appear in a hierarchy of natural systems are of course the subject matter of scientific disciplines. Though the ultimate ideal of its practical philosophy is reductionist explanation, most scientific work takes as given the wholes existing at some level in the natural hierarchy and tries to describe them and the laws which govern their behaviour. Most science is therefore concerned with the behaviour of particular systems, even though interest is not usually centred on system hierarchies or on the property of 'wholeness'. Neither does most science use systems language; but some does.

As an example of a natural science describing a particular kind of system it is instructive to take an example from chemical thermodynamics. The prime question which chemistry faces is the question: Does A react with B? In general, the natural world will contain in close proximity only substances which do not react together under normal conditions. The air could not contain oxygen and nitrogen if they reacted easily with each other. Beaches would not exist if salt water reacted with sand. On the other hand zinc and hydrochloric acid, if brought together at room temperature, do react spontaneously, producing hydrogen gas. The science of chemistry tackles the problem: In a system consisting of substances A and B, what decides whether or not they react together to produce C? Thermodynamics has answered this question. Initially it might be thought that such a system would suffer a spontaneous change if the product of reaction, C, represented a lower energy state. This is a good way towards an answer, in fact, in that in very many instances A reacts with B to give a reaction product and to release energy in the form of heat. But this cannot be the whole answer because in some cases a spontaneous reaction is accompanied by an absorption of heat. Iodine and chlorine form iodine chloride, for example, in a reaction of this kind. It is now known that the behaviour of a system consisting of A + B, as regards its propensity to produce C, is determined by a combination of both energy and entropy considerations (the latter being a measure of the degree of disorder in a system). The system's behaviour is described, beautifully succinctly, by a now well-tested equation

$$\triangle G = \triangle H - T\triangle S$$

where $\triangle G$ measures the propensity to react, $\triangle H$ is the heat change in the reaction, T is the temperature, and $\triangle S$ is the entropy change in the reaction. If $\triangle G$ is negative for the system then A + B will react to form C. (Other physico-chemical considerations, ignored here, determine the rate at which C will be formed.) The equation above (itself literally a designed abstract system) is a description of the behaviour of a natural system consisting of substances A, B, and C in close proximity. Chemical thermodynamics is a science of systems of this particular kind, though it is not usually recognized as such.

As an example of a science which now makes an explicit use of both systems ideas and systems language we may take physical geography. There has been increasing discussion of systems ideas in geography in recent years (e.g. Harvey, 1969; Chorley and Kennedy, 1971; Davies, 1972; Wilson, 1973; Chisholm, 1975; Chapman, 1977; Bennett and Chorley, 1978). According to Harvey:

> The history of systems thinking in geography is very much bound up with the functional approach, with the organismic analogy, with the concept of regions as complex interrelated wholes, and with the ecological approach to geography ... . In the same way that considerations of logic in functional analysis (and cause-and-effect analysis) lead to the concept of system, so, it seems, the various pathways of geographic thought lead inevitably to systems thinking.

Concentrating on physical, rather than human and economic geography (which will involve more than natural systems) we find an interesting example of systems thinking by Chorley and Kennedy (1971). They regard physical geography as concerned primarily with systems of four kinds: morphological systems (for example, a beach); cascading systems, composed of sub-systems 'linked by a cascade of mass or energy' (for example, weathering cascades in which chemical and biological constituents cascade through soil and rock layers); process–response systems, which 'represent the linkage of at least one morphological and one cascading system' (for example, the system in which scree accumulates at the foot of a cliff); and finally 'control systems', which are 'process–response systems in which the key components are controlled by some intelligence' (for example, man-induced changes in sand movement on a beach as a result of the construction of groynes). They examine the structures and processes of these systems, and determine their behaviour in terms of measurable variables. This is a particularly clear example of a scientific discipline studying some natural systems, and of course ecology provides a similar example.

In all such cases in which the object of study is a natural system, the scientist, no less than when he studies a natural phenomenon such as magnetism, is in the position of an outside observer. He hopes to give a convincing account of the object of study which can be tested experimentally and can be repeated by other observers, thus turning his account into public knowledge. There is a very different situation, however, when the observer's object of study is a system—such as a human activity system—which could be other than it is; here the whole question of establishing public knowledge is much more complex.

Human Activity Systems

We see in the world many examples of sets of human activities related to each other so that they can be viewed as a whole. Often the fact that they form an entity is emphasized by the existence of other systems (often designed systems) which are associated with them: the activities which make British Rail a human activity system, for example, are associated with the designed physical system which is the railway network, with its stations, track, engine depots, etc. Even if there are no closely associated systems to emphasize the grouping of the activities, as in the example above of 'the eating habits of the octogenarians of Basingstoke', it is difficult to deny the right of an observer to choose to view a set of activites as a system if he wishes to do so. What is less obvious, and needs to be argued, perhaps, is that human activity systems (and, for that matter, designed systems) are fundamentally different in kind from natural systems.

The difference lies in the fact that such systems could be very different from how they are, whereas natural systems, without human intervention, could not. And the origin of this difference is the special characteristics which distinguish the human being from other natural systems.

There are many different ideas about what makes man a unique animal. Engineers, for example, like to see man essentially as a tool-maker; communication experts cite the sophistication of man's language as a distinguishing feature, and there are many other possibilities including religious awareness and artistic activity. But against the view of man as tool-maker we might quote the example of woodpecker finches using cactus spines as tools to poke into crevices in the bark of trees in order to expel the insects they feed on. And against the image of man-the-creator-of-language we could draw attention to the sophistication of the dance language of bees by means of which the returning bee can indicate the direction, distance, and quality of a source of food. Thorpe (1974) even argues from experimental evidence that bird song is not random but indicates the glimmerings of artistic awareness. It is unlikely that there will be agreement on the degree of importance we give to the various ways in which man is clearly different from the other animals, at least in degree. But probably all the different arguments advanced for the uniqueness of man include the same common factor: *self-consciousness*. In Thorpe's well-documented survey of experimental evidence bearing on the difference between 'animal nature' and 'human nature' (1974) he not only emphasizes the importance of self-consciousness as a distinguishing characteristic of the human animal but also goes on to deal with a consequence of self-consciousness which is extremely important to an understanding of the nature of human activity systems and justifies their special position in this systems typology. The consequence of self-consciousness is that the human being is irreducibly free; he has genuine freedom of choice in selecting his actions. Thorpe sets out arguments for this which derive from Popper (who shows that 'even a computing machine of unlimited capacity would be unable completely to predict the future of a physical system of which it was itself a part') and, especially, MacKay (1967, 1970). The arguments have already been mentioned briefly in Chapter 3, since they are relevant to the special difficulties which face social science if it tries to match the methods of natural science; but at the risk of a little repetition it is worth dealing rather carefully with a part of the argument which in my experience is found difficult.

The argument as a whole rests on the experimental findings of neurophysiologists that all conscious human activity and experience, including 'believing', has a correlate in electrical activity in the brain. This might be thought to indicate that man is a machine, since *in principle* it would be possible to describe precisely the electrical state of the brain and hence, given relationships between brain state, the content of consciousness, and consequent actions, to make predictions which the individual has no choice but to act out. Ironically, following through the argument proves the reverse of this: that the individual is free to choose his actions.

Mackay considers the case of an observer obtaining an accurate description of the state of the brain of a subject called Joe. (Never mind that this is not at present practicable. This is an argument about what is possible in principle. To be as fair as possible to those who think that human beings are nothing but machines, assume it is possible.) Suppose the observer uses this description to predict Joe's next action. If, before making the action, Joe confirms that the description is correct, then clearly the prediction is now of no interest, since Joe's belief in the correctness of the description will have changed the state of his brain, and made the observer's description obsolete. But what about the case in which the observer's predictions are kept secret? Mackay (1970) writes:

> To test the strength of the argument, let us take the most favourable case imaginable by the determinist. Suppose that from our observations of Joe's brain and its environment we can write down, well in advance, a whole series of predictions that we keep secret until after the events, and then triumphantly produce to Joe as a proof of our success . . . we convince him beyond doubt that our mechanistic theory of his brain is correct. Would not this show . . . that he was mistaken in believing that he was facing genuinely *open* possibilities?

> It would not. It would show, of course, that the outcome was *predictable by us*. What it would not show is that it was *inevitable for him*. It cannot do so, for it cannot produce a specification of the outcome that Joe would have been unconditionally correct to accept before he made up his mind. In this sense, no matter how many detached observers could predict the outcome, Joe—and you and I—are *free* in choosing.

It is this last part of the argument which causes difficulty, and is worth amplification.

Consider two groups of observers observing, in a scientific spirit, two experiments. The first group are watching an experiment in elementary physics. They note that when the north-seeking poles of two magnets are brought close together they repel each other. This is the repeatable finding, and of course any observer may check it for himself. For our present purposes the important aspect is that there is absolutely no restriction whatsoever on membership of the group of observers. The observation that two north-seeking poles repel each other could in principle be made by *anyone,* hence it attains the status of a scientific fact. This is a part of the very definition of what we mean by 'scientific fact'.

The second group are observing the actions of an amazing machine which measures the precise electrical state of Joe's brain and hence, from known relationships between brain-state and consciousness, predicts his next action. They note that every time a prediction is made Joe does in fact do what the machine predicted. (Remember that we are assuming this to be possible in practice in order to create 'the most favourable case imaginable' for the determinist.) Here also we have a repeatable finding which the observers in the group can check by working the machine for themselves. Suppose that they do this, and confirm that they can without fail predict Joe's next action. Is this situation similar to the first one? It is not, for the reason that this time there is a restriction on the membership of the group of observers. Joe himself cannot join it. Were he to do so he would continually spoil the experiment. Every time he knew of the account of the content of his brain, upon which the prediction is based, he would thereby make the description worthless because it would now be out of date. Even if the machine were sophisticated enough to include Joe's knowledge of the description as part of the description itself, Joe's knowing of his knowledge of the description prior to his action would still make this augmented description obsolete. The mechanistic prediction of Joe's action will forever be inaccessible to Joe. He can never join the group of observers, and he has no alternative but to make his next action in a state of freedom of choice: its predictability by observers does not make it inevitable for him because he is forever barred from joining the group of observers.

Both of the outcomes of this 'thought experiment' on the hypothesis that human beings might be merely ultra-sophisticated machines are extremely important for systems thinking. Firstly, the restriction on membership of the group of observers of Joe's actions means that whatever they observe cannot acquire the full status of public knowledge. Hence there cannot in principle be a strict science of human activity exactly similar to a science of a natural phenomenon like magnetism. Secondly, the irreducible freedom of Joe (and every other) human actor means that there can never be accounts of human activity systems similar to, and having the same logical status as accounts of natural systems.

These differences between natural and human activity systems justify separating them in the systems typology. Different kinds of investigation will be appropriate for the two kinds of system. The well-established methods of science will be entirely appropriate for the study of natural systems, perhaps with the addition of attempts to generalize accounts of specific examples by using systems terminology. In the case of human activity systems the way to proceed is less obvious. The research on real-world problem-solving to be described later, which has been much concerned with systems of this type, suggests that it is essential always to include with a description of human activity system an account of the observer and *the point of view from which his observations are made*. The observers of the phenomonena of the natural world can usually be taken for granted; this is much less the case with systems which could be other than they are, such as human activity systems, as we shall see in later chapters which describe the research.

Designed Systems

We could if we wished use a piece of rock, a natural system, to knock nails into pieces of wood to make a hen coop. But we would perform the task better if we used a hammer, a physical system designed with *fitness for purpose* in mind. Clearly, once it has been designed and made, the hammer has much in common with natural systems: it simply exists, implacably. But in spite of this similarity it is justifiable to distinguish designed physical systems from natural systems because of their different origins and because hammers are possible in many different forms which can be changed at will depending upon the precise intentions of the designer. Designed physical systems exist because a need for them in some human activity system—such as the construction of hen coops—has been identified. Man as designer is able to create physical artefacts to fulfil particular defined purposes. And similarly he may create structured sets of thoughts, the so-called 'designed abstract systems'.

Man as designer is a teleological being, able to create means of enabling ends to be pursued, and to do so on the basis of conscious selection between alternatives. It is proper to restrict the word 'teleological' to use in this sense, involving human will, and not to apply it casually to natural systems. Many natural systems are of course apparently 'designed' to fulfil a purpose efficiently but 'design' is here the result of the operation of blind evolutionary forces over a long time period, and ought to be distinguished from purposeful design by a human being. The neutral word based on the notion of 'serving a purpose' is 'teleonomy'; it is a pity that 'teleology', with its metaphysical overtones, is too often confused with it in the systems literature.

This distinction suggests that at a more everyday level of discourse there is also a need to distinguish carefully between activities (or systems) which simply serve a purpose and activities (or systems) which are the result of willed choice by human beings. Here dictionaries and common usage do not readily provide a solution but in this book I use the word 'purposive' when the meaning is the neutral 'serving a purpose', and 'purposeful' when conscious human action is involved. Thus the escapement is a *purposive* system of a clock; telling the time by reading the dial of a clock is a *purposeful* action by a human being. So also are all the actions involved in monitoring and recording our mental activity, enabling plans to be argued over, communicated, opposed, carried out, subverted, and/or modified in a purposeful fashion by groups of human beings. This purposeful activity distinguishes our species from the stimuli-provoked, goal-seeking, merely purposive behaviour of the other animals.

'Social Systems'

There is a problem for the typological map whose discussion gives a useful indication of some of the likely difficulties in research on human activity systems; it is summarized in the question: Which of the classes of systems contains 'social systems'?

In everyday language social systems are probably thought of as groupings of people who are aware of and acknowledge their membership of the group. They accept various responsibilities as a result of their membership and have certain expectations of other members. This covers a wide range of such 'systems' from ethnic groups and families to drinking clubs, the Girl Guides, political parties, trade unions, and industrial firms. Much of what we experience as ordinary members of society is due to our multi-membership of such groups. If we accept this everyday definition of a social system then clearly we would expect each such grouping to be characterized by a particular set of human activities. This would seem to make such 'systems' describable as human activity systems in the sense in which I am using the name here. On the other hand, what we are most aware of in our everyday membership of 'social systems' is the texture of the interpersonal relationships involved, the extent to which our membership engages our emotions as individual personalities. This is most marked in a grouping such as the family, less so in associations into which we enter voluntarily, such as an industrial firm. But even in such a professional context anyone who has worked in a factory, a school or an office will know that such groupings develop something of the characteristics of a family: tensions develop, alliances form and re-form, and emotions colour what in principle could be objective professional relationships. Now these characteristics, typical of the tribe and the family, presumably have their origin in the nature of man as a gregarious animal, one who has a basic need for the support provided by his fellows in the community. Hence it might be argued that our everyday life 'social systems' are properly located in the typology as natural systems.

In the first account of this systems map I suggested that Social Systems should be placed astride the boundary between human activity and natural systems, to mark their equivocal nature. The activities associated with a social system, and the connections between them are certainly amenable to rational design; but any actual manifestation of such a system involving a group of real people will exhibit properties due to the natural characteristics of man the social animal. I subsequently discovered in the work of the sociologist Ferdinand Tönnies a sharp expression of the distinction I was groping for.

Tönnies, writing towards the end of the 19th century, was centrally concerned with the problem which was 'the very *raison d'être* for the emergence of sociology: namely the disruption of the old traditional order of society and the explosive development of a new commercialism and industrial capitalism' (Fletcher, 1971b). His approach was to construct a typology for use as an analytical tool in illuminating and understanding the major transformation of society to which Fletcher refers. Tönnies's main work, published in 1887, is *Germeinschaft und Gesellschaft*, translated (Loomis, 1955) as *Community and Association*; in it Tönnies presents models of two types of social system: 'Gemeinschaft' (community), which is natural, and 'Gesellschaft' (society, or association) which is contrived. The two types are based on the kind of willing, choosing or deliberating which characterizes the human relationships in the two kinds of social system. Fletcher, in arguing that 'community' and 'society' are the best translation of the German words, describes the distinction thus:

> ... we think of ourselves (and others) as members of a *natural order of relationships*—as members of a family, relatives among kinsfolk, neighbours in a particular area ... (and) we think of ourselves (and others) as sometimes, and for some particular purposes, having to associate formally with others—not as whole persons but partially—in order to accomplish a certain end.

In Tönnies's own words 'All intimate private and exclusive living together is understood as life in Gemeinschaft (community). Gesellschaft (society) is public life. In Gemeinschaft with one's family, one lives from birth on bound to it in weal and woe. One goes into Gesellschaft as one goes into a strange country.'

This distinction is very much the one I was trying to make in arguing that any actual social system observed in the world will be a mixture of a rational assembly of linked activities (a human activity system) and a set of relationships such as occur in a community (i.e. a natural system). In practical work in the real world it will be necessary to take both aspects into account. A purely behavioural approach based upon the idea of man as a gregarious animal will neglect the power and influence of rational design; but an approach which assumes human beings to be rational automata and ignores the cultural dimension will also pass the problems by.

Basic Systems Thinking: Conclusion

Some very rudimentary systems thinking has been expounded. Such thinking starts with an observer/describer of the world outside ourselves who for some reason of his own wishes to describe it 'holistically', that is to say in terms of whole entities linked in hierarchies with other wholes. This leads to the most basic prescription of what the observer's description will contain: his purpose, the system(s) selected, and various system properties such as boundaries, inputs and outputs, components, structure, the means by which the system retains its integrity, and the coherency principle which makes it defensible to describe the system as a system.

Boulding's hierarchy and Jordan's taxonomy are examples of broad attempts to survey the whole of the real world in systems terms, and the typological map similarly provides concepts by means of which holistic analysis might be started. In the latter case the intention was to provide a base in systems thinking for research work which attempted to find out whether and/or how systems ideas might help in tackling the kind of unstructured real-world problems which defeat the reductionism of the method of science.

The systems map suggests that the absolute minimum number of systems classes needed to describe the whole of reality is four: natural, designed physical, designed abstract, and human activity systems. Some properties of the four classes have been discussed.

It is important to note that the typological map is itself a designed abstract system. It provides not so much an account of reality, rather a set of conceptual types to be used in systems-based descriptions of reality. There may in practice be no argument about a foxglove *being* a natural system; and few would dispute that a hen coop *is* a designed physical system (rather than it being useful to take it to be so). But the case of what in everyday language are called 'social systems' shows that real-world entities may well not fit easily into one class; in particular it may not be easy to obtain descriptions upon which all observers agree. Nevertheless the gradual development of tested conceptual models of the four classes of system, with the logical, structural, and regulatory entailments worked out, should make simpler the interpretation and holistic analysis of complex reality.

Part 2

SYSTEMS PRACTICE—Action Research to Establish the Use of Systems Concepts in Problem Solving

Chapter 5

'Hard' Systems Thinking—The Engineers' Contribution

The idea of 'systems practice' implies a desire to find out how to use systems concepts in trying to solve problems. Such an endeavour might expect at least some general guidance from the systems way of regarding the world. In terms of that outlook described in the previous chapter, in which the world is taken to be a complex of systems describable as 'natural', 'designed physical', 'designed abstract', or 'human activity' systems, we may surmise that each type of system will be relevant in a different way to the would-be problem solver. Firstly, there will be much to *learn* from natural systems. Study of the energy flows of the biosphere, for example, the nitrogen cycle or the control mechanisms by which the body maintains its temperature constant, will teach us much about the dynamics of systems and the means by which their integrity is maintained. Secondly, the problem solver is free to *use* designed systems, whether physical or abstract, to achieve his ends. And thirdly, we may anticipate, since human beings are purposeful, that we may seek to 'engineer' human activity systems, using the word 'engineer' in its broadest sense.

This latter idea suggests a possible approach to systems practice aimed at real-world problem-solving: it might be a useful hypothesis that problems of this kind, which will certainly relate to human activities, can be tackled by *identifying, designing, and implementing* human activity systems.

These words suggest that the exploration of this hypothesis, and the testing of it, will involve work of a kind similar to that of professional engineers. Hence it makes sense to start an examination of systems practice by examining briefly the evolution of the thinking and practice of engineering as a professional activity. In fact the generalizations arising from the practice of engineers constitute an important second strand of systems thinking, one which complements that already examined, the one arising in science. (We have already seen the importance of the engineering concept of 'control', in Chapter 3.) This chapter examines the difference between science and engineering or technology, reviews the development of thinking about the engineering of systems rather than merely components of systems, and describes the 'hard' systems thinking which was the basis of the research into real-world problem-solving described in subsequent chapters.

Science compared with Engineering and Technology

The intention behind the activity of science is to establish well-founded knowledge about the world and our place in it. Its method, the carrying out of reductionist, repeatable experiments which aim to test hypotheses to destruction, has been very successful, and much knowledge of a special kind—public knowledge—has been established by the use of it. In our civilization the *application* of scientific knowledge through technology has so dominated the man-made world that we may forget the crucial differences between science and technology, and between the aims and methods of professional scientists and engineers or technologists.

All human communities have faced problems arising from man's basic need for food, shelter and, from an early stage of cultural development, transport; and all societies have developed technologies to solve the problem of meeting these needs. In our civilization the incorporation of science into technological problem-solving has produced a uniquely powerful cultural force, so much so that some writers (for example Bunge, 1966) would describe mud hut building, or ancient Chinese paper making as 'crafts', restricting the label 'technology' to the combination of a craft with science. But however the word is used, there is no doubt that its Greek origin *techne* has to do with objects produced artificially by man, and it is in this that the crucial distinction between science and technology (or engineering) lies.

Any purposeful human activity implies commitment to a particular ranking of values. Science implies the belief that the highest value attaches to the advancement of knowledge. Engineering and technology, on the other hand, prize most highly the efficient accomplishment of some defined purpose. Where the scientists ask: 'have we learnt anything?' the engineer and the technologist ask 'does it work?'

This difference of aim and motivation is on the whole reflected in statements about science and technology in the rather sparse literature concerned with the philosophy of technology (Durbin, 1978, provides a useful review of this area; see also Bugliarello and Doner, 1979). According to Wisdom (1967): 'The difference is not *in rerum natura* but in aim: the one to understand structure, the other to create a structure for a certain purpose'. Skolimowski (1966), discussing the structure of thinking in technology, summarizes the distinction in the sentence: 'In short, science concerns itself with what *is*, technology with what *is to be*', while Jarvie (1966) in commenting on Skolimowski's paper argues that the epistemological distinction is that 'Scientific statements are posed to solve scientific problems; technological statements allow that certain devices are not impossible'.

These theoretical discussions, which I would summarize in the crucial difference between the value systems of science and technology, match well my own experiences as a Research and Development manager in science-based industry. In the 1960s, for example, the largest team in my group, ten science or technology graduates under the leadership of a very able scientist-turned-technologist, were trying to define a synthetic shoe upper material and a process to make it economically. The technological base was our knowledge of polymers and our expertise in making synthetic fibres, and the composite leather-like product developed was what has been called a 'third generation' product in that industry (Thompson, 1967). In the experiments carried out we were not at all interested in refuting hypotheses, in gaining knowledge for its own sake, only in gaining *useful* knowledge in the shape of *rules* for obtaining a product which performed satisfactorily in shoe uppers. Such technological rules are very different from the laws which science seeks, being circumscribed by a range of considerations specific to a particular situation and purpose: we were trying to develop a new material within a certain time, at a certain cost, to meet a specific opportunity believed to exist in a particular market. Such considerations are determining factors in technology and engineering but are not fundamentally relevant to science. An individual scientist may be in a hurry to make the discovery which will gain him a Nobel prize, but the timing is intrinsically unimportant except to those who believe that new scientific knowledge is welcome for its own sake, and welcome early rather than late. In our technological projects we and a number of competitors were trying to meet a market need which, we believed, would exist in the mid-1970s as a result of a world shortage of natural hides. As it happens the market forecasts were wrong, but that does not affect the fact that the timing was not just a factor in our objectives but a determining factor.

One other experience in that project also illustrated the difference between a scientific and a technological approach. Many groups with polymer and fibre-making skills were interested in the possibility of a 'synthetic leather'. In one research department which got left behind in the race, a research physicist was asked to report on the possibility of a project to make such a material. He assumed this to be a scientific question, one concerning knowledge, and his negative report was based on the following argument: 'the three-dimensional matrix of natural leather is so complex that it cannot at present be accurately described; therefore we cannot hope to simulate it'. Had he assumed the question to be a technological one he would have asked, not 'can we copy leather' but 'can we imagine a material which will perform satisfactorily in the *end uses* in which natural leather is now used?' Where science is interested in new knowledge, and whether it is true or false, technology and engineering are interested in action directed to a defined end, and whether it is successful or unsuccessful.

The fact that this thinking underlying technology and engineering is directed towards action makes it a very direct and explicit guide to exploring the possibilities of 'engineering' human activity systems. This strength eventually turns out to be also a weakness, as we shall see, but as a first basis for systems practice it is useful.

Systems Engineering

It is not surprising that engineering as a professional activity attracts action-oriented people who value practical achievement above all else. A result of this is that engineers (and technologists) are impatient with theorizing; after a good design has been successfully realized in practice they are little inclined to analyse the way they went about achieving it. As a result the literature of engineering methodology is not extensive, in spite of the fact that public speculation on the role of engineers usually embodies rather grand visions: thus Sporn, 1964:

> The engineer is the key figure in the material progress of the world. It is his engineering that makes a reality of the potential value of science by translating scientific knowledge into tools, resources, energy and labour to bring them into the service of man . . . the engineer requires the imagination to visualise the needs of society and to appreciate what is possible as well as the technological and hard social understanding to bring his vision to reality.

Engineers have not in fact developed the broad systems approach which is implied by clarion calls of this kind, but nevertheless there is one part of engineering in which methodological prescriptions are common. This is that area of the subject concerned with the engineering not of components but of systems (both physical and organizational) which involve the mutual interactions of many components: the engineering of the telephone network, for example, rather than the telephone instrument or the switching equipment in an exchange.

It is obvious that engineers throughout history have faced tasks of this kind, from the pyramid builders to the NASA engineers who worked on the moon-landing mission, although it is only in relatively recent times that the methodological principles for carrying out such projects have been defined. It is characteristic of the engineering world that principles should be learnt from experience and grasped intuitively long before they are codified and expounded. Eli Whitney in the 1790s in America contracted to produce 10,000 muskets by using jigs, thus enabling unskilled workers to make large numbers of interchangeable parts; he must have had an intuitive grasp of what is now known as 'industrial engineering'. And in the 1830s in Lancashire James Nasmyth produced standardized machine tools in a factory which was a pioneer of assembly lines. By 1913 the knowledge was available to enable Henry Ford to set up a moving assembly line on which a chassis was produced in 1 hour 33 minutes compared with the previous 12 hours 28 minutes; in 1914 he produced more than 250,000 cars (Armytage, 1961). Similarly, the history of the development of electrical engineering, including the provision of power, lighting, and telecommunication systems abounds in examples of 'systems engineering' even though the phrase gets no mention in a standard history of the subject, that by Dunsheath (1961).

The principles of work organization began to be codified around the turn of the century, with the work of Frederick Taylor, and the Gilbreth's, Frank and Lillian. Exposition of methodology for the total engineering of complex man–machine systems involving sophisticated machines as well as logistic support systems came rather later. It dates only from the 1950s, although since then a number of versions of this kind of 'hard' systems thinking have been described. The following dozen references illustrate the steady stream of discussion of 'systems engineering' over the last twenty years: Goode and Machol, 1957; Eckman, 1961; Williams, 1961; Gosling, 1962; Hall, 1962; Chestnut, 1965 and 1967; Miles, 1973; de Neufville and Stafford, 1971; Chase, 1974; Daenzer, 1976; Wymore, 1976. It will serve my purpose in describing the background to the action research discussed later to review briefly some of the systems engineering methodologies advocated.

Both Goode and Machol, and Gosling, identify the increased complexity of human requirements, and hence of the engineer's task, as the prime reason for the development of 'systems engineering' as a new set of considerations which engineers must take into account. Gosling makes a good *systemic* point when he writes;

> The system engineer must also be capable of predicting the emergent properties of the system, those properties that is, which are possessed by the system but not its parts

but in general the emphasis in both books is on a *systematic* approach to engineering design by means of model building and model optimization. For Goode and Machol the computer is 'the basic tool of interior system design', and their overall approach comprises: deciding what system has to be designed and suggesting some possible designs; mathematical and experimental evaluation of potential designs according to some defined 'measure of effectiveness'; principal design, a phase which 'may last from 1 year to 10 years'; prototype construction; and testing, training, and evaluation, the purpose of the latter being 'to decide whether the design accomplishes its objectives'. In the case of large-scale systems designed and constructed over a number of years, evaluation in the field is not really possible, and in any case the problem itself may have altered. Thus 'evaluation at the end of the system-design process overlaps evaluation at the beginning of the system-design process'.

This picture of systems engineering as the total task of conceiving, designing, evaluating, and implementing a system to meet some defined need—the carrying out, in other words, of an engineering *project*—is the one which persists throughout accounts of this activity; and, from the 1950s on, many engineers and project managers in large organizations were consciously formulating the procedures needed to make such projects successful, including the necessary sequencing of activities as well as approaches to the problem of co-ordinating the efforts of numerous specialists. Williams (1961) gives an account of the approach adopted within the Monsanto company; but the leaders were probably Bell Telephone Laboratories (Hunt, 1954) and the classic account of systems engineering methodology, that of Hall (1962, 1969) is the result of Bell Telephone experience.

Hall sees systems engineering as part of 'organised creative technology' in which new research knowledge is translated into applications meeting human needs through a sequence of plans, projects, and 'whole programs of projects'. He goes on:

Thus systems engineering operates in the space between research and business, and assumes the attitudes of both. For those projects which it finds most worthwhile for development, it formulates the operational, performance and economic objectives, and the broad technical plan to be followed.

Hall envisages, and illustrates, a sequence of steps in the systems engineering process, these being 'generalised from case histories' rather than developed theoretically. This is an important indication of the spirit behind this approach.

The problem-solving sequence in Hall is as follows

Problem Definition   (essentially definition of a need)
          ↓
Choice of Objectives   (a definition of physical needs and of the value system within which they must be met)

          ↓
Systems Synthesis   (creation of possible alternative systems)
          ↓
Systems Analysis   (analysis of the hypothetical systems in the light of objectives)

          ↓
Systems Selection   (selection of the most promising alternative)
          ↓
System Development   (up to the prototype stage)
          ↓
Current Engineering   (system realization beyond prototype stage and including monitoring, modifying, and feeding back information to design).

As an initial example Hall quotes from Engstrom (1957) an account of the development in America of colour television. The need for such a system was examined, together with its technical and economic feasibility. These considerations introduced an important constraint into the statement of objectives: because of the existing investment in black-and-white receivers the systems engineers concluded that the new system must be compatible with the existing service. Systems synthesis to meet the objectives required some technical inventions to be made; system selection related to apparatus design, practical operation under normal broadcast service conditions, and Federal approval of signal specifications. Then came the problems of establishing the service, which involved developing new studios, connecting broadcasting networks, installing transmitters, and creating programme production groups. Finally the technical performance and public reaction to the new system had to be measured: all this Engstrom sees as 'a "text book" example of the systems concept in action', with a systems project team 'working toward the single defined objective'. Hall's own major case history describes the fourteen years of work on establishing and improving a continent-wide microwave radio relay system. Again there is careful emphasis on defining a range of precise objectives so that expected performance can be stated and the criteria named by which performance will be measured.

The pattern of Hall's exposition is recognizable in most later accounts of systems engineering. Chestnut (1967) in an account broadly similar to Hall's emphasizes that the systems engineering process is itself a system which has to be 'engineered':

> ... the overall problem of systems engineering [is] composed of two parts, one being the systems engineering associated with the way that the operating system itself works and the other with the systematic process of performing the engineering and associated work in producing the operating system.

The 'operating system' itself, in Chestnut's account, is still a physical entity of the kind which has traditionally been the concern of the engineering profession, though the overall system will also include activity systems. More recent accounts have emphasized that logically the thinking applies equally to systems of the kind which professional *planners* are now expected to design—for example health care or public water systems. Thus de Neufville and Stafford (1971) address their account to managers as well as engineers, and, like Hall include a summary of welfare economics as one tool relevant to a systems study. Their emphasis is on the need for multiple criteria in evaluating public projects, and on the theory of welfare economics as a source of the various elements which will make up a function which measures 'social welfare'. They accept that the decision on how objectives should be traded off against each other in social projects is taken by means of the political process, and that the role of the systems study is to feed that process, improving the quality and range of information available to decision makers:

> It is important that engineers, planners, and economists recognise not only their incapacity to determine a social welfare function, but also the legitimacy of the political process to decide social priorities.

It is their inclusion of social projects, presumably, which causes their account to be restricted to the analysis phase of the whole systems engineering sequence. Within that phase they advocate five steps which match those in the accounts already discussed: definition of objectives; formulation of measures of effectiveness; generation of alternatives; evaluation of alternatives; selection of the proposed system.

A recent European account (Daenzer *et al.*, 1976) emphasizes that the starting point for the systems engineering process is often only a feeling of unease, an awareness that things could be better than they are, a point made earlier by Jenkins (1969). The assumption is that improvements will follow the modification or restructuring of a system, and the recommended iterative problem-solving cycle begins with the 'Zielformulierung' phase, the formulation of objectives which are neutral with regard to possible solutions, complete, precise and intelligible, and realistic. This is followed by the search for possible solutions and the choice between alternatives, on the usual pattern.

Finally, in a recently published account of 'systems engineering methodology for interdisciplinary teams' (Wymore, 1976) we have an account which, at least in the claims made for it, goes beyond earlier versions. Wymore claims as within the scope of the team's efforts just those considerations which de Neufville and Stafford accept as 'political' and hence, in their view, outside the boundaries of the systems study as such. The interdisciplinary team is said by Wymore to be concerned with 'the analysis and design of large-scale, complex, man/machine systems'. The remarkable claim made is that this category may include not only such obvious examples as communication, transportation and manufacturing systems but also education systems, 'health delivery systems' and law enforcement systems!

The methodology Wymore expounds is said to be 'powerful enough to state *any* system design or analysis problem' and to be based on an earlier mathematical theory of systems (Wymore, 1967) which is itself 'based rigorously on set theory'. The underlying structure of the methodology in fact fits the pattern which runs through earlier versions: define the proposed performance the system is required to achieve, generate alternative possibilities, and select one on the basis of defined criteria. The specific form of it involves a definition of what is required in terms of a set of input trajectories (varying over time) and a set of output trajectories. A potential *system* is an arrangement which matches an input trajectory to an output trajectory:

> ... a way to look at an input/output specification is as a formal description of the total range of possible performances by the system to be designed, from the possibility of abysmal failure to the possibility of perfect performance ... it is the function of the interdisciplinary system design team eventually to choose one possible (input/output) matching for the purposes of design and development of a system ...

Having generated a set of input/output matchings which meet the ultimate requirements—that is, having carefully specified what is *desirable*—the team now proceed to define the set of potential systems (i.e. input–output matchings) which are *feasible* with available technology. Any system which is a member of both the desirable set and the feasible set is a candidate for implementation, and selection is done on the basis of criteria such as: the input/output merit ordering, the technology merit ordering, profit, performance, quality/reliability, cost/benefit, etc.

Wymore's account sets out in detail the logic involved in synthesizing sub-systems into systems, and so designing the latter that objectives are met and constraints accommodated. It is the most elaborate exploration of the logic inherent in the kind of engineering process of which Hall gives the classic account. Whether or not we should accept the extravagent claim that 'soft' human activity systems, such as education or health care systems, can be engineered by this process unfortunately cannot be decided by a reading of the book. This restricts itself to spelling out the logic of the systems engineering process and claiming that it is universally applicable. This claim is really a version of the view which may be vulgarly expressed as the idea that there is a 'technological fix' for every problem. Early in the book Wymore points out that there are many models available which purport to describe and predict various aspects of human behaviour; he advocates using them, since 'we must be able to predict (the human being's) behaviour as a component in a system' and claims, without alas illustrating the assertion, that

> Extant insights from the behavioural sciences *are* sufficient to enable the development of system-theoretic models of human behaviour in a restricted environment.

His advice is to bring in at the appropriate time the expert with the appropriate model, as when in discussing a notional example of an education system whose objective is to produce a person specializing in engineering he writes:

> Do not worry too much at this stage about how to measure social contributions or family contributions, for example, or the student's

competence in self-education. We will enlist experts to serve on the interdisciplinary team who can help to design state-of-the-art symbols for levels of states of these various attributes and who can help to design instruments for assessing these levels or states . . .

Only specific case histories could dispel the Utopian ring of all this, and these we are denied. Certainly the outlook expressed here takes a view of human activity systems very different from that I have argued in Chapter 4. Wymore's account was published after the completion of most of the action research described in the next two chapters, and I shall return in Chapter 8 to the issues raised by the claims he makes. At this stage the relevant point is that here is the latest in the long line of accounts of how to go about the engineering of systems which has been appearing over the last two decades and which ought in principle to provide guidelines for attempts to solve real-world problems by seeking to 'engineer' human activity systems.

Another source of such guidelines is the tradition of 'systems analysis', which parallels and overlaps that of 'systems engineering'

Systems Analysis

Simultaneously with the development of systems engineering in the 1950s there emerged the strand of methodological thinking known as 'systems analysis', a development associated especially with the RAND Corporation, a non-profit-making corporation in the advice-giving business. Though it is not easy to find explicit accounts of the methodological thinking which gradually evolved in the course of actual studies made by RAND professionals, the outlines of the methodology are quite clear, and what is being practised and developed at the International Institute of Applied Systems Analysis (IIASA) in the 1970s is undoubtedly RAND-style analysis (Quade *et al.*, 1976).

The development which made possible the emergence of institutions like RAND Corporation was the involvement of scientifically-trained civilians in the planning of military operations during the Second World War. The success of operational research teams established in military thinking the value of scientific analyses of the kind OR could provide (Morse, 1970). In the now-classic wartime OR studies the concern was usually tactical (Waddington, 1973). Early applications included studies of anti-submarine tactics and the co-ordination of the use of radar with anti-aircraft guns and interceptor aircraft. In the immediate post-war years scientifically-based advisory work shifted to broader issues and there emerged what a historian of RAND Corporation describes as 'a broader and more refined sister discipline—"systems analysis"'. Actually, there is a strong case for describing systems analysis as *less* refined than OR! But we can agree that it was 'less quantitative in method and more oriented toward the analysis of broad strategic and policy questions' (B. L. R. Smith, 1966).

In 1944 and 1945, discussions among officials of the War Department and the Office of Scientific Research and Development in Washington led to the idea of 'a contract with a private organisation to assist in military planning, and particularly in coordinating planning with research and development decisions. Thus the concept of Project RAND began to emerge in nascent form in mid-summer 1945' (Smith, 1966). RAND is an acronym for 'research and development', and *Project* RAND was a contract with the Douglas Aircraft Company for a study of 'intercontinental warfare, other than surface, with the object of advising the Army Air Forces on devices and techniques'. Work on the contract began in 1946, the initial four employees being located in the main Douglas building in Santa Monica. The first report carried the title 'Preliminary Design of an Experimental World-Circling Spaceship', and it is essentially a contribution to a systems engineering study. By 1947 Douglas felt that Project RAND was a commercial liability, in that the Air Force, in placing other contracts, tried hard to avoid the appearance of giving preferential treatment to Douglas; Douglas lost some contracts they expected to get. The result was that in 1948 the RAND Corporation separated from Douglas and was established as an independent non-profit advisory corporation funded initially by the Ford Foundation and some San Francisco banks.

The emphasis on systems engineering in early RAND studies was quickly replaced by an emphasis on cost and strategic considerations. In later years RAND's President described its role to a congressional committee in the following terms:

> RAND is engaged primarily in long-range research and analyses . . . as an aid to strategic and technical planning and operations. We have no laboratories . . . we do not manufacture hardware . . . we do not act as systems engineers as that term is usually used in industry (Smith, 1966).

During the 1950s the pattern of RAND-style 'systems analysis' became clearer. The work done consisted of broad economic appraisal of all the costs and consequences of various alternative means of meeting a defined end. It was a refinement of the kind of cost–benefit analysis which had been developing in government since the 1930s and of the 'requirements approach' especially associated with the Department of Defense, an approach characterized by definition, by officials, of a 'requirement' whose provision will solve a problem. The requirement might be a task, a piece of equipment or a complete system. Feasibility and performance characteristics of alternatives are then checked, and the analysis is passed to government decision makers for a decision on whether the necessary budget can be obtained. In one of the first book-length accounts of systems analysis, McKean (1958) of RAND Corporation takes the 'requirement approach' as given, and argues that systems analysis extends it by putting more emphasis on cost estimation during the course of the analysis.

Systems analysis asks, he says: 'what are the pay-offs *and the costs* of alternative programs?' He argues that there is a need for formal quantitative cost-benefit analysis in all aspects of government, not merely in defence, because government spending lacks any 'natural' mechanism promoting efficiency, such as a free-market price mechanism might provide. His discussion centres on the problems of choosing performance criteria, selecting alternatives to be compared, dealing with intangibles and uncertainties, and taking into account the fact that time is an important aspect of both gains and costs. The emphasis throughout is on the search for economic efficiency in government-funded activities, and the same theme was developed in more detail for defence spending by Hitch and McKean (1960). They regard systems analysis as '*a way of looking at* military problems' which they insist can be treated as problems of economics in the sense that 'economics is concerned with allocating resources—choosing doctrines and techniques—so as to get the most out of available resources'. The search for economic efficiency can be helped by 'increased reliance on systematic quantitative analysis to determine the most efficient alternative allocations and methods'. By 1967, a book entitled *Defense Management* (edited by Enke) could claim that the adoption of the RAND approach was 'a revolution' which was now in the past. That book could now address itself, in an article by McKean, to 'remaining difficulties'. The revolution referred to was the introduction of RAND-style systems analysis/cost–benefit analysis/programme budgeting into the Pentagon by Secretary McNamara in the 1960s. McNamara selected Charles Hitch as his Comptroller, and by 1965 there was an 'Assistant Secretary of Defence for Systems Analysis'.

The formal methodology of systems analysis was described briefly by many RAND authors during the 1950s (Optner, 1973, edits a useful collection of papers). In a RAND report Hitch (1955) gives an account which has many similarities with systems engineering (and OR) methodology which was emerging at the same time. The essential elements are described as:

1. An objective or objectives we desire to accomplish.
2. Alternative techniques or instrumentalities (or 'systems') by which the objective may be accomplished.
3. The 'costs' or resources required by each system.
4. A mathematical model or models; i.e. the mathematical or logical framework or set of equations showing the interdependence of the objectives, the techniques and instrumentalities, the environment, and the resources.
5. A criterion, relating objectives and costs or resources for choosing the preferred or optimal alternative.

These are the elements in the approach. The making use of them, says

Hitch is 'shot through with intuition and judgement', systems analysis is 'a framework which permits the judgement of experts in numerous sub-fields to be combined'. It is clear that the word 'system' in systems analysis has two connotations. It is used in the same sense as in the phrase 'systems engineering', and this derives from the fact that from the 1940s defence requirements were usually expressed in terms of a total complex of equipment, personnel and procedures, rather than simply as a requirement for a specific piece of equipment. And the word is also used to indicate that the analyst tries to be comprehensive, to take into account many of the factors—financial, technical, political, strategic—which will affect decision on an important problem. The flavour of a RAND systems analysis is best described by Quade and Boucher (1968):

> One strives to look at the entire problem, as a whole, in context, and to compare alternative choices in the light of their possible outcomes. Three sorts of enquiry are required, any of which can modify the others as the work proceeds. There is a need, first of all, for a systematic investigation of the decision-makers objectives and of the relevant criteria for deciding among the alternatives that promise to achieve these objectives. Next, the alternatives need to be identified, examined for feasibility, and then compared in terms of their effectiveness and cost, taking time and risk into account. Finally an attempt must be made to design better alternatives and select other goals if those previously examined are found wanting.

Given the establishment of systems analysis as a way of tackling complex problems of resource allocation in defence, it is inevitable that it should be advocated as a methodology for business managers, who face problems of a similar kind. We have already noted that de Neufville and Stafford (1971) address their account of what they call 'engineering systems analysis' to managers as well as engineers, and there are a number of accounts of systems analysis as an approach to business and industrial management (for example, Optner, 1965, 1975; Lee, 1970), as well as systems-oriented accounts of management and organizational studies (for example, Johnson et al., 1963; Litterer, 1963; Beckett, 1971; Schoderbek et al., 1975). In the business and management application area there is some confusion of 'systems analysis' in the broad, RAND, sense with the more limited kind of computer systems analysis which must precede the installation of computers, as when Schoderbek et al., in a systems-based account of the management process, nevertheless define systems analysis as

> the organised step-by-step study of the detailed procedures for the collection, manipulation and evaluation of data about an organisation for the purpose not only of determining what must be done. but also of ascertaining the best way to improve the functioning of the system.

However, when the broader sense of the term is used, we find the RAND methodology being urged upon managers. Lee, for example, advocates analysis and comparison of alternative solutions to the identified problem followed by the allocation of resources to 'a usable system, plan or method based upon the "best" solution'. Advocacy of this kind tends to be more common then detailed accounts of specific case histories which we might hope would discuss the strengths and weaknesses of systems analysis in this area of application. Nevertheless the general concepts of RAND-style systems analysis are now part of the knowledge of any competent manager.

The Nature of Systems Engineering and Systems Analysis

Systems engineering comprises the set of activities which together lead to the creation of a complex man-made entity and/or the procedures and information flows associated with its operation. Systems analysis is the systematic appraisal of the costs and other implications of meeting a defined requirement in various ways. Both are 'research strategies' rather than methods or techniques (Fisher, 1971), and both require 'art' from the practitioner while making use of scientific methods wherever possible. It is obvious that the two overlap: SE is the totality of an engineering project in the broadest sense of that term; SA is a type of appraisal relevant both to the decision-making which ought to precede the setting up of any engineering project and to the early stages of such a project once it is started. Hall (1962) acknowledges this in his statement that 'the RAND Corporation . . . developed a useful philosophy . . . similar to what we shall call later the first phase of systems engineering'. Both activities use the word 'system' to indicate their nature. Systems engineering is systemic in the control engineers' sense of that word, while both SE and SA are *systematic* in the sense that they proceed by rational and well-ordered steps. Their similarities stem from this commitment to a systematic approach, and behind this, at the core of both SE and SA, is the single idea which links them, the idea that an important class of real-world problems can be formulated in the following way: there is a desired state, S_1, and a present state, S_0, and alternative ways of getting from S_0 to S_1. 'Problem-solving', according to this view, consists of defining S_1 and S_0 and selecting the best means of reducing the difference between them. Thus, in SE, $(S_1 - S_0)$ defines 'the need', or the objective to be attained, and SA provides an ordered way of selecting the best among the alternative systems which could fulfil that need. *The belief that real-world problems can be formulated in this way is the distinguishing characteristic of all 'hard' systems thinking*, whether it emerges as SE or SA. Faith in this belief has been strong, which explains why the literature of systems methodology has been insisting since the 1950s that at the start of a systems study it is necessary to define the need, the aim to be achieved, the system which when engineered will meet the need, the mission to be accomplished, etc. Table 4 collects some representative statements made in the period 1955–76 about the initial phase of a systems study, whether it is called SE, SA or the 'systems approach'. The words differ somewhat but the thought is always the same, that at the start of the study it is essential to know, and to state, what end we want to achieve, where we want to go. Given that definition, the systems thinking then enables us to select a means of achieving the desired end which is efficient, if possible *economically* efficient.

It is not surprising that this outlook underlies 'hard' systems thinking, it is inevitable, given the historical origins of systems engineering and systems analysis in the world of engineering and engineering economics. Hard systems thinking makes use of the kind of thinking which is natural to design engineers, whose role is to provide an efficient means of meeting a defined need. The design engineer exercises his professionalism in a situation in which *what* is required has been defined, and he must examine *how* it can be provided. His skill and flair are directed to providing ingenious possible answers to the question: How? The best design engineer is the man who generates the cheapest, most efficient and ingenious alternatives, the man, for example, who first designs a bridge with arch and keystone where previous beam-based structures have required closely spaced support pillars, thereby achieving a cheaper, more elegant and more efficient means of meeting a defined need. Much skill is called for in engineering design, and much resolution in converting design into realized artefact, but the relevant point here is that the design engineer's problem is a *structured* one: there is a gap to be bridged between the desired future state and the present state; *how* to bridge it is the problem. For the engineer as a professional, the need and objective-defining are taken as given at the start of his problem-solving, and we find this carried over into 'hard' systems methodology together with the structured model of problem-solving which objective-defining implies (Checkland, 1972).

The ultimate identity of 'hard' systems thinking with engineering and economics is seen in the intermittent debate which goes on about whether 'systems engineering' is not simply 'good engineering'. Thus, where Gibson (1960), for example, sees systems engineering as 'a branch of the art [of engineering] with problems, methods and objectives peculiar to itself', it is not surprising that *chemical* engineers fail to see any distinctive attribute which is not already present in their discipline. They are used to modelling complex designed physical systems, and to estimating the contribution of individual components to an overall economic measure of performance. Sargent (1972) sees the coining of the term 'systems engineering' as signalling

a determined take-over bid from the communications and control engineers, who saw this as a natural extension of their own field, with

Table 4   The initial phases in a dozen accounts of 'hard' systems methodology published 1955–1976 (Checkland, 1977a)

Author	Date	Name given to the systems activity	Initial phases
Hitch	1955	Systems analysis	The first element is 'an objective which we desire to accomplish'
Hall	1962	Systems engineering	'Problem definition is isolating, possibly quantifying, and relating that set of factors which will define the system and its environment . . . a problem is an outward expression of an unsatisfied need . . .'
Quade	1963	Military systems analysis	'Systems analysis is undertaken primarily to suggest or recommend a course of action. This action has an aim or objective. Policies or strategies, forces or equipment are examined and compared on the basis of how well and cheaply they can accomplish this aim . . .'
Machol	1965	Systems engineering	'In the first go around the general outlines of the system and one-significant-digit estimates of its performance can be drawn up.'
Chestnut	1967	Systems engineering	'Establish the value or need for the system'
Jenkins	1969	Systems approach	'1.1 Recognition and formulation of the problem 1.2 Organization of the systems project 1.3 Definition of the system . . .'
Lee	1970	Systems analysis	'Recognition that a problem or a challenge exists. . . . Research to find possible solutions . . . analysis and comparison of these alternatives to determine the "best" one . . .'
de Neufville and Stafford	1971	Engineering systems analysis	'1. Definition of objectives. . . . The ultimate purpose . . . is to develop an application for the relative effectiveness with which selected alternatives meet some set of goals'
Miles	1973	Systems approach	'1. Goal definition or problem statement'
Chase	1974	Systems engineering	'Describe mission or use requirements'
Daenzer	1976	Systems engineering	The concrete and problem-oriented work begins with the formulation of objectives. 'What is involved . . . is systematic thinking concerning the formulation of objectives'
Wymore	1976	Systems engineering	'What is the system supposed to do. basically?'

perhaps a nervous look over their shoulders at the proponents of 'cybernetics'.

He concludes, fairly enough, that

the basic systems approach is seen to be nothing more than the traditional engineering approach. Indeed chemical engineering owes its emergence as a separate discipline to the application of precisely this philosophy to analysis of complex chemical processes. The recognition that these processes were built up from a basic set of 'unit operations' would nowadays be hailed as a brilliant 'systems' concept.

The Application of 'Hard' Systems Thinking to 'Soft' Problems

Systems engineering and systems analysis have undoubtedly been successful in introducing systematic rationality into one important area of human decision-making, that in which the problem is to select from among a number of alternatives an efficient means of achieving an end we know we wish to reach. This success makes it inevitable that the methods will be tried out on problems of a different kind; and indeed, attempts at 'technology transfer' of this kind are not to be deplored if they are carried out in a spirit of experimentation.

There have in fact been many attempts to use the 'hard' concept of systems engineering and systems analysis in problems a good deal 'softer' than those of engineering and defence economics in which they were developed, as well as discussions of the methodological implications of doing this. These latter range from careful discussion of the implications of applying systems analysis to 'public decisions' (Quade, 1975) or of the requirements of a methodology in what have come to be called the 'policy sciences' (Lasswell and Lerner, 1951; Dror, 1968, 1971) to polemical attacks on the use in public policy-making of methods which might even be seen as 'a force undermining the very form of government prescribed by the (American) Constitution' (Hoos, 1972; see also Hoos, 1976).

In California in the early 1960s there occurred an important experiment in the application of the systems analysis methods developed in the defence and aerospace industries to problems of public policy. Hoos (1976) drily suggests that the motivation for the experiment was not unconnected with the fact that retrenchment in the defence industry had produced a surplus of engineers in the aerospace firms; but whatever the motivation, Governor Brown's decision to use systems analysis as a vehicle for the design of systems of criminal justice, state information handling, waste management, and mass transportation provided an excellent test of the limits of systematic analysis when applied to problems much softer than those for which it had been developed. Here was an example of the professional bureaucracy examining methods of achieving managerial efficiency in government, and doing it with the enthusiastic support of the politicians. Senator Hubert Humphrey had suggested to representatives of the aerospace industries that

> there is so much that your systems approach, your experienced management approach can bring to solving problems of transportation, air and water pollution, transit, communications, education, neighborhood development, and crime control. In short, you can make this environment here on Earth a better place to live (Kelleher, 1970).

Hoos is very scathing about the California experiment, and, indeed about every attempt to use systems analysis in formulating public policy. Her argument—not unfairly dubbed 'a diatribe' by one of her critics (Pollock, 1972)—refuses to admit that there could be a useful systems analysis carried out by a careful practitioner in a spirit of earnest enquiry, and in her 1976 paper she quotes approvingly an Assistant Secretary of the Navy who replies to a charge that he is writing only about *bad* systems engineering, by stating that he seldom sees any other kind! But of course the systems approach as a hypothesis is unaffected by the consequences of misapplying it, and we are more likely to learn about the limitations of hard systems thinking in soft problems from a rather more sympathetic appraisal of the California projects. Churchman (1968a) usefully reviews the State information system study in this spirit. It provides a good example of the application of systems analysis to a soft problem.

The 'need' in the study was taken to be a management information system for the state of California, one which included information transmission both between the public and government agencies and from one agency to another. Hard systems thinking requires at the outset, as we have seen, a clear definition of the objectives of such a system. But in systems of this kind a definition which is *operationally useful* is extremely difficult to obtain. Also, as Churchman points out, the propaganda version is hardly helpful—it would be something like the following: 'to provide the public and the civil servants with the right kind of information at the right time and with the right precision and in the right form, to meet the needs of the State of California'. We may feel some sympathy for the analysts who faced this kind of dilemma at the very start of their study. Their solution is an interesting one, and one which reflects the methodology of systems analysis. Faced with the methodological requirement of an explicit statement of the objective to be attained which could form the basis of systems design, they reduced the problem to one in which their objective was to provide those information flows which *now exist* between public and agencies, and between agencies, but to do so by means of a single computer-based system which was at least as quick as present arrangements and no more costly. This is exactly the kind of specification envisaged in the methodology of systems engineering and systems analysis, and its definition of the criteria by which the system will be judged opens the way to the examination of alternative versions of the system and the selection of the best one for implementation. In this instance major alternatives included a central collection of all relevant state information versus agency-based collections with a central file recording what information was available in which location. Much effort went into the evaluation of existing computer hardware in order to suggest a configuration within the budget allowed by the state. All of this, of course, eliminates from the study many of the most interesting questions, especially those concerning the purposes served by the existing flows of information and the desirability of others. Ruled out of the study from the start was any consideration of the *meaning* of the information flows in relation to decision-making at the state level. A few years after the study was completed, and before its recommendations were implemented, Churchman records, a new governor, 'efficiency minded', was advocating a policy of across-the-board cuts! Churchman remarks that even if the study's recommendations had been fully implemented at this time, the computerized information system would not have helped very much in evaluating the efficiency policy of the new governor. He is right, of course. The analysts had defined the problem at a level which enabled it to fit the procedures of hard systems analysis as they understood them, but precluded consideration of whether or not the existing information flows were sensible.

Churchman's criticism of the study centres on the initial decision to define a limited system whose objectives and measures of performance could be precisely defined but which ignored wider systems, including those having political implications. He is critical of the failure to include in the study any examination of the political support for the new system. That this was not asked for is, he contends, no excuse for its omission if the analysts are purporting to use a systems approach: 'One can scarcely say that a systems approach has been taken if a large part of the design is bound to die on the vine for lack of political fertiliser.' To say this is to claim that adopting a systems approach—which presumably is something one would expect to find enshrined in 'systems analysis'—necessarily entails moving outside the class of problem for which systems engineering and RAND-style systems analysis were created, namely that in which selection is made among alternative means of reaching a known end. It is to claim that 'a systems approach' entails wider considerations than were taken into account in the 1960s by those who developed systems engineering and systems analysis methodology.

The position regarding the application of this 'hard' systems thinking to 'soft' problems can be stated something like this:

1. The traditional thinking of engineers envisages that the activity starts with the engineer's acceptance of a *specification* of what it is he is required

to create. His ingenuity then goes into answering the question. How?; the question What? is taken as already answered elsewhere.

2. Similarly the traditional thinking of early systems analysis, adopting what McKean (1958) calls 'the requirements approach' takes as given the definition of a *need* and undertakes the systematic analysis of the economic and other costs and benefits of alternative ways of meeting the requirement. Again the question faced is How? rather than What?

3. Both traditions are aware of the need to question assumptions. Thus Hall (1962) pays considerable attention to investigating the values which underlie a given choice of objective, and Quade (1975) emphasizes that the stages of a systems analysis are not sharply separable, and do not lead step by step to some 'optimum' solution, especially in applications to public policy. Nevertheless, the 'technology' of both traditions implies that objectives will be defined and that efficient means of achieving them will be found and compared.

4. This assumption, justified within the fields of engineering and defence procurement in which systems engineering and systems analysis were developed, is less applicable when the methods of analysis are applied more broadly to a wider range of management problems. Selecting a means to achieve a defined objective constitutes only a small part of managerial decision-making. Most management problems, and especially those in the public sector, cannot be formulated as hard problems in this way, and it would be surprising if methodology devised for problems which are 'hard' in this sense survived intact the transfer to soft problems in which the inability to define precise objectives is at the core of the problem situation.

5. In the 1970s this difficulty for systems analysis has been recognized, but not resolved (see, for example, Kelleher, 1970; Dror, 1971; Quade, 1975). In the opening essay of the collection edited by Kelleher which examines the 'challenge' to systems analysis of 'public policy and social change', Root suggests that the original role of the analyst was to supply a quantitative base for management decision 'without, consciously or unconsciously, gradually encroaching on the judgement function'. Root and other contributors accept that tackling social problems *will* involve encroaching upon the judgement function in some way. Several advocate a sternly scientific approach: Engel suggests that we 'observe repetitive quantifiable operations involving men and machines', and even suggests

> we need to put all (sic) of our social scientists, sociologists psychologists, all our people-oriented people to work on these problems.

More realistically, Quade describes a flexible use of systems analysis procedures as guidelines in an iterative exploration of the issues involved in making public policy. Dror, discussing the possibility of the design of a 'policy science' incorporating 'systematic knowledge, structured rationality and organised creativity' points out that present methods cannot deal with the *primary uncertainty* involved in the definition of overall goals or objectives.

To write out the five-point argument above is to state rationally a problem which has also been the subject of much emotional discussion. I refer to the anti-technology school of thought. While systems engineers, systems analysts and, for that matter, operational researchers, have been coming to realize the limitations of their kind of systematic rationality in the problems of the real world, there has been something of an outcry against the whole notion of technology, centred on the idea that the imperatives of scientific thinking, embodied in the technology of the developed countries, serve to diminish our humanity and to subordinate our personalities to technology. If we are to improve systems analysis and to prevent its misuse, we can no doubt learn from this school of thought, which attacks as anti-human the whole notion of applying scientific thinking in human affairs (Checkland, 1970; Mingers, 1980).

'Anti-technology' overlaps with many other concerns and areas of debate, including the environmental conservation movement and the development of the 'counter culture' (Roszak, 1970, 1973; he provides useful, if eccentric, annotated bibliographies). For my purposes here, I may restrict attention to Ellul's (1965) *The Technological Society*, which Roszak describes as 'the best theoretical statement' on the technocracy. Ellul's pessimistic thesis is that our civilization progressively and irreversibly reduces areas of human activity to technique, and that life is not happy in a world dominated by technique, a world in which human freedom is gradually lost. In the words of Merton, in his introduction to the book:

> ... Ellul means that the ever-expanding and irreversible rule of technique is extended to all domains of life. It is a civilisation committed to *the quest for improved means to carelessly examined ends. ... technique turns means into ends* [my italics].

And Wilkinson, Ellul's translator (the book was published in France as long ago as 1954), referring to economic and social model builders as 'assiduous technocratic apes' says that although they may point out that all sciences have to specify a universe of discourse,

> It remains unfortunately true, however, that such specification proceeds by way of elimination of the human,

The strength of feeling behind Ellul's expression of his thesis, as well as the passion of the less coherent proponents of the theme, provide a useful reminder to those of us who would try to increase the part played by reason in human affairs that autonomous human beings are of a complexity far beyond the present reach of our intellectual tools. When allied to the more objective observation that hard systems thinking has not so far been noticeably successful in ill-structured problems which involved human beings, the thought ought to make us circumspect as we try to develop systems thinking for use in such problems. But it ought not to make us give up the attempt. Anyone who accepts the picture of scientific method developed in Chapter 2—that it is a learning system which establishes only provisional findings which may then be replaced by later learning—and who subscribes to the value system underlying this (according to which, finding things out is intrinsically 'good') ought to be ready to learn from the relative failure of systems analysis so far and to try to make the next incremental step, knowing that it too will lead only to provisional knowledge which will itself be replaced eventually.

The Basis of the Action Research on 'Soft' Systems Methodology

The next two chapters describe a research programme which aimed at developing ways in which systems ideas could be used in tackling soft, ill-structured problems. The programme assumed that the concept of a human activity system would be relevant to such problems, and hence its aims also included that of finding out more about descriptions of this kind. Its method was to tackle actual problems facing real-world managers; its criterion of success was that the people concerned felt that the problem had been 'solved' or that the problem situation had been 'improved' or that insights had been gained.

Its starting point was one of the versions of hard systems thinking listed in Table 4, that of Jenkins (1969.) This methodology starts from an organizational definition of 'system' as a complex grouping of human beings and machines for which there is an overall objective. The procedure is then to select that system, the engineering of which will solve the problem whose existence, perceived by a manager, initiates the activity. The system selected is placed in a systems hierarchy, objectives and measures of performance are defined, and the chosen system is designed (via model building, simulation, and optimization) implemented and reappraised in operation.

The core idea in this methodology is that businesses and industries are systems. This idea, as a basis for problem-solving, seems first to have been expressed by Optner in 1965, in his book *Systems Analysis for Business and Industrial Problem Solving*. Optner set out 'to examine business problem solving from all sides, all levels, and all angles', and to do so via 'the concept of business as a system'. He argued that 'systems concepts make it possible to extract both the general and the special properties of a problem' and one of his objects was to persuade business executives not to treat solutions to problems as special cases but to look upon their normal professional activity as that of setting up, running, and maintaining systems. He gives four reasons for using the concept 'systems'. Firstly, many problems are seen to recur when they are looked at as problems of systems; hence solutions may be transferred. (Note that this is also a core assumption of OR, discussed in Chapter 3.) Secondly, the systems view enables problem-solving to concentrate on the *processes* by which things are done, rather than on (*ad hoc*) 'final outcomes'. Thirdly, 'systems may provide the objective standard by which problems can be organised for solution. . . . From objective standards we may be able to gain greater insight to generalise on business phenomena.' Fourthly, although OR has made important mathematical contributions to solving quantitative problems, many business problems are 'mixed', that is they have both qualitative and quantitative attributes. 'Systems analysis', Optner concludes, 'is the newest technique to be brought to these mixed problems', and it seems to be the method most likely to do the three things he asks of a problem-solving methodology for business and industry:

1. Prescribe a system that functionally organises a general problem-solving process.
2. Stipulate parameters that provide the format necessary for the solution of problems.
3. Describe system models and capabilities that provide the means for the iteration of alternative outputs in the problem-solving process (Optner, 1965).

This was an admirable programme to propound in 1965. Nearly fifteen years later, we know that the anticipated generalities pertaining to business and industrial systems have barely begun to emerge; but at least we begin to understand why this is so. In 1969, when the work to be described began, I took as my starting point the methodological cycle shown in

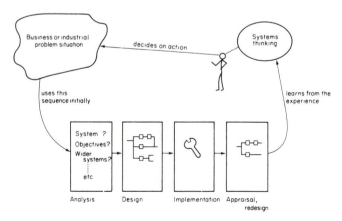

Figure 5. The methodological cycle at the start of the action research

Figure 5, and started by loosening up the phrasing of the methodology as Jenkins had expressed it, so that, for example 'Definition of overall economic criterion' became 'Definition of measures of performance'. Beyond that, the intention was simply to try to apply the hard methodology to soft problems and to observe how the methodology had to adapt or change if successful problem-solving were to be achieved.

The work itself, which produced the degree of surprise which is usually associated with useful learning, is the subject matter of Chapters 6 and 7.

Chapter 6

The Development of 'Soft' Systems Thinking

In the previous chapter was discussed the version of 'systems analysis' which emerged within the engineering discipline and profession. This was demonstrably successful within its own field, that of the provision of complex hardware, and given this success, together with the glamour which in the 1960s attached to computers and 'Space Age' technology, it was probably inevitable that systems analysis of this kind should be extended to social systems and to civilian problems, and should be seized upon by professional politicians as something they could use for their own (legitimate) purposes (Hoos, 1972). The results were equivocal, as we have seen, and this was probably inevitable also, given the difference between the problem of meeting a precise hardware need which is described in a detailed specification and the problem of *defining*, never mind *meeting*, a social need. The core of all the versions of systems thinking discussed in the previous chapter is that they are goal-oriented. Their implicit assumption is that the problem which the systems analyst faces can be expressed in the form: How can we provide an efficient means to meet the following objective ...? Hoos describes systems analysis of this kind as 'Utopian' precisely because it works from a supposedly 'optimum objective' for a given system and tries 'so to order the organisation of the components and their interactions as to achieve a desired and presumably desirable goal'. 'Hard' systems thinking is goal-directed, in the sense that a particular study begins with the definition of the desirable goal to be achieved.

The application of such methods to management problems, to 'soft' problems in social systems where goals are often obscure, was the subject of the research work to be described in this and subsequent chapters. The research had as its subject one of the kinds of system which make up the systems map discussed in Chapter 4, namely the *human activity system*. It sought to answer three questions:

What are the special characteristics of this kind of system?
Can such systems be improved, modified or designed?
If so, how?

and to do so by working within real-world manifestations of human activity systems in which something was perceived to be 'a problem'. Trying to solve the problem(s) would, it was hoped, reveal crucial aspects of sets of interacting human activities viewed as systems.

After experience as a manager in an innovating industry—man-made fibres—I was interested to see to what extent 'hard' systems thinking could be applied both to the kind of fuzzy problems which managers face and to social problems which are even less well defined. I did not imagine that methods suitable for tackling 'hard' engineering problems would survive unscathed their transfer to 'soft' problem situations; on the other hand there seemed no justification for postulating at the start some novel methodology hopefully suitable for ill-defined problems in social systems. If the work started from the well-established methods of goal-directed systems analysis and consisted of trying to use them in ill-defined problems, then it would be possible both to cling on to the known as far as possible and to mark out the areas in which the known failed. Hopefully, it was thought at the start, it would be possible to learn *how*, *why*, and *to what extent* the existing hard systems thinking failed. Hopefully, the research would end up with new systems principles which could be defined rather closely in terms of their departure from the existing methods developed in the engineering tradition. In particular, the work took as its starting point the account of goal-directed methodology given by Jenkins (1969).

A main outcome of the work is a way of using systems ideas in problem-solving which is very different from goal-directed methodology. It emerges from the research experiences as a systems-based means of *structuring a debate*, rather than as a recipe for guaranteed efficient achievement; but its departures from the starting position can all be plotted and described in terms of the particular experiences which forced the modifications to be made. Without the well-defined base of 'hard' systems thinking to serve as an initial hypothesis to be tested, the research would have been difficult to define, control or describe. Such conviction as the work carries must lie in the story of its journey from its base in 'hard' systems thinking to its end point in 'soft' systems thinking. The course followed was not a theoretical pathway but the result of a particular set of experiences in actual problem situations. These were rather small in scale. Unlike the major problems which Governor Brown of California defined for the systems experts he consulted in the early 1960s—involving a state information system, transportation, crime, and sanitation — this research tackled small-scale problems in well-defined social systems of modest size. Unlike the Californian experiment its aim was not to achieve social (or political) ends, but rather to ascertain to what extent systems concepts could be used in a helpful and coherent way to tackle problems which reside in social systems and are of their nature difficult to define.

Achieving that was then to reveal something of the fundamental nature of 'human activity systems'.

The Research Context and Method: Action Research

The research work to be described was carried out as part of an on-going action research programme in a university Department of Systems. The programme is an unusual one compared with most university research, and because the outcome of the work was undoubtedly influenced by its context it is necessary to describe this briefly; a fuller account is given in Checkland and Jenkins (1974).

The Department teaches mainly at postgraduate level, and offers a one year Master's Course in 'Systems in Management'. The average age of students has usually been in the late twenties, and most of the Master's students have been mature postgraduates from numerate disciplines who after typically five years' experience in employment wish for one reason or another to change their career direction or broaden their prospects. Typical student motivations include: a desire to move into line management from 'management services' or vice versa; a desire to move into employment in the private sector from the public sector or vice versa; and, a desire to move into general management from a specialist function or vice versa. Students using the course as a means of moving into the computer world, for example, frequently meet on the course fellow students using it as a route out of the computer world. Now, given mature postgraduates of this kind, when he established the course in 1966 (it was then mainly concerned with 'hard' systems) G. M. Jenkins's bold concept was that it would be possible to explore systems concepts by entering actual problem situations with a team consisting of a faculty member, a full-time mature postgraduate Master's (or Doctoral) student and a member of a client group who posed a problem the sponsors wished to see solved. Systems ideas would be the basis of project thinking and the systems projects themselves, as well as seeking practical improvements in the problem situation, would provide a growing body of experience which ought to be a source both of insight into systems ideas and methodology for using systems concepts. A methodology derived tentatively from a number of studies could hopefully be further tested and refined in later ones: that was the pattern to be followed. Later, a university-owned consultancy company with its own full-time employees was established and this also provided a source of systems studies from which lessons could be learned. The work to be described made use of a flow of studies associated with the Master's Course or Doctoral work, as well as some carried out by the consultancy company. The former had both advantages and disadvantages which stemmed from the link with the course. Because of that link they were restricted to a five-month duration. While this could be restrictive on occasion, the urgent need to reach by a certain date a point in the work at which a significant report could be written provided both a spur to effort and a bridle on irrelevant speculation. The limited duration of the projects and the one-year cycle of the course also provided a rapid means of testing methodological changes.

This university context is one reason why the research has taken the form it has. The other determining factor has been the fact that it is 'action research', and before describing the outcome of the work it is necessary to discuss the special nature of 'action research' and the special problems which it brings.

The concept of action research arises in the behavioural sciences and is obviously applicable to an examination of human activity systems carried out through the process of attempting to solve problems. Its core is the idea that the researcher does not remain an observer outside the subject of investigation but becomes a participant in the relevant human group. The researcher becomes a participant in the action, and the process of change itself becomes the subject of research. In action research the roles 'researcher' and 'subject' are obviously not fixed:

> the roles of the subject and the practitioner are sometimes switched: the subjects become researchers . . . and the researchers become men of action (Clark, 1972).

The theoretical point here is that it is taken for granted that the researcher is himself part of the field of study and since he cannot be separated from it it is better to accept his involvement as itself part of the subject to be researched (Blum, 1955; Susman and Evered, 1978; Warmington, 1980; Hult and Lennung, 1980).

The origin of action research is usually taken to be Kurt Lewin's view of 'the limitations of studying complex real social events in a laboratory, the artificiality of splitting out single behavioural elements from an integrated system' (Foster, 1972). This outlook obviously denotes a systems thinker, though Lewin did not overtly identify himself as such—he was in fact a psychologist of the Berlin Gestalt group who worked in America from 1933. A central idea of the group was that psychological phenomena should be regarded as existing in a 'field': 'as part of a system of coexisting and mutually interdependent factors having certain properties as a system that are not deducible from knowledge of the isolated elements of the system' (Deutsch and Krauss, 1965, quoted in Sofer, 1972). Lewin's field theory regarded any observed behaviour as the outcome of the operation of a large number of factors which it was the task of the social scientist to identify and analyse. The idea of action research follows from this, and it may be regarded as one particular response to the problem of the special nature of social phenomena which was raised in Chapter 3.

Clark (1972) places action research as one extreme in a typology of social science research which includes Pure Basic (concerned with a theoretical problem), Basic Objective (tackling a general practical problem), Evaluation (assessing some aspect of performance), Applied (aiming to solve a problem by applying appropriate knowledge—i.e. not aiming to enlarge the stock of knowledge), and Action Research which he sees as 'one strategy for influencing the stock of knowledge of the sponsoring enterprise . . . also . . . of scientists'. Foster (1972) gives a formal definition based upon one by R. N. Rapaport:

> A type of applied social research differing from other varieties in the immediacy of the researcher's involvement in the action process and the intention of the parties, although with different roles, to be involved in a change process of the system itself. It aims to contribute both to the practical concerns of people in an immediate problematic situation and to the goals of social science by joint collaboration within a mutually acceptable ethical framework.

In these terms, these definitions do fit the systems studies carried out in the research to be described. The intention, although it was not always realized, was always to be involved in 'a change process in the system itself' as a means to both practical action in solving a problem and experience relevant to the wider research aim of developing systems concepts.

The problem with action research arises from the fact that it cannot be wholly planned and directed down particular paths. The essential nature of it is revealed by asking: 'What would action research in physics be like?'. It is, of course, impossible. You cannot do action research on magnetism because the researcher has no alternative but to accept the role of outside observer of a phenomenon which he must take to be unalterably following a fixed pattern which he can discover. But when the phenomena under study are social interactions the researcher will find it almost impossible to stay outside them. If he accepts wholeheartedly that he cannot remain aloof—which is what he does in the intervention in purposeful systems which is action research—then he may express his research aims as hopes but cannot with certainty design them into his 'experiments'. He has to be prepared to react to whatever happens in the research situation; he has to follow wherever the situation leads him or stop the research.

In the present work an attempt was made to tackle ill-defined problems by using systems concepts, to explore the difficulties encountered and to propose and test ways in which the systems concepts could be used. The problems were diverse in content but for the wider purposes of the research were treated as a single group. Lessons from the ones first tackled modified the approach used in later ones; but whatever the happenings in any individual study this did not deflect the research from its aims since each individual study only represented one more example of 'a problem of a human activity system' to be tackled using systems concepts. This remained true whatever the content of a particular study. This use of action research projects as a vehicle for the wider research is thus legitimate because, and perhaps only because, the research aim is the very general one of developing a methodology for tackling 'soft' ill-structured problems and using that experience as a source of insight into the special properties of social systems.

The Problem of 'Problems' and 'Problem-Solving'

Experience at the start of the research quickly showed that it was not possible to take for granted the concept of 'a problem' and the activity of trying to solve it. This is obvious enough in retrospect but was a mild surprise at the time. Apparently, at the start, all the research needed to get under way was a perceived problem which was in or related to human activity systems (the research not being concerned with *objective* problems such as 'Design a better mousetrap') and a desire on the part of a client or sponsor to see that problem solved. Responses to this idea quickly showed that a more subtle view of this starting position had to be taken: What is 'a problem'? itself became a part of the research.

Two extreme but typical responses were quickly noted. Some potential clients in industrial firms took the view: 'We have no problems; things operate fairly smoothly; over a period of time we've organised ourselves in a way which works, so we have no problems as such'. This attitude assumes a rather dramatic definition of 'problem' as a clearly discernible, somewhat cataclysmic turn of events. On this definition a manager in charge of a plant filling milk bottles 'has a problem' if he finds the floor covered in broken glass and swilling with milk; a community in the path of an advancing hurricane 'has a problem' in the same sense. Clearly the research assumed a broader, less dramatic definition of what constitutes a problem. But the other extreme response, again not uncommon, showed that there were difficulties in the other direction too. This response was the one which said 'We think we've got problems but are unsure what they are; if we could tell you what they are we would get on with solving them ourselves!'

These difficulties drew attention to the need to recognize problems of two kinds—*structured* problems which can be explicitly stated in a language which implies that a theory concerning their solution is available (for example: how can we transport X from A to B at minimum cost?) and *unstructured* problems which are manifest in a feeling of unease but which cannot be explicitly stated without this appearing to oversimplify the situation (for example: What should we be doing about inner-city schools?) It became clear that structured problems are what 'hard' systems thinking and most operational research are concerned with. In these terms, the distinct approach of OR, for example, as discussed in Chapter 3, is to argue that those problem types which recur have a particular structure; given such a structure techniques may be developed which are applicable to any problem of that form whatever its specific content. Within that tradition, in fact, it has been argued that *all* problems may be reduced to a single form, that which we have seen (Chapter 5) underlies both the RAND Corporation style of systems analysis and the 'hard' systems engineering methodologies. Ackoff (1957), for example, argues that

> All problems ultimately reduce to the evaluation of the efficiency of alternative means for a designated set of objectives.

It became clear that the present research was to be concerned not with problems as such but with *problem situations* in which there are felt to be *unstructured* problems, ones in which the designation of objectives is itself problematic.

It also became clear that the language used initially implied a time-independent situation in which the sequence of events was: recognition of problem: definition of problem: action to solve problem: problem solved. This adequately describes the tackling of many structured problems but is inadequate for unstructured problems for two reasons. Firstly, such problems, though 'recognizable', cannot be 'defined'. Secondly, in problems in human activity systems history always changes the agenda! The contents of such systems are so multivarious, and the influences to which they are subject so numerous that the passage of time always modifies the perception of the problem (such problems really do sometimes 'go away'!). Such perceptions of problems are always subjective, and they change with time. This is something which the research had to take into account. In fact a number of studies have been completed which are successful in the sense that they are judged so by both client and systems analyst but in which 'the problem' was never defined throughout the whole course of the work.

In formal terms the research proceeds on the basis of the following definition of the word 'problem'.

> A problem relating to real-world manifestations of human activity systems is a condition characterised by a sense of mismatch, which eludes precise definition, between what is perceived to be actuality and what is perceived might become actuality.

In the early stages of the research it was accepted that whereas the definition of structured problems implies what will be accepted as 'a solution', unstructured problems—the concern of the research—must not be pressed into a structured form but must somehow be tackled in the absence of any firm definition of them. They are conditions to be alleviated rather than problems to be solved.

Two Project Experiences

During the period 1969–71 nine systems studies in problem situations which were unstructured in the sense discussed above enabled a basic methodology for the use of systems ideas in such situations to be developed (Checkland, 1972). The methodology was tested and refined in later studies and so far more than a hundred studies have contributed to this process (Checkland, 1970, 1972, 1975, 1976a, 1977a,b, 1979c,d). It would be tedious to present a chronological story of the wandering course of the work. Instead there will now be described some of the most significant aspects of two of the early studies which were crucial in determining the shape of the methodology as it departed from 'hard' systems practice. An account of the methodology is then given, followed by the detailed story of a particular study which illustrates all its main features. Chapter 7 describes later experiences which have contributed significantly to the work.

The first study was in a small textile firm of 1000 employees, here to be referred to as Airedale Textile Company Ltd. Its business was the manufacture of a wide range of yarns, twines and ropes, as well as some fabricated products, from a range of fibres, natural and man-made. The work began as a five-month study carried out by myself and a postgraduate student, R. Griffin, but this eventually led to further work with the same firm over the following two years. The work overall made a contribution to solving the cataclysmic problems of a company which at the start had recently failed to pay a dividend to its shareholders for the first time in its history and was in danger of going out of business. By the end of the three-year period recovery was well advanced, but of course by then the whole problem context had changed (with the takeover of a group of smaller firms bringing in a new managing director, among other changes) so that it is not possible to define clearly the specific contribution of the systems study. But the lessons relevant at this point are those which arose from the very first stages of the work and by chance provided a dramatic example of the inadequacy of a goal-directed approach in a very 'soft' situation. (At a later stage the work also provided useful lessons concerning information systems—Checkland and Griffin, 1970.)

The initial contact was a conversation with the Marketing Director of the Company. He was newly appointed, having had previous experience in the marketing function of a large textile corporation. He and a new Finance Director had both been recruited by Airedale at Board level as a response to declining performance. This was a significant step for Airedale to have taken, its tradition being that all its managers joined the Company on leaving school. It was thus an unsophisticated Company, unversed in modern management methods, and inbred. Nevertheless the recruitment of the two new Directors from outside showed a readiness for change, as did the fact that Airedale had recently spent many thousands of pounds which it could ill afford on management consultant reports. The consultants had attempted to introduce what for them was an 'off the shelf' production scheduling system but this system was not at all understood by anyone in Airedale, nothing had happened in practice, and the consultants' reports remained in the Production Director's bookcase. He did show us, on our first day in the Company, the regular reports he received on production operations and said 'Look at those, they tell me nothing'. The Managing Director was also surprisingly ready to admit that he and the other Directors did not feel themselves to be in charge of the enterprise, and he encouraged us to range widely in the Company in deciding what work we could most usefully do. The Marketing Director who had brought us into the Airedale Textile Co. Ltd, busily introducing a small-scale version of the marketing organization he had known in his previous job, also did not wish to push us into any particular definition of the problem to be tackled, although he personally felt that the real problem was the Production Department's inability to provide him with the yarns he wanted at the appropriate time, cost, and quality—a not unusual Marketing Department view.

At first we were naively pleased to have no definition of the problem provided by the client: here was the desired 'unstructured problem'; but puzzlement increased as investigation showed that virtually every aspect of Airedale's activity was a candidate for our problem. Production planning and production scheduling were certainly poor; quality control was virtually non-existent and the opinion of the Production Director was that a clear definition by Marketing of product quality related to end use would enable most of the problems to be solved; relations between the area-based sales representatives and the headquarters manufacturing site were not good, with representatives running what were virtually private businesses from regional stores; much basic financial information about production and selling costs was not available, but the new Finance Director assured us that he was reorganizing all the accounting systems; customer complaints were frequent; there was no planning at Company level and the Managing Director had no concept in his head of planning as an on-going activity; management information flows at all levels were sparse and random; morale throughout the Company was low. The situation was so bad that tackling any of these perceived problems could be expected to produce improvement but the existence of so many candidates inhibited the choice of any one. What use would the elimination of any one obvious inadequacy be if the problem nexus as a whole forced the Company out of business? On the other hand if the conglomeration of problems forced us to consider the Company as a whole as 'the system' on which our study was to concentrate, then the intellectual weapon available to us as systems analysts, namely 'hard', goal-directed systems methodology seemed singularly inappropriate. It required us to define the objectives of this system as a precursor to model building and the choice between alternative means, not an operationally useful thing to do because definition of the obvious objective—'Company survival'—took us no further forward. (We did try asking the Managing Director his objective for Airedale and received the expected unhelpful reply: 'Start making profits again'.) At this stage we were ruefully remembering Wittgenstein's remark: 'The methods pass the problem by'.

Our personal dilemma, something of an intellectual crisis in fact, was resolved six weeks into the study. We blundered into doing what with hindsight can be seen as *choosing to view the Company as a whole in a particular way which seemed relevant to the multivarious problems, and working out the logical systemic consequences of that view*. This amounted to taking 'the system' of the study to be not simply 'the Company', which implies a uniquely describable single entity—a concept which does not engage with the problems—but the Company from a particular viewpoint which did seem specifically relevant to the core of the problems. *The system* for purposes of the study was taken to be a notional system which generates and accepts customer orders for a defined range of products and uses its expertise to meet them sufficiently quickly and efficiently to ensure an increasing flow of them. This encapsulated the view that because the Company was in danger of going out of business it was most important to engineer some improvement within a matter of months to a human activity system fundamental to its operation. Of course the Company *is* such an order-generating-and-processing system, but it is many other things as well, all of which were ignored at this stage. In subsequent studies production planning and scheduling, company planning, and distribution systems were all given further attention. But what was done six weeks into the first study was to make a logical conceptual model of an order-generating-and-processing system in order to see where present activity of that kind could be improved. The sequence of *decisions which recur* in any such system was isolated and converted into a systems model by postulating a system component to take each decision. (Decisions were of the kind: Do we accept this order? Is this product in stock? etc.) The logic of the decision sequence led to a system structure and this enabled the purpose, source, content, and recipient of the necessary information flows within the notional system to be defined.

The concept led to the creation of two units for order processing, one customer-oriented, the other production-oriented, and the subsequent projects sought to bring actual Airedale activities *gradually* nearer to those of the systems model.

The second systems study, which was instrumental in shaping the general methodology which eventually emerged from the sequence of action researches, was of a very different kind. It was carried out in a large and highly sophisticated engineering company which designed and constructed a small number of very large and very complex objects. It employed a very large number of skilled technical graduates, and, as is sometimes the case in the British engineering industry, its self-image had more to do with maintaining technical excellence than with being successful in a competitive business. This was reflected in the organization structure which consisted of specialist functions within which professional experts worked alongside their peers. The senior managers were largely men who had performed outstandingly in specialist functions and had been promoted because of it: in the value system of this organization technological excellence came first.

A perspicacious director of the Company (which will here be referred to as Cordia Engineering Company Limited) concerned about the difficulties of integrating highly specialist expertise into the enterprise as a whole, had organized several residential seminars for groups of middle managers from various functions in which, in a country house away from the factory, various Company problems were discussed. In order to stimulate discussion the organizing director suggested that the groups examine the problem of computers in Cordia, not because he necessarily considered this the crucial problem but because the provision of computer services in the Company was an activity which cut across functional (departmental) boundaries, and because it was a topic on which all managers had opinions as well as experience. Various outside speakers provided additional stimulus to the seminars, the admirable purpose of which was to help free the thinking of middle managers and to create more flexible attitudes towards change in the Company.

After lecturing on systems concepts as one of the outside speakers, I was asked to discuss with the seminar chairmen a possible five-month project. A study was set up and was carried out by myself and a postgraduate student, D. I. Thomas, who had had experience of advanced technology while working in the aircraft industry in America, experience which was reasonably relevant to the problems facing Cordia. The study theme was to be 'information flow for decision making', and again this seemed usefully vague, given our desire to work on unstructured problems.

At an early meeting the representatives from the seminars made a specific suggestion that the problem be taken to be the difficulties of information flow at the departmental interface between design and production planning. Not surprisingly there were differences of opinion in the two departments concerning the extent, quality and timeliness of the information which passed from one to the other. In our systems enthusiasm, and—as I would now see it—our arrogance, we argued that that particular interface, like all internal organizational boundaries, was arbitrary, the demarcation which the Company happened to have chosen for the time being, and hence was not necessarily relevant to the more fundamental problems which the Company faced. The interface problem was not pressed and we were left to define our own problem.

At that stage our picture of the overall problem situation included the following elements: a remarkably complex functional organization within which decision-making was diffuse and which contained a number of project 'task forces', put together in the past because of particular crises, which had lived on beyond the end of their intended life; a 'project' organization which cut across the functional organization but which was purely a reporting, not a managing function; a senior management which continually became involved in relatively low-level engineering problems and obviously enjoyed it; and a middle management remarkably dedicated to the technical success of the total enterprise, showing much more commitment to it, in fact, than I had observed in my own experience in the process industries. Because of middle manager dedication, procedures which to an outsider seemed obviously unworkable, were made to work.

Again the systems problem was that the choice of 'the system' was itself part of the problem. 'Hard' systems thinking assumes that 'the system' (together with its sub-systems and the wider systems of which it is itself a part) are not problematic; they are 'obvious', they can be taken as given, and the problem then resides in defining their objectives and examining alternative ways of meeting them. In vague problem situations, it was again apparent, no systems hierarchy relevant to the problem could be taken as given. Problem definition again depended upon the particular view adopted and again it seemed necessary to *make that viewpoint explicit* and work out the systemic consequences of adopting it.

In this particular instance it was decided to take as the system for the study a notional human activity system which carries out the gigantic task of converting physical and abstract resources into a specified complex engineered object within a certain time, at a certain cost and while meeting various technical and safety constraints. A system to do this was envisaged, initially without reference to what was currently going on in Cordia, and the resulting model was then compared with Cordia's functional organization and project administration. General conclusions were drawn and these were then tested and illustrated by examining a specific historical example. This phase involved finding out exactly how one specific part of the engineered product *had actually* been provided within existing procedures and comparing this with how it *would have* been provided by the conceptualized system. Such an illustration shows in microcosm the operation of the system as a whole and hence, in systems terms, potentially reveals more than could be revealed, for example, by the examination of information flow across a particular (arbitrary) organization boundary. In fact this study revealed rather more than the Company was prepared for, and the needed changes defined by the systems analysis went far beyond anything the Board of Cordia were then prepared to contemplate. In terms of achieving any practical outcome within the period of the work, the project was a failure, but in contributing to the development of 'soft' systems methodology it was an extremely valuable experience, not only because of the lessons concerning problem definition which are relevant to this stage of the argument, but also in other respects which will be taken up later in this chapter.

Both of the studies described briefly here, as well as others among the first nine carried out, presented situations in which the would-be problem solvers were faced not with defining the goals of systems which could be taken as given, but with selecting what was to be taken as 'the system' out of a very large number of possibilities. The situations forced a choice to be made and a need for coherence dictated that that choice should be made explicit. What then seemed useful was to work out the consequences of the choice by making a conceptual model of the human activity system implied by that choice—an order-generating-and-processing system in the case of Airedale, a specified-task-carrying-out system in the case of Cordia. The systems models were then set against what was actually observed to be happening in the problem situations so that comparison might suggest and justify useful changes.

These early studies laid down this outline for the 'soft' methodology which is described and illustrated in the sections which follows.

A Main Outcome of the Research: Systems Methodology for Tackling Unstructured Problems

1. *Methodology in General*

The systems studies which were the content of the research tackled small problems and larger problems, problems of detail and problems of principle; they were based in organizations both small and large, and in organizations which are public-supported, such as hospitals or the Civil Service, as well as in ones which are user-supported, such as industrial firms or Electricity Boards.† The factor which unites them all into a single group is the fact that all were vehicles for the same thing: the development of principles concerning the use of systems ideas in problem-solving in real-world situations. All the studies had this in common: systems methodology.

By 'methodology' I do not mean 'method'. The word does derive from the Greek word for method and this, according to Kotarbinski (1966), originally meant the path of a person pursuing another, then came to mean generally a path, then a way of doing something, and later expert behaviour in formulating one's thoughts. As a result of this history Kotarbinski distinguishes three current conceptions of methodology, which he calls *praxiological*—'the science of ... ways of expert procedures', *logical*—'the study of methods of using one's mind', and *epistemological*—'the study of sciences as historical products and processes'. My sense of the word here is that the outcome of the research is not *a method* but a set of *principles of method* which in any particular situation have to be reduced to a method uniquely suitable to that

†The distinction between public-supported and user-supported organizations. more useful than public sector/private sector. is taken from Vickers (1965).

particular situation. I believe this point to be an important one and am prepared to labour it. In attempting to work in the real-world we face an astounding variety and richness. If 'soft' systems thinking is reduced to method (or technique) then I believe it will fail because it will eliminate too much of the munificent variety we find in real life, just as the generalized algorithms of management science have tended to lose contact with the uniqueness of each individual management problem. My use of the word is thus nearest to Kotarbinski's 'praxiological' version: methodology not as 'ways of expert procedures' but *the science of* [such] procedures'. (I would want to eliminate Kotarbinski's 'expert', trying to make systems thinking a conscious, generally accessible way of looking at things, not the stock in trade of a breed of experts.) I take a methodology to be intermediate in status between a philosophy, using that word in a general rather than a professional sense, and a technique or method. A philosophy will be a broad non-specific guideline for action: it might be held as a *philosophy*, in this everyday sense, that, say, 'political action should aim at a redistribution of wealth in society', or that 'industrial expansion should be carefully balanced against environmental degradation'. At the other extreme a *technique* is a precise specific programme of action which will produce a standard result: if you learn the appropriate technique and execute it adequately you can, with certainty, solve a pair of simultaneous equations or serve a tennis ball so that it swerves in mid-air. A methodology will lack the precision of a technique but will be a firmer guide to action than a philosophy. Where a technique tells you 'how' and a philosophy tells you 'what', a methodology will contain elements of both 'what' and 'how'. In this sense the research programme sought a methodology for using systems concepts which would have four characteristics: it should be capable of being *used* in actual problem situations; it should be *not vague* in the sense that it should provide a greater spur to action than a general everyday philosophy; it should be *not precise*, like a technique, but should allow insights which precision might exclude; it should be such that any developments in 'systems science' could be included in the methodology and could be used if appropriate in a particular situation.

2. *The Methodology in Outline*

For brevity the methodology is here expressed in the form of a diagram. This section gives a dense account of it which will then be expanded. Figure 6 represents a chronological sequence and is to be read from 1 to 7, a logical sequence which is most suitable for describing it but which does not have to be followed in using it! Recent work (see Chapter 7) has provided proof that it is possible wholeheartedly to start a project at stage 4, for example, and in principle a start can be made anywhere. Backtracking and iteration are also essential; in fact the most effective users of the methodology have been able to use it as a framework into which to place purposeful activity during a systems study, rather than as a cookery book recipe. In an actual study the most effective systems thinker will be working simultaneously, at different levels of detail, on several stages. This has to be so because the methodology is itself a system—of the Designed Abstract type discussed in Chapter 4—and a change in any one stage affects all the others.

The methodology contains two kinds of activity. Stages 1, 2, 5, 6, and 7 are 'real-world' activities necessarily involving people in the problem situation; stages 3, 4, 4a and 4b are 'systems thinking' activities which may or may not involve those in the problem situation, depending upon the individual circumstances of the study. In general, the language of the former stages will be whatever is the normal language of the problem situation, that of 3, 4, 4a and 4b will be the language of systems, for it is in these stages that real-world complexity is unravelled and understood as a result of translation into the higher level language (or meta-language) of systems.

Stages 1 and 2 are an 'expression' phase during which an attempt is made to build up the richest possible picture, not of 'the problem' but of the *situation* in which there is perceived to be a problem. The most useful guideline here—in the interest of assembling a picture without, as far as

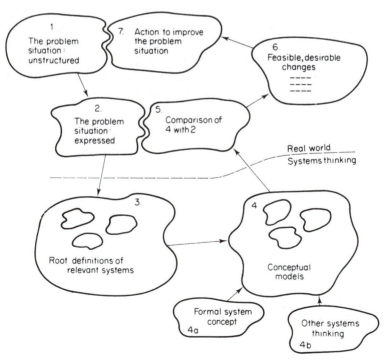

Figure 6. The methodology in summary (after Checkland, 1975).

possible, imposing a particular structure on it—has been found to be that this initial analysis should be done by recording elements of slow-to-change *structure* within the situation and elements of continuously-changing *process*, and forming a view of how structure and process relate to each other within the situation being investigated. Stage 3 then involves naming some systems which look as though they might be relevant to the putative problem and preparing concise definitions of what these systems *are*—as opposed to what they *do*. The object is to get a carefully phrased explicit statement of the nature of some systems which will subsequently be seen to be relevant to improving the problem situation. This cannot be guaranteed, of course, but the formulation can always be modified in later iterations as understanding deepens. These definitions in stage 3 are termed 'root definitions', which is intended to indicate that they encapsulate the fundamental nature of the systems chosen—'specified-task-carrying-out', for example, in the case of Cordia Engineering. Given this definition, or, better, these definitions, for it is always better to explore several possibilities, stage 4 consists of making conceptual models of the human activity systems named and defined in the root definitions. The model-building language is very simple, but emerges as a subtle and powerful one: it is simply all the verbs in the English language! A structured set of verbs is assembled which describes the minimum necessary activities required in a human activity system which is that described by the root definition. Model building is fed by stages 4a and 4b: 4a is the use of a general model of any human activity system which can be used to check that the models built are not fundamentally deficient; 4b consists of modifying or transforming the model, if desired, into any other form which may be considered suitable in a particular problem. For example it may be thought appropriate to re-express it in the language of systems dynamics (Forrester, 1961, 1969) or re-structure it as a Tavistock-style 'socio-technical system' (Emergy and Trist, 1960). Whether or not this kind of transformation takes place, the models from stage 4 are then, in stage 5, 'brought into the real world' and set against the perceptions of what exists there. The purpose of this 'comparison' is to generate a debate with concerned people in the problem situation which, in stage 6, will define possible changes which simultaneously meet two criteria: that they are arguably *desirable* and at the same time *feasible* given prevailing attitudes and power structures, and having regard to the history of the situation under examination. Stage 7 then involves taking action based on stage 6 to improve the problem situation. This in fact defines 'a new problem' and it too may now be tackled with the help of the methodology.

In the next sections this brief account of the whole will be amplified as each of the stages is discussed in more detail.

3. *Stages 1 and 2: Expression*

Experience so far has shown that although the intention of the first two stages, to find out about the problem situation while trying not to impose a particular structure on it, is usually clearly understood by people using the methodology, these are in practice difficult stages. There is a marked reluctance to pause and reflect over the initial expression, and this is perhaps not helped by the use of the word 'problem'; people find it difficult to interpret the word in the loose way described above, and often show an over-urgent desire for action. But the best studies have been characterized by a holding back in stages 1 and 2, by a readiness to collect

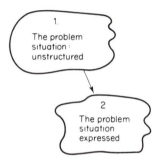

as many perceptions of the problem as possible from a wide range of people with roles in the problem situation, and by a determination not to press the analysis in systems terms at all. In 'hard' systems analysis the concept is that there is a system to be engineered and that this occupies an unequivocal place in a manifest hierarchy of systems. In 'soft' systems—which include most human activity systems considered at a level higher than that of physical operations—there will always be many possible versions of 'the system to be engineered or improved' and system boundaries and objectives may well be impossible to define. Vickers (1968, 1970) has argued cogently against taking social systems to be goal-seeking, pointing out that 'relationship-maintaining' is often a better description of their purpose, and this work endorses that view. It has been found most useful to make the initial expression a building up of *the richest possible picture* of the situation being studied. Such a picture then enables selection to be made of a viewpoint (or viewpoints) from which to study further the problem situation. Once that selection is made, of course, one or more particular systems, which will be part of a hierarchy of systems, are being defined as *relevant to problem-solving*. Indeed, stages 3 and 4 consist of a systems-oriented exploration or 'design' of it, but the spirit in which this is done is one of entertaining the idea that 'this one is a relevant system', in the full knowledge that other choices are possible and might be more insightful. Happily, the initial selection is not being made once and for all.

Suppose that the problems of a public library were the subject of the systems study. It might well be thought appropriate in the particular circumstances to take a public library itself to be a system. But what kind of a system? Various possibilities are apparent: Is it to be taken as a local authority amenity system, one among others? Or as a system which is part of a wider education system in the locality concerned? Or will it be useful to define it functionally as a system which seeks to maximize the exposure of a range of stored material to a particular population? Any of these might be fruitful in a particular situation in which people in particular roles perceive there to be 'problems' of a particular kind. The function of stages 1 and 2 is *to display the situation so that a range of possible and, hopefully, relevant choices can be revealed*, and that is the only function of those stages.

It is in achieving as neutral a display as possible that the concepts of 'structure', 'process', and 'the relation between structure and process' have been found useful. 'Structure' may be examined in terms of physical layout, power hierarchy, reporting structure, and the pattern of communications both formal and informal. 'Process' may frequently be examined in terms of the basic activities of deciding to do something, doing it, monitoring both how well it is done and its external effects, and taking appropriate corrective action. The relationship between structure and process, the 'climate' of the situation, has frequently been found to be a core characteristic of situations in which problems are perceived. In Cordia Engineering the technology-based functional structure was ill-matched with process operations of a task-carrying-out system but managed to survive given the attitudes of middle managers. This was taken to be the core characteristic of this particular problem, the end point of the analysis phase on the first iteration.

4. Stage 3: Root Definitions of Relevant Systems

At the end of the expression stage we answer the question, not: What system needs to be engineered or improved? but: What are the names of notional systems which from the analysis phase seem relevant to the problem? It is essential to answer the question carefully and explicitly, writing out and discussing openly a rather precise account of the nature of the system or systems chosen. The choice will represent a particular outlook on the problem situation and the purpose of naming the system

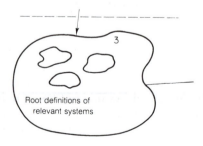

Root definitions of
relevant systems

carefully is both to make that outlook explicit and to provide a base from which the implications of taking that view can be developed. The choice of what I have called 'a Root Definition of a relevant system' is not ultimately committing, in the sense that if later stages reveal the choice to be lacking insight, irrelevant or infertile then other viewpoints may be tested. In fact the best systems thinkers will at this stage be quickly testing out various possibilities by looking ahead to stages 4, 5, and 6 and seeing what *kind* of model will follow from the root definitions entertained and what *kind* of changes will be likely to emerge when the models are examined alongside what presently exists in the real world. Root definitions thus have the status of hypotheses concerning the eventual improvement of the problem situation by means of implemented changes which seem to both systems analyst and problem owners to be likely to be both 'feasible and desirable'. To propose a particular definition is to assert that, in the view of the analyst, taking *this* to be a relevant system, making a conceptual model of the system, and comparing it with present realities is likely to lead to illumination of the problems and hence to their solution or alleviation. 'Relevant' does not here imply that the system selected is necessarily desirable, certainly not that it is the system which *ought* to be designed and implemented in the real world. I have tried to avoid such Utopian connotations. Thus it might well be useful for a priest making a systems analysis of certain problems in the Church to take as a basis for a root definition the famous epigram in Karl Marx's essay of 1844 (Bottomore and Rubel, 1956):

> Religion is the sigh of the oppressed creature, the sentiment of a heartless world, and the soul of soulless conditions. It is the opium of the people.

The priest would presumably not agree with this view, but it could well provide the basis of an insightful analysis of his problems.

A root definition should thus be a concise description of a human activity system which captures a particular view of it. In trying to use the methodology a number of people have been dismayed by their inability to think up 'brilliant' definitions, but a root definition does not have to be noticeably clever to be useful. It is not necessary to describe NASA as, say, 'a show business system' to feel that a satisfactory root definition has been achieved. The question is: Given the picture of the problem situation and the perceptions of 'the problem' by people in it, does the suggested root definition seem to have a chance of being useful? And that can be answered only by testing some possible definitions, even if they seem commonplace.

In conveying the idea behind the formulation of root definitions, however, it is easier to use rather more dramatic examples. Here is one more. One of the new social phenomena of recent years in the Western world has been the pop festival. If we were making a systems study of this phenomenon we might well view this particular kind of human activity system as if it were a conventional commercial enterprise (and this might be useful even for the 'free' festivals for which no admission is charged). That would be one possible view. Another, which might well provide more insight, would start from the idea that many thousands of young people do not travel long distances to live in tents in muddy fields in rather squalid conditions simply to be the customers of pop music sellers; such festivals display 'the alternative culture'. A systems study of pop festivals might well take as a root definition 'a system to celebrate a particular life style using pop music as an emblem of the sub-culture concerned'.

Here finally are a few actual root definitions which have been used in studies carried out by myself or my colleagues and associates in recent years.

In an early study (by J. K. Denmead and C. Driver) it was found useful to take part of a Blood Transfusion Service to be the operation of *a transfer system*, one concerned with locating a particular commodity, blood, in a particular place, namely the veins of potential donors, and removing it from there by some appropriate technology to a new location for storage.

In a study of the role of a community centre which serves a deprived area of a northern city and is largely funded by a local industrialist, R. H. Anderton and P. Thomas took as a root definition:

> An institution encouraging and helping community action aimed at development of the community's own resources.

The most obvious characteristic here is that this definition clearly expresses a particular *Weltanschauung*, that the centre is concerned to develop self-help by the community rather than simply to distribute charitable benefits.

In a study of part of the work of the Social Services Department of a Local Authority M. R. Jackson and R. Douglas used the following portmanteau definition:

> A department to employ social workers and associated staff to build and maintain residential and other treatment facilities and to control and develop the use of these resources so that those social and physical needs of the deprived sections of the community which Government statute determines or allows, to the extent to which County Council, as guided by its professional advisers, decides is appropriate, are met within the annual capital and revenue constraints imposed by the Government and Council.

It is often useful to include a number of constraints in the definition, as is done here; their effect can always be tested subsequently by relaxing them and seeing how the model of the system then has to change. And it is also clear here that a carefully elaborate definition of this kind is of great help when moving on to making a model of the system named in the definition.

5. *Stage 4: Making and Testing Conceptual Models*

As we saw in Chapter 4, given the concept of a whole which we choose to regard as 'a system', we have two possible approaches to the task of describing it. It may be described in terms of its 'state' by describing the elements which comprise it, their current condition, their relationships with external elements which affect the system, and the condition of those external elements. This method of description is very appropriate to many physical systems, whether natural or man-made. Alternatively we may

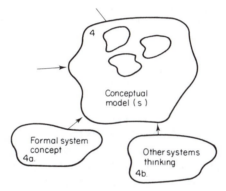

provide a systems description by regarding a system as an entity which receives some inputs and produces some outputs; the system itself *transforms* the inputs into the outputs. This kind of description can also be used successfully with many physical systems; in the case of the class of system with which this research was concerned—human activity systems—it seemed the only feasible descriptive mode. Assembling a state description of a human activity system seemed too daunting a task, whereas describing purposeful activity according to the scheme: input — transformation—output did seem possible.

Now, any root definition may be looked at as a description of a set of purposeful human activities conceived as a transformation process. What is now done in stage 4 is to make a model of the activity system needed to achieve the transformation described in the definition. We now build the model which will accomplish what is defined in the root definition. The definition is an account of what the system *is*; the conceptual model is an account of the activities which the system must *do* in order to *be* the system named in the definition.

Definitions are formulated without thinking: 'this system ought to be engineered'. And note that the resulting model, when complete, is not a state description of any actual human activity system. It is in no sense a description of any part of the real world; it is simply the structured set of activities which logic requires in a notional system which is to be that defined in the root definition. This is a hard point to grasp, and once conceptual model building starts there is a noticeable tendency for it to slide into becoming a description of actual activity systems known to exist in the real world. This needs to be resisted because it negates the whole purpose of the approach, which is to generate radical thought by selecting some views of a problem situation as possibly relevant to improving it, working out the implications of those views in conceptual models and comparing those models with what exists in the real-world situation. If descriptions of the real world slip into the model then in the comparison stage we shall be comparing like with like, and novel possibilities are unlikely to emerge. (If the conceptual model derives properly from the root definition but still leads to an unexciting comparison then of course the root definition itself was not sufficiently radical, and another version should be tried.)

The step from root definition to conceptual model is the most rigorous in the whole methodology, the nearest to being 'technique'. That modelling 'technique' will now be described, but first an illustration will help to make the above points clear. (Checkland, 1979c, gives a detailed example of conceptual model building. See Appendix 1.)

In the case of the Blood Transfusion Service problem situation mentioned above it was decided that it might be relevant to look at the operation of a blood-collection unit as that of *a transfer system*. This process of transfer was the core of the root definition; viewed as a transformation process it entails as input 'blood in the veins of potential donors' and as output 'blood stored in a location from which it is available for medical use'. The necessary model is that of a notional system which brings about the transformation of this input into this output. To construct such a model we do *not* examine the actual operations of blood collecting units and blood banks within the National Health Service: to include them in the model would be positively to invite a poverty-stricken comparison stage. The fact is of course that prior participation in an analysis phase in which the actual operations of that part of the Health Service have been examined will in a subtle way facilitate model building; but it ought not to direct it, and nothing ought to be included in the model which cannot be justified by reference to the root definition. Modelling thus becomes a question of asking: What activities, in what sequence, have to occur in order to do the transfer?

Because the conceptual model is a model of an *activity* system its elements will be *verbs*. The 'technique' of modelling is to assemble the *minimum* list of verbs covering the activities which are *necessary* in a system defined in the root definition, and to structure the verbs in a sequence according to logic—for example 'define potential donors' would have to precede 'locate potential donors'. The main verb in this model, covering the core of the transformation would be 'transfer' (blood from collection point to storage point).

That the model building is not quite a technique, in the sense that a technique is a procedure which, properly applied, will produce a guaranteed result, is shown by the fact there are always arguable issues about whether one person's model is as adequate a representation of a root definition as another's. In this example, for instance, it might be argued that the inclusion of the verb 'classify' (blood according to group) could be justified on the grounds that this is fundamental to any system which collects and stores this particular commodity. Or it might be argued that the root definition as given is concerned only with collecting and transferring to store and hence the model should exclude 'classify'. What is clear is that if the root definition were more explicit, and contained a phrase such as 'transfer, type, and store' then it would be unarguable that the model should contain a 'classify' sub-system.

Experience has shown that it is best to begin conceptual model building by writing down no more than about half a dozen verbs which cover the main activities implied in the root definitions. Sometimes, as for example in the case of the Social Services Department quoted in the previous section, a portmanteau definition virtually outlines the main activities and their relationship to each other, and hence the structure of the model. Whether or not this is the case it has been found best always to complete a model at a low 'resolution level' (little detail) and then to expand each major activity at a higher level of resolution. Thus 'classify blood according to group' could itself define a sub-system which at a higher resolution level contained all the verbs necessary to cover the more detailed actions involved in determining the blood group of a series of examples. The art in model building of this kind lies, in fact, in keeping separate the major activities of the system and, in a given model, in maintaining consistency of resolution level.

For a complete example of a simple conceptual model used during this research we may take that developed during the work in the Airedale Textile Company. Here a 'relevant system' was taken to be one concerned with generating customer orders for a range of textile products and operating a technology to enable those orders to be met expeditiously, the intention being to provide good service at a cost the Company could afford. (This was the equivalent of what would now be called a root definition although that concept had not been formulated at the time of this study.) What was done then was to ask what decisions would necessarily recur in any such system—'decisions which recur' later being generalized to 'activities which are necessary'. Six basic activities were listed as describing the system envisaged, and logic dictated the structure given in Figure 7 (based on Checkland and Griffin, 1970).

Modelling at a higher resolution level was carried out by concentrating on the information flows needed if this linked set of activities were to be

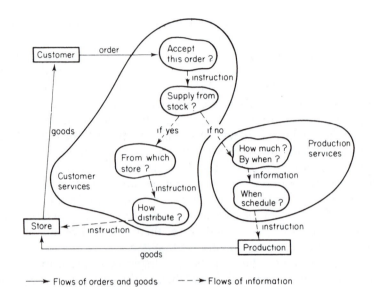

→ Flows of orders and goods   — — → Flows of information

Figure 7.   A conceptual model of the order-processing system
(after Checkland and Griffin, 1970)

carried out efficiently. For each activity (or 'decision' as they were here called) the question was asked: What information must the decision maker have in order to take this decision (i.e. 'carry out this activity'), what is its content, source, and frequency? From the detailed information system model resulting from this questioning it was clear that the system could be structured as an operational system consisting of two 'doing' systems, one concerned with making yarns, the other with storing them, these being served respectively by two instruction-issuing systems which were respectively production-oriented and customer-oriented, as in Figure 8.

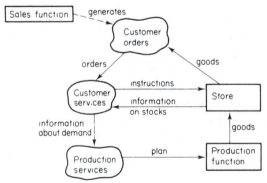

Figure 8.   Conceptual   model   of   the
order-generating-and-processing system — two
'planning' and two 'doing' systems (after
Checkland and Griffin, 1970)

The system would seek efficiency by learning to meet as many orders as possible direct from store without incurring unacceptable storage costs.

Once a conceptual model of the kind described has been built, it would be reassuring to be able to establish its validity, just as many chemical engineers' 'hard' models of chemical reactors can be validated by showing that the performance of the model on a computer simulates the observed performance of the reactor itself. Such validation is not possible with conceptual models based on root definitions. There are not valid models and invalid ones, only defensible conceptual models and ones which are less defensible! But at least it is possible to check that conceptual models are not fundamentally deficient, and this is done in stage 4a of Figure 6 by checking the model against a general model of any human activity system which I have called the 'formal system' model. This model is not descriptive of actual real-world manifestations of human activity systems, still less prescriptive. It is a formal construct aimed at helping the building of conceptual models which are themselves formal: they are not accounts of what *ought* to exist in the real world, for it is absolutely not the intention of the methodology to diminish the freedom of actual human activity systems to be, if they wish, irrational or inefficient. But the formal system model, though not normative, is nevertheless related to experience, as will be seen. The model is a compilation of 'management' components which arguably have to be present if a set of activities is to comprise a system capable of purposeful activity. The model extends the 'summary of properties of systems' which Jenkins (1969) proposed for systems defined as groupings of men and machines with an overall objective and characterized by an economic criterion which measures performance; and it follows the 'Anatomy of Systems Teleology' which Churchman (1971) presents as a definition of that sub-class which are 'teleological things, i.e. things some of whose properties are functional'. My formal system model draws on both sources but contains only components whose absence or inefficiency in actual problem situations has turned out to be crucial to the existence of something perceived to be a problem. This is what makes the model, though not prescriptive, practical. Thus, while a strong case can be made for 'negative entropy' being a characteristic of any purposeful human activity system, this is not included in the formal system model because it has never emerged as a crucial characteristic in any of the hundred odd systems studies in which the methodology has been used. It is therefore not included in what is intended as a formal but practical tool. The components of the model are as follows. S is a 'formal system' if, and only if:

(i) S has an on-going purpose or mission. In the case of a 'soft' system this might be a continuing pursuit of something which can never be finally achieved—something such as 'maintaining relationships'. In harder systems this is what sharpens up into 'objectives' or 'goals' characterized by being achievable at a moment in time. It is meaningful to describe objectives as 'not yet achieved', 'achieved', 'abandoned' or 'changed' but this is not so with the less precise 'purpose' associated with soft systems. Families, and many organizations, do not have objectives in this sense, but they do have purposes or missions which serve to cohere and link their activities. For a university, for example, which as a system might be regarded as 'hard' in some characteristics and 'soft' in others, this characteristic might be taken to be 'the discovery, preservation, and transfer of knowledge'.

(ii) S has a measure of performance. This is the measure which signals progress or regress in pursuing purposes or trying to achieve objectives.

(iii) S contains a decision-taking process—notionally 'a decision taker', as long as this is taken to be not a person but a role which many people in a given system may occupy. Via the decision-taking process the system may take regulatory action in the light of (ii) and (i).

(iv) S has components which are themselves systems having all the properties of S.

(v) S has components which interact, which show a degree of connectivity (which may be physical, or may be flows of energy, materials, information, or influence) such that effects and actions can be transmitted through the system.

(vi) S exists in wider systems and/or environments with which it interacts.

(vii) S has a boundary, separating it from (vi), which is formally defined by the area within which the decision-taking process has power to cause action to be taken—as opposed to hopefully influencing the environment.

(viii) S has resources, physical and, through human participants, abstract which are at the disposal of the decision-taking process.

(ix) S has some guarantee of continuity, is not ephemeral, has 'long-term stability', will recover stability after some degree of disturbance. This might be helped from outside the system; it might derive internally from participants' commitment to (i).

Note that if the analysis is pressed to lower levels in greater detail, then below sub-systems and sub-sub-systems, etc. will eventually be found items which, from the analyst's point of view, are not systems at all but only system components. Similarly, analysis in the other direction will eventually reach larger entities which *in the analyst's judgement* have to be taken as environments rather than systems, the distinction being that an environment may hopefully be influenced but cannot be 'engineered' whereas a wider system can, at least in principle, be 'engineered'. The classic example would be an economy. In the West the performance of post-1945 governments in their dealings with state economies (and maybe also ideology) would probably persuade an analyst to take the economy of an industrial country to be an environment within which industry creates wealth; but an East European would probably regard an economy as a system which makes known what it requires of the industrial firms which would be regarded as its sub-systems. But this is a matter of judgement, there are no absolute definitions. This is not a weakness because the systems analyst is not saying: 'This is what *is*', he is saying 'This is what I shall (temporarily) take things to be in my analysis'.

The representation of components (i) – (viii) in a single system is best done by using the kind of diagram which social scientists sometimes use to illustrate the interactions among qualitative factors. This is done in Figure 9.

The value of the formal system model is that it enables questions to be framed which, when asked of the conceptual model, reveal inadequacies

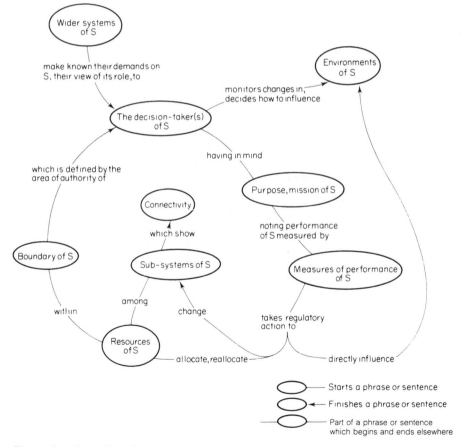

Figure 9. A model of the concept 'human activity system' (from the point of view: taking purposeful action in pursuit of a purpose or mission)

either in it or in the root definition which underlies it. Typical questions might be: Is the measure of performance in this model explicit, and what would constitute 'good' and 'bad' performance according to it? What are the sub-systems in this model and are the influences on them of their environments taken into account in the activities of the system? Are the system boundaries well defined?

Although the use of the formal system model cannot ensure that conceptual models are 'valid', it can at least ensure that they are not so sloppily constructed as to be useless when set against real-world activities in the comparison of phase 5. But there is something else which can be done before moving to the comparison, and that is to examine the models for validity in terms of any other systems thinking which the analyst reveres. Because the methodology was developed at the level of 'principles of method', rather than as a technique, it was essential that it should not in principle exclude any systems thinking being developed elsewhere. This is the stage at which those other veins may be mined. This is the point at which the conceptual models may be inspected alongside any systems theory which is relevant to human activity systems. For example if the analyst finds useful the Tavistock Institute concept of a task-system which is inseparably and simultaneously both a social system and a technological system (Emery and Trist, 1960) then the conceptual model may be extended or recast to include this particular insight. Or the analyst working on a conceptual model which happens to be that of an institution as a whole might wish to incorporate Beer's model of organization, or to make sure that his model is compatible with Beer's. Beer (1972) sees an industrial organization as 'a viable system which tends to survive', as do organic systems like the human body, and his 1972 book 'continuously compares the unfolding story of corporate regulation in the body with its manifestations in the firm'. From this emerges a hierarchical model of organization which exhibits autonomous control at its various levels. Now, if this analogy appeals (and to be fair to Beer he actually regards it as a description of how things are rather than as an analogy) then the analyst may recast his conceptual model into the five sub-systems of Beer's model. Other possibilities would be to check the model against Ackoff's (1971) compendium of systems concepts, against Churchman's (1971) concepts or against one of the models of adaptive control systems such as that developed by Gomez et al. (1975). I personally have found it useful, after making my choice of the models which seem to provide a useful way of structuring the examination of the problem, to ask how they do or could embody Vickers's (1965, 1968, 1970, 1973) concept of an 'appreciative system' namely 'the interconnected set of largely tacit standards of judgement by which we both order and value our experience'. Most important of all, however, is the fact that it will be possible in the 1980s to make use of whatever systems concepts have by then been developed in order to obtain further reassurance that the conceptual models are, if not strictly 'valid', at least defensible.

6. *Stage 5: Comparing Conceptual Models with Reality*

It is a matter of judgement as to when to stop conceptual model building and move on to a real-world comparison between what exists there and what is in, or is suggested by, the models of systems thought to be relevant

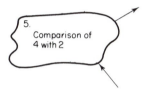

to the problem.The temptation is always to indulge in prolonged and elaborate model building: it is a more comfortable activity than bringing the models out into the chill wind of reality and engaging once more with the difficulties of the problem situation itself! On the whole, though, it is better to move fairly quickly to the 'comparison' stage, even if models subsequently have to be refined in a return to conceptualization.

The 'comparison' stage is so-called because in it parts of the problem situation analysed in stage 2 are examined alongside the conceptual models: this should be done together with concerned participants in the problem situation with the object of generating a debate about possible changes which might be introduced in order to alleviate the problem condition.

My colleague R. H. Anderton has pointed out that this stage is not in fact a proper comparison of like with like, and he is correct. A number of experiences have led to a more subtle understanding of what exactly is being done at stage 5, and this will be discussed in Chapter 7. Here will be described the rationale of this stage in the methodology and the four different ways of doing it which have emerged from the studies carried out during the research.

Whenever we set out consciously to do some serious thinking we are aware of a number of operations of our minds: perceiving, predicating, and *comparing* images, pictures or models. In the methodology this is somewhat formalized: the perception of a problem situation is recorded in the first two stages of analysis; root definitions and conceptual models then use systems ideas to predicate certain selected features of it; these predications, in the form of systems models, are then compared with the perceived realities in the problem situation itself. The comparison is the point at which intuitive perceptions of the problem are brought together with the systems constructs which the systems thinker asserts provide an epistemologically deeper and more general account of the reality beneath surface appearances; it is the comparison stage which embodies the basic systems hypothesis that systems concepts provide a means of teasing out the complexities of 'reality'.

Studies of different kinds have seemed to call for different ways of carrying out the comparison, and in a variety of experiences four ways of doing it can be identified.

In Airedale Textiles it was obvious that what was going on in the Company was very different indeed from what was in the conceptual model of an order-generating-and-processing system. The latter is not normative, and it would be very foolish to argue that the analyst, as a result of his brief use of systems thinking, is now ready to tell these particular people, whose attitudes will have been moulded by living through the particular history of this unique situation, what *ought* now to be done. Rather, the need is to use the systems models to open up debate about change. In Airedale this was done by using the models as a source of questions to ask of the existing situation. These were written down and answered systematically, and it was these answers which provided illumination of the problems for the Airedale participants. (Eventually, in Airedale, a central chart room was created in which were taken, with appropriate information available, the kind of decisions included in the notional 'Customer Services' and 'Production Services' sub-systems. But the decision to do this came from the Managing Director as a result of the discussions the systems work had initiated, rather than as a direct immediate outcome of the first study.) This method of using the conceptual models as a base for *ordered questioning* in the problem situation has subsequently been much used in many later studies. It will always be one possible way to proceed in any study.

In Cordia Engineering, as has been mentioned earlier, a different method of comparison was used. The comparison was done by reconstructing a sequence of events in the past (the provision of a relatively complex part of the final engineered product) and comparing what had happened in producing it with what would have happened if the relevant conceptual models had actually been implemented. This is in logic a satisfactory way of exhibiting the meaning of the models, and maybe the inadequacies of the actual procedures, but it is a method to be used delicately because it can easily be interpreted by participants as offensive recrimination concerning their past performance. It was most successful in a study by D. H. Brown and myself which was done for a consultant who wanted to know why one of his studies for a client had been a spectacular failure (see Chapter 7). In that case the whole content of the study was history, and the analysis compared the story as remembered and recorded at the time by participants, with a systems model of consultant/client interaction.

In a number of studies the conceptualization stage raises major strategic questions about present activities rather than detailed queries about procedures, questions of the kind: Why do this at all? rather than: Is this done well? In such cases, one of which is described below in the illustration of the methodology as a whole, it is usually appropriate to make the comparison of stage 5 a general one, asking what features of the conceptual models are especially different from present reality and why. In the case to be described, a day was spent on this stage in discussion with the manager who sponsored the study, and at the end of it we had a chart listing six major differences which opened up possibilities for change.

Finally, in a study of our own problems arising from the overlapping of a university department and a university-owned consultancy company, a fourth method of doing stage 5 was used. In carrying out this study Su Lau and myself were very aware of the 'physician-heal-thyself' problem, and we decided on extreme—even flat-footed—methodological rigour, with each step in the thinking made explicit. For the comparison, after completing conceptualization based on the chosen root definition, we made a second model, this time of 'what exists'. The second model had as near as possible the same *form* as the conceptual model, the aim being to re-draw that model, changing it only where the reality differed from the conceptual model. With this method, direct overlay of one model on the other then starkly revealed the mismatch which is the source of discussion of change. With this method, piquancy may be added to the comparison by asking of the model of what exists: What root definition is implied by this 'system'? How does it compare with the one which was the basis of conceptualization in stage 4?

This 'model overlay' method of making the comparison was used very successfully by my colleague B. Wilson in a study in an engineering firm which (rather unusually) was consciously seeking a more rational organization structure (Wilson, 1979). Here the senior management agreed a root definition appropriate to the Company as a whole and the activities required by it were assembled in functional sub-systems; these were then compared, by 'model overlay', with the activities actually going on in the various departments and sections. Where a mismatch was revealed a conscious decision was then taken either to shift an activity to a more appropriate location or to modify the responsibility of the manager responsible for a given section or department.

All four methods help ensure that the comparison stage is conscious, coherent, and defensible. In any particular study it may be useful to adopt any one of them or to carry out several comparisons using different methods.

In a 'greenfield' situation in which a new human activity system is being designed the comparison is not in principle different from that described above, although the comparison cannot be with what exists, only with some defined *expectation*. It is likely in such cases that the comparison, though it might reveal some basic omissions, will be less fruitful than attempted implementation, which will quickly reveal design inadequacy. As Popper (1945) points out we have little certain knowledge on which to base this kind of 'social engineering'; incrementalism and trial and error are the wisest approach.

7. *Stages 6 and 7: Implementing 'Feasible and Desirable' Changes*

The purpose of the comparison stage is to generate debate about possible changes which might be made within the perceived problem situation. In practice, initial work on this stage frequently draws attention to inadequacies in the initial analysis or in root definitions, and further work is required there. Eventually, however, if necessary after several iterations,

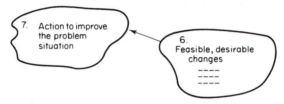

the comparison will lead to discussion of possible changes. These are of several kinds, and any combination may be appropriate in a particular situation.

Note that in 'hard' systems work the 'change' envisaged is the creation and implementation of a *system*. This is sometimes the case in 'soft' problems also. Sometimes, appropriate action might entail, say, the implementation of a planning system, or an information system to serve existing functions. But in general in these more nebulous problem situations, the eventual action is likely to be less than the implementation of a system; it is more likely to be the introduction of a more modest change.

Changes of three kinds are possible: changes in structure, in procedures, in 'attitudes'. Structural changes are changes made to those parts of reality which in the short term, in the on-going run of things, do not change. Structural changes may be to organisational groupings, reporting structures, or structures of functional responsibility. Procedural changes are changes to the dynamic elements: the processes of reporting and informing, verbally or on paper, all the activities which go on within the (relatively) static structures. Changes of both these kinds are easy to specify and relatively easy to implement, at least by those having authority or influence. Once made, of course, such changes may bring about other effects which were not anticipated, but at least the act of implementation itself is a definite one and can be designed. This is not the case (and we may think this a good thing if we are to remain human) in the case of changes of the third kind, changes in 'attitude'. This I take to include not only changes in the attitudes such as may be sampled in the 'attitude surveys' beloved of behavioural scientists, but also many other crucial, but intangible characteristics which reside in the individual and collective consciousness of human beings in groups. The term is intended to include such things as changes in influence, and changes in the expectations which people have of the behaviour appropriate to various roles, as well as changes in the readiness to rate certain kinds of behaviour 'good' or 'bad' relative to others—changes, in fact, in what Vickers terms 'an appreciative system'. Such changes will occur steadily as a result of shared experiences lived through by people in human groups, and they will also be affected by deliberate changes made to structures and procedures. It is possible in principle deliberately to try to bring about changes of this third kind, although it is difficult in practice to achieve exactly the result anticipated. Whether or not this is attempted, the main essential is continuously to *monitor* 'attitudes' in the sense discussed here if changes are to be made in situations perceived as problems so that concerned actors in the situation agree that 'improvement' has been achieved.

The purpose of stage 6 is to use the comparison between conceptual models and 'what is' to generate discussion of changes of any or all of the three kinds discussed above. The discussion should be with people in the problem situation who care about the perceived problem and want to do something about it. Of course, a 'concerned actor' in the problem situation may actually be the 'systems analyst' himself, who may be making a systems study of one of his own problems, but it is useful to differentiate between the two roles, 'concerned actor' and 'analyst', even if the same person occupies both. If an individual is examining problems with which he is himself concerned, it is still very important to distinguish between the 'below-the-dotted-line' activity in Figure 6 (stages 3 and 4) where all the thinking is explicit and clinical, and the activity above that line (stages 1, 2, 5, 6, 7) where the analyst can again behave like a human being: sagacious, rational, calm, and charitable, or, equally, blinkered, obtuse, impulsive, irrational. Below-the-line, systems concepts are used to penetrate beneath surface characteristics; above-the-line we are in the real world with all the untidiness that that implies; and it is not a characteristic of this approach to pretend that real life is tidier than it is, still less is it the intention to force it into a more rational form.

The debate about change, then, carried out in the real world of the problem with 'concerned actors', aims at defining changes which meet two criteria. They must be arguably systemically *desirable* as a result of the insight gained from selection of root definitions and conceptual model building, and they must also be culturally *feasible* given the characteristics of the situation, the people in it, their shared experiences and their prejudices. It is not easy to find changes, of whatever kind, which do meet both criteria.

In the case of Cordia Engineering, for example, after the comparison stage we could argue with some cogency for major changes in the management of a project which was making fitful progress as a result of the functional specialization of departments, but this concept was an unbridgeable gap away from what was feasible, given the prevailing attitudes and the unwillingness to contemplate major change. In fact, this failure helped to teach me the importance of moving quickly and lightly through all the methodological stages, several times if necessary, in order to engineer a bridgeable gap between 'what is' and 'what might be', a lesson which other studies have reinforced. In another study in the engineering industry, by J. H. Collins and P. E. Gillett, which was concerned with the creation of management information systems, the implications of the work were that structural changes to the organization were required. Getting wind of this, the Managing Director decreed that proposed changes must be procedural only. It was only at this stage appreciated that for the Managing Director his ability to decide the organization structure was a crucial aspect of his power as chief executive, something he was not prepared to concede to outsiders. Such happenings cause a reappraisal of root definitions, which may have to incorporate 'root constraints' if the study is to end in action.

In another study which provided useful illumination of these later stages, carried out by J. K. Denmead and J. K. Mackley, a meeting with management was held at which a concept of a quality control system was expounded. Such a system was conceived as 'a system to balance the cost of achieving a certain desirable quality in the manufactured product concerned, against the cost of lost sales if adequate quality were not achieved'. In our opinion the company needed to set up such a system. Considerable management interest in the concept was expressed, but in the event the change which was not only systemically desirable but also culturally feasible was the creation of a system for dealing with customer complaints. The introduction of this system, one of the sub-systems of the complete quality control system, was the biggest change which, in the opinion of the people concerned, could be absorbed at that time.

Once changes have been agreed, implementation of them may be straightforward. Or their introduction may change the situation so that although the originally perceived problem has been eliminated, new problems emerge. Or the activity of implementing changes may itself be problematic—and this new problem may also be tackled by means of the methodology. In a study concerned with a sophisticated engineering problem by L. Watson and N. Jarman, the work included development of a particular way of carrying out a complex task. Implementation became the new problem of creating a temporary system to carry out the task under the supervision of the analysts, followed by a transition to the operation of that system on a permanent basis with supervision by managers in the company.

The methodology has in fact not emerged as a once-and-for-all approach to something sharply defined as 'a problem' but as a general way of carrying out purposeful activity which gains from the power of some formal systems thinking but at the same time does not require individual human beings to behave as if they were rational automata—which latter has been the unfortunate implication of much management science in the 1960s and 1970s.

The Methodology Illustrated: Structural Change in a Publishing Company

The study now to be described is one which illustrates all the stages of the methodology discussed above. It was commissioned by R. H. Anderton, then in charge of a corporate planning unit in a publishing and printing company, now a university colleague. His knowledge of systems concepts and his interest in their use in problem-solving made it possible at all stages to discuss the course of the study without inserting a translation from systems language into management language, and this made it an ideal vehicle by means of which R. W. Keen and myself could test the methodology which had emerged from the early action research experiences, including Airedale Textiles and Cordia Engineering. At the time of this study, I felt ready to try out the whole methodology in a fairly explicit manner on an unstructured problem.

This description of the twenty-week study, which at the end helped bring about structural change in the company, will follow the stages sequentially, as described in previous sections. This is in the interests of clarity, and it should not be imagined that there were no iterations or that the complete rich picture of the problem situation was gained at the start, before any of systems thinking in stages 3 and 4 began. Nevertheless, in this instance, such retracing of steps was very limited, and an account from stage 1 to stage 7 is unusually close to what actually happened.

The study was located in R. H. Anderton's corporate planning group, they being the survivors of what had once been a much larger management science team. The corporate planners responded organizationally to the Secretary of the Company but substantively to a group consisting of the Managing Director and the heads of the operating divisions. The planners were engaged in introducing corporate planning into the Company (here to be called Index Publishing and Printing Company) and they had adopted the sensible strategy of concentrating on producing a Company *plan* for the next period, the idea being that working on a specific plan with the managers concerned would eventually lead to the establishment of corporate *planning* as a normal activity within Index. We were not to be concerned with the embryonic Company plan; our presence was to provide an opportunity to examine some of the ill-defined but nevertheless persistent problems which were felt to dog the Company. We were not in fact offered any definition of a problem at all, rather we were given statements of the kind: 'We find it very difficult to define, for example, what our mix of printing facilities should consist of over the next five years. We are not asking you to answer that question, we are asking you if you can do anything to improve a situation in which such questions cannot be readily answered'. This seemed an admirably vague indication of a problem, or rather a *problématique* of connected problems, and we set about building up a picture of the situation perceived to be problematic. We were asked to confine our attention to the publishing and printing of consumer magazines, this being an area of Index's activities which was felt to exemplify the problems.

Within the Company itself a study by well-known management consultants had led to a structure of operating divisions which were intended to be profit centres. We were concerned with Magazine Division and Printing Division, both a part of the consumer magazine business, although some Index magazines were printed by outside printers, while Printing Division printed some magazines for outside publishers. In arguing for the divisional structure the consultants had said of the divisional chief executives 'Thus, they can be held truly accountable for their own profit contribution or cost', but the next sentence of their report tarnished this bold concept by remarking

> However, they cannot be entirely free to take action in isolation from the needs of the Company or in competition with each other. Before setting out on a major course of action, directors must seek approval and coordination from higher management.

Unfortunately no way of defining 'major' or of providing 'coordination' was suggested, and we were not surprised to find that a committee set up to define internal prices between divisions had made little progress in more than a year.

The magazine publishing market as a whole was at that time experiencing considerable difficulties. The industry had equipped itself with expensive but inflexible high-quality photogravure printing machines, ideal for printing 3 million identical copies of a woman's magazine but not very suitable for coping with the fragmentation of the market. Women's weekly magazines, in particular, illustrated the dfficulties. Thus, according to Ferguson (1974) sales of five national women's weeklies declined by 1.6 million in nine years to 1973, with the magazine *Woman*, for example, 'Originally and successfully aimed at a cohesive classless, second-sex market, designed to be all things to all women aged 16–60' losing 35% of its circulation between 1966 and 1973.

As we learned about the magazine business, and about the various technologies of printing, we also noted—when relating the 'structures' of Index to its 'processes'—that publishers and printers lived in very different worlds, though both were involved in the same business operation. The self-image of magazine editors, for example, seemed to be that they were 'media people', almost a part of 'showbiz'; whereas printers behaved not so much as operators of a 20th-century technology, more as if they were guardians of a craft skill. Editors clearly felt themselves to be creative people, with an almost mystical rapport with a particular audience, with whose hopes and dreams they could identify. (The industry was full of praise for a lady who had spotted, correctly, that there was a new affluent audience of adolescent girls too old for comics but too young to be interested in the recipes and knitting patterns of traditional women's weeklies. Magazines were created to meet that new market.) Printers, on the other hand, did not have any particular allegiance to an individual magazine title, but organized their activity as in a job shop.

We decided, as the picture was built up, that a 'relevant system' would be a notional unified system which edits, prints, and publishes a single consumer magazine. We were aware that this was very different from the way in which Index was organized, but we decided to make a root definition of such a system and see how its implications illuminated Index's problems. The considerations which went into the formulation of the root definition were: that both the editorial side and the printing side are necessarily involved together in a business operation (that is, it is user-supported and must earn a surplus); and that the crux of the operation is the identification of an audience need in the 'information, entertainment, education' area and the provision of a succession of images relevant to that need. The root definition was expressed as a diagram, and is represented in Figure 10, our notional system being one example of what is generically expressed in the root definition. The conceptual model deriving from it contained the minimum necessary activities for a system in which editors and printers regularly create a magazine for a carefully defined audience

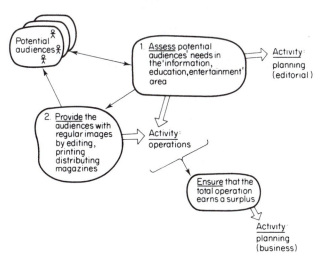

Figure 10. Diagrammatic representation of a root definition of a system relevant to consumer magazine publishing (from Checkland, 1972)

186

(the model is given in Checkland, 1972). The system as a whole generates £x of sales and advertising revenue by selling *n* copies, and this is a measure of performance of this operational system. Notionally this performance must be the concern of a wider system in which the results for this one title are compared with those for others owned by the same wider system. This is the system which would allocate resources by deciding to continue or to close down individual titles. It is obviously a very important one in any organization owning more than one title, and so it too was modelled; part of this wider system model is shown in Figure 11. This system is assumed to be carrying out resource management for a notional owner outside the system, referred to as 'Corporation'. The system appraises the available resources and in the light of the reported performance of individual titles makes plans for the total use of resources, both in publishing and printing. The core of its activity is that in the central portion of Figure 11. Individual titles are compared in terms of the audience sought, the *concept* of the magazine (say, for example, 'provision of technically accurate information about a practical hobby', or 'selling social reassurance to an ill-educated non-liberated female audience', etc.,), the technology by which the magazine is to be printed together with the location of printing, and an economic assessment of the chance of profitable business. This is in fact a control sub-system for a system owning both publishing and printing expertise and facilities. It would take decisions on resource allocation in the light of the total deployment of resources.

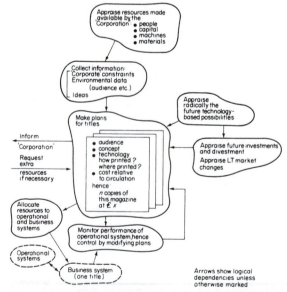

Figure 11. Part of the conceptual model of the system which allocates resources among magazine titles (from Checkland, 1972)

As previously mentioned the comparison stage in this study was carried out by spending a day making a general comparison between the modes of operation in the conceptual models and actual happenings in Index. There were about half a dozen areas in which it was fruitful to debate the differences. For example, in the models, decisions concerning environment forecasts, magazine specification, resource acquisition and allocation, and the decision where to print, were all taken within one system which assembled information relevant to those decisions; in Index these decisions were split between Magazine Publishing and Printing Divisions. In the models the summation of individual plans was related to total resources available; in Index separate negotiations took place between the two Divisions on printing contracts for individual magazine titles. One Company document, which was typical, recorded:

> Title . . . is printed at . . . in photogravure. It can no longer afford the process, and it is hoped to switch to a web-offset printer to restore some reasonable profitability. Printing Division are unhappy about the transfer; negotiations are in progress.

In the models the printer's activity was explicitly linked to the publishing business, whereas Index printers regarded themselves as technology-based jobbing printers, sometimes making across-the-board price changes which might or might not be in the corporate interest, even if they were in the interest of the Division as an autonomous profit centre. Research and development activity in the models was linked to the business activity, whereas in Index research was mainly confined to technological changes within Printing Division.

Discussion of these many differences between what Index was doing and what was implied would be done in formal systems selected as 'relevant' to Index's problems, enabled a range of possible specific changes to be assembled. Firstly, two directions of change were possible. It would be possible to split the business up into small totally autonomous operations with complete freedom to compete with each other; alternatively, continued attempts could be made to manage Index Publishing and Printing Company as a single entity. The arguments in favour of this latter course were rehearsed: the alternative would be a gesture of despair; Index as a whole was a profitable company and had been for many years; there would not be much chance of Printing Division being successful on its own, and no ready buyer for its expertise and assets; finally, no person we had talked to in Index had advocated this kind of fragmentation. These arguments suggested that further attempts should be made to manage Index as a single enterprise. Assuming this to be desirable, seven possible significant changes which could help alleviate Index's problems as revealed in this study were proposed. They ranged from a modest possibility that the responsibilities of existing managers with respect to links with other organizational groups should be explicitly defined, to the radical possibility that a central management committee, with supporting functions, should run the whole operation from the centre. The changes were displayed in a table in order from least to most radical, their implications for Index were recorded, and in a third column was listed our estimate of the readiness of Index management to accept them. In our judgement middle managers were not complacent about the present state of affairs and were ready for significant change without wishing to see Index centrally controlled as if it were a process industry. Our own choice of a change which met our criteria of being both desirable and feasible was one which envisaged a new decision-taking body which had control over some of the matters affecting both publishing and printing, in particular the decision on 'where to print?'. Our comment on this possibility, in our final report, was as follows:

> This body's role begins to approach that of the planning body in the conceptualised system of section 3. To work, it would have to be seen to have authority, and to be distinct from existing Divisional groups. Its formation would accord well with much of the thinking we encountered among middle managers. Thus, we were given a number of instances of decisions being taken by individual divisions which affected other divisions without reference to what would be best for ['Index'] as a whole. (Some see the transfer price system achieving this but others are pessimistic.) This body would be a logical one through which to carry out work, such as forecasting, which is relevant to groups of titles rather than individual titles.

The head of the group with whom we were working, R. H. Anderton, embodied this proposal in a paper he put to the Managing Director and the Divisional Chief Executives entitled 'Corporate Influences on the Publisher-Printer Problem—A Proposal'. His proposal was accepted and a new unit within Index was formed, headed by one of the directors. Writing of the new unit in the Index house magazine, the Managing Director wrote:

> Primarily the unit is concerned with trying to develop a more effective relationship between our publishers and printers. Systems will be developed so that decisions relating to pricing, placing and scheduling of work, and to investment, can be taken more effectively by the divisions to the benefit of ['Index'] than they are now.

The Index study illustrates the 'soft' systems methodology in a very straightforward fashion, and that is its main purpose here. I am aware in writing the account of it that no such account does justice to the rich, floundering confusion which actually characterizes all systems studies, even methodologically straightforward ones like the Index study. As the picture of the problem situation is built up, and impressions and judgements change, the direction of what will only later be recognized as 'progress' is difficult to see. With hindsight, some of the confusions melt away, and no doubt to the reader the course of the thinking, and the final outcome seem 'obvious', where to the analyst at the time all was benighted struggle. Ironically, this *post hoc* rationalisation—which is not simply a deliberate attempt to make things clearer for the reader—is revealing of the nature of attempts to solve problems in human activity systems. The point is that if coherent corporate action is to be taken (as opposed to merely talking about taking action) in a human activity system then *that action must seem to be 'obvious' to people in the system*, it must fit in with the state of their perceptions of their situation and their valuations of what constitutes 'good' or 'bad' activity relative to on-going purposes. In Index the idea of the new unit was not a startling new notion such as no one in the Company had ever thought of. What the systems thinking provided was an argument for change which, because it was convincingly 'obvious', actually helped significant change to take place.

Finally, the Index study helped to teach me the role of the methodology for unstructured studies which it set out to test. Although in this instance there was relatively little backtracking in moving from stage 1 to stage 7, it was during the confusions of finding out about what was for me a new technology in a new culture, that I appreciated that the role of the methodology is to provide an ordering framework for problem-solving. Its stages provide what Hoag (1956), in discussing RAND systems analysis calls '. . . a division . . . useful in order to talk about some conceptual issues one at a time'. At the same time, the linking of the stages makes the whole methodology into a single system—a learning system. The matter is best put this way: 'Do what you like, in the order you like, but be in the position that if tapped on the shoulder and asked what you are about, you can produce a reply of the kind: "Now I am reading reports to gather information about the structures and processes in this situation"; "Now I am making a conceptual model, i.e., I am not describing *what is*"; "Now I am interviewing to find out why the changes I consider desirable are not considered feasible"; or "Now I am thinking up a new relevant system because the first one led to an unbridgeable gap between my model and the way reality is perceived in the problem situation"'.'

'Hard' and 'Soft' Systems Thinking Compared

The action research which led to the methodology described above began by attempting to use the ideas of 'hard' systems thinking in 'soft' problems, moving away from the 'hard' engineering tradition when forced to do so by the difficulties of actual situations. It was stated earlier that it was not imagined that methods suitable for tackling well-defined engineering

Table 5 'Soft' and 'hard' systems methodology compared

'Soft' systems methodology	Jenkins (1969)	RAND Corporation (1950s)
(a) Start: an urge to bring about improvement in a social system in which there is felt to be an ill-defined problem situation	Start: an urge to solve a relatively well-defined problem which the analyst may, to a large extent, take as 'given', once a client requiring help is identified	
(b) Express by examining elements of 'structure' and 'process' and their mutual relationship. Tentative definition of systems relevant to improving the problem situation	Analysis by naming the system, is objectives, etc., and its place in a hierarchy of systems	Analysis by examining the decision-makers' objectives as expressed in the stated need for the required system with a specified performance
(c) Formulate root definitions of relevant systems and build conceptual models of those systems	Design the system by quantitative model building and simulation	Identify alternative systems for meeting the defined need and compare them by modelling using the performance criteria
(d) Improve the conceptual models using the formal system model and other systems thinking	Optimize the design, using the defined (economic) performance criterion	Select the alternative which best meets the need and is feasible
(e) Compare the conceptual models with 'what is' in the real situation, and use the comparison to define desirable, feasible *changes* in the real world	No equivalent stage: both approaches know from the start what change is needed	
(f) Implement the agreed changes	Implement the designed system	

problems would survive unscathed their transfer to fuzzy problem situations set in social systems, and they have not. Now that an approach to 'soft' problems has emerged and been tested, it is interesting to compare it with the earlier forms of systems analysis which derive from engineering.

Table 5 compares the author's approach with both the RAND Corporation version of systems analysis and with the account given by Jenkins (1969) from which this work started.

The main difference between 'hard' and 'soft' approaches is that where the former can start by asking 'What system has to be engineered to solve this problem?' or 'What system will meet this need?' and can take the problem or the need as 'given'; the latter has to allow completely unexpected answers to emerge at later stages. This difference forces the 'soft' methodology to include the comparison stage, which has no equivalent in the 'harder' approaches. In this stage, systems thinking provides a structure for a debate about change which hopefully will be of good quality as a result of the insight captured in root definitions. By this stage the 'hard' approaches are busy preparing to implement the designed system. A RAND-type systems analysis of a weapon system will always produce a definition of a weapon system. Analysis using the 'soft' methodology might suggest disarmament, turning the other cheek, or political negotiation.

Apart from this main difference, the 'soft' methodology is seen to be the general case of which 'hard' methodologies are special cases. Thus *conceptualization* becomes, if the problem is sufficiently well defined, systems *design*. 'Improving a conceptual model' sharpens up into 'optimization of a quantitative model'. Implementing some variety of change becomes implementing a designed system.

The methodology can thus be seen as a general problem-solving approach appropriate to human activity systems. When problems can be expressed unequivocally it may simplify into one of the systems engineering approaches. When problems cannot be stated clearly and unambiguously, it becomes a means of exploring that ambiguity, and contains an additional stage which uses systems analysis as a means of orchestrating debate about change. This additional stage (stage 5: 'comparison') is a reflection of the main characteristic of human activity systems as revealed in this research. This characteristic is that, in contrast to other types of system, human activity systems can never be described (or 'modelled') in a single account which will be either generally acceptable or sufficient. For a system of this kind there may well be as many descriptions of it as there are people who are not completely indifferent to it. This is the characteristic of the real world which forces the methodology to become a means of organizing discussion, debate, and argument rather than a means of engineering efficient 'solutions'.

Chapter 7

The Systems Methodology in Action

The main outcome of the 'systems practice' of this book has been the methodology described in the previous chapter, together with the problem-solving experiences which were tests of it. As a research outcome, this is both curious and unusual. It is curious because of the ambiguous status of methodology as, in its own right, 'science' (in the 'public knowledge' sense) and it is unusual in that the literature reveals that methodological prescriptions are, alas, frequently offered to the reader before they have been tested in practice. It is deplorable but true that the literature of management science, and, especially, of social science, sags under a heavy load of methodological assertions and conceptual models which have not been tried out. Authors would better keep their models and methodologies to themselves until they can demonstrate a problem solved by the use of them, since conceptual models and methodologies (the latter being themselves only conceptual models of systems to enquire and learn) are intrinsically of little value, a harsh fact which is itself a consequence of the curious scientific, or rather, lack of scientific status of methodology.

In developing this particular systems methodology for tackling unstructured problems it was hoped to steer a course between precise technique and vague 'philosophy', between what Boguslaw (1965), in his study of modern systems designers as the new Utopians, calls 'handbook truths' and 'unbridled intuition'. But even if success in that endeavour is achieved, the end product of the research is not quite 'public knowledge'.

It was argued in Chapter 2 that science is characterized by reductionist hypothesis-testing experiments whose results are repeatable, and that progress can be made by the refutation of the hypotheses. A methodology, as such, cannot measure up to this definition. In the case of the 'soft' systems methodology, we cannot obtain or expect precisely repeatable results in purposeful systems, and the idea of making progress by refutation is equally inapplicable. In 'testing' methodology the best we can do is to ask a question which is always difficult to answer, and especially so when applied to social situations, the question being 'Was the problem solved?' (The situation is analogous to the problem of the irrefutability of philosophical theories, Popper, 1963.) If, over a period of time and a number of experiences, problems are solved, in the sense that things 'improve' as measured by some agreed criteria, or that concerned people in the situation themselves feel that insight has been gained or useful changes made, then confidence in the methodology may grow, and we may gradually come to feel that it has been tested and found useful.

In the present case, more than a hundred serious attempts to use it in a wide variety of problems of several different types suggest that that point has been reached, even though, given the nature of methodology, that assertion cannot be proved! A number of studies have made a special contribution to the testing process, either because they concerned a new type of problem or because they led to increased understanding of the use of systems ideas as a way of tackling unstructured problems. In this chapter the different types of study carried out using the methodology are described, then accounts are given of six particular studies which led to useful learning. Finally, a number of general lessons, to which many experiences have contributed, are described. These lead, in the following chapter, to the discussion of the implications of this systems practice for systems thinking.

Five Types of Systems Study

Some of the studies carried out since 1970 have been posed as problems of systems design. 'Help us create a corporate planning system,' 'Install an order-processing system,' 'Design a system to cost non-standard products'; problem definitions of this kind imply that the problem situation is not obscure, and that the task is purely one of systems design and implementation. In a number of situations this has been the case, or it has been necessary, at least initially, to take this to be so, and the studies have differed from 'hard' systems engineering only in that, since the systems to be engineered are manifestations of human activity systems, it has not been possible to build quantitative models and optimize them using some numerical measure of performance. In studies of this kind, conceptual model building itself becomes the basis of systems design, with root definition replaced by system specification, and the comparison stage becoming a comparison of models with expectations which various people have of the proposed system.

Most of the studies, however, have been of the kind described in Chapter 6, in which in a problem situation which defies sharp definition there is nevertheless a desire to take action to bring about improvements. The Airedale Textiles, Cordia Engineering, and Index Printing and Publishing studies were all of this kind, as were most of the studies mentioned briefly in the previous chapter. But though the main purpose of the action research was to learn about human activity systems by tackling this kind of problem, the experiences have included three other types of study.

In Cordia Engineering part of the work consisted of examining how one part of the final product had been produced historically and comparing this with how it would have been produced according to a particular conceptual model. This examination of past happenings became the main content of a remarkable study in which an engineering consultant asked us to examine one of his completed projects which he regarded as a disaster, and tell him why the failure had occurred. This project is described below; it revealed 'soft' systems analysis as a way of doing historical research.

A fourth kind of study is one in which the aim is not to take action but to make a survey of a particular area of concern. Many government reports are of course of this nature. The methodology provides a coherent way of going about such reporting, and an example of this kind is also described briefly below.

Finally, the methodology was used in a fifth mode, in a purely theoretical problem. The study was done for a government committee and had as its aim the clarification of the concept 'terotechnology'. It too is described briefly below.

In summary, then, the hundred-odd attempts to use the methodology have included studies of five different kinds having five different aims: systems design; action to improve an ill-defined problem situation; historical analysis; survey of an area of concern; and clarification of concepts.

Six studies which were important in defining these last three ways of using the methodology, or which provided other significant lessons, are summarized below.

Six Systems Studies

1. *Systems Analysis of Past Events*

The consultant who asked us to do a retrospective study of one of his failures, here called Mr Cliff, was a mining engineer. His general approach to mining problems started from the proposition that mining was concerned first with a natural physical system, the geology and geomorphology of a mineral-bearing area, then with a designed system, the shafts, roadways, extraction equipment, etc., and finally with the human activity system which operates the mine economically under various constraints concerned with safety and working conditions. His theme was that the mine design system interacts with the geological system and that both interact with the mine operation system. His systems approach to mining problems, which we obviously applauded, argued that the design and operation of a mine had to be considered as a total system. His dilemma in the present instance was that in spite of this, a study he had carried out for a client, a manufacturer of a major item of mining equipment, had been a disastrous failure. He asked us to tell him why.

D. H. Brown and I carried out a four-week study for Mr Cliff, spread over a period of several months. He gave us access to his files on his study for 'Mining Support Ltd', recounted his memories of it, and arranged with his former clients that we would also be able to visit them in order to hear their version of the unhappy events.

The first phase of the work saw us examining the files and gathering accounts of the project both from Mr Cliff and from Mining Support. We established the chronology of events in the project and the content of the main exchanges which had taken place between Cliff and Mining Support while they were his clients. We also learnt that the two parties had rather different views of the nature and purpose of Cliff's study. As Mr Cliff had understood it, he was approached by Mining Support Ltd following difficulties they had been experiencing with their equipment in American coal mines. They were to supply him with details of installations which had failed, he was eventually to give them advice which would eliminate such failures in the future. Mining Support, however, claimed that they had always been clear that what they looked for was advice on the next generation of their equipment. They had wanted, they said, a report which would be of value to their 'field engineers' who made on-site examinations and defined what equipment would be suitable in a particular location. Not surprisingly, Cliff's report when presented was discussed at a meeting which both sides remembered for its acrimony. Further work by Cliff led to a second report, mainly dealing with the mining parameters which affect Mining Support's equipment in mine situations, and pointing out the role of these considerations in total mine design. This report was no better received than the first one and after a further bad-tempered meeting the parties separated. Mr Cliff did not invoice Mining Support for the outstanding part of his fee.

Clearly in a study of this kind any outsider could shed light on the controversy simply by talking in turn to the two sides. Equally, it was almost certainly true by definition that problems arose because the two sides had different views of the nature of Cliff's study and its expected outcome. However, that is only another way of stating that the two parties were in conflict; and in any case we were not brought in to make 'common-sense' observations of that kind. We had been approached specifically because we would make a *systems* study. It therefore seemed appropriate in this instance to use systems ideas as a means of understanding the structure of the Cliff/Mining Support Ltd dispute, the structure underlying the detailed happenings and the personality clashes. We made a systems reconstruction of the outline of the project. In these terms, Mr Cliff's consultancy is 'an advice-giving system'; its inputs are his know-how together with a request for advice, its output satisfactory advice to the client who requests it in exchange for an economic fee. Mining Support on the other hand may be viewed as one system among others in a wider 'design-and-operation-of-mines system'. Their input is a need for one particular function in a mine to be fulfilled, their output commercially successful equipment to meet that need. Mining Support's asking Cliff for advice creates a *temporary* system; its inputs are Cliff's expertise and the Company's request for advice; its output is satisfactory advice transferred from Cliff to Mining Support. We decided that this was the 'relevant system' in the given problem, and made the simple conceptual model of it shown in Figure 12, using the input–output formulation as its root definition.

This system arises because party A (in this instance Mining Support) calls for expert advice from party B (Cliff). The activities of this (temporary) system are those minimally necessary to ensure that satisfactory advice passes from B to A in return for an agreed fee. The six main activities are those shown in the diagram, the monitoring and redefinition of remit ensuring that the system can operate to achieve the advice transfer satisfactorily as measured by its measure of performance, which is A's readiness to deem the advice satisfactory. Given the root definition, the decision-taking procedure in the system will have to involve both A and B; the activities then ensure that B knows exactly what kind of advice is required and A knows the changing context of the work, thus enabling A to up-date its notion of 'satisfactory'. The main activity 'Carry out the work according to the (re)defined remit' is expanded, in Figure 12, into three sub-activities. These embody the idea that B has expertise to offer and will exercise professional judgement in

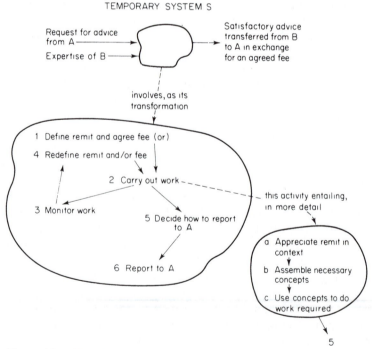

Figure 12. Model of the consultant/client 'advice transfer' system

using appropriate concepts. The system as conceptualized here ceases to exist when, outside the system, B hands over the fee in return for the advice. The model of Figure 12 was satisfactorily coherent in terms of the characteristics of the formal system model (see Chapter 6) and hence was used for a comparison with the actual history of the Cliff/Mining Support collaboration.

We had already assembled a detailed historical account of the exchanges between the client and the consultant, and the stage 5 comparison consisted of framing questions derived from the model in Figure 12 and answering them from the historical narrative. It was obvious from the start, of course, that the story was one of an *inadequate* temporary system in which Mining Support and Mr Cliff were in the roles A and B respectively. But it was interesting to specify precisely what the inadequacies were. There were four of them. Firstly, the remit was never clear between the two parties (main activities 1,4). Analysis of letters and working papers had revealed a gradual widening of the remit for the study. First it was concerned with the technical specification of the item of equipment which Mining Support Ltd manufacture. Further discussion led to Cliff's introducing interaction with 'mine environment factors' and finally the 'economics of mine operation' as well. This was clearly the result of Cliff's systems approach to mine design; but it was also clear that in this project the remit discussions had gone on until each side had read into the remit what they wished to find there. Secondly, the concepts used by Cliff in the study (main activity 2, expanded) were inadequate in that he failed to note that Mining Support do not see themselves as mine designers, only as suppliers of a particular piece of equipment. Thirdly, the monitoring of the project (main activity 3) was intermittent, with Mining Support not kept up-to-date with progress. Fourthly, once work was completed, Cliff had not thought consciously about the audience for his report and how best to address it (main activity 5). We had asked Mining Support to comment on the relevance of the two reports chapter by chapter, and this had revealed much material which, far from enhancing the message, had actually irritated senior managers in the company. One comment had been 'We didn't want all this stuff we've already read in his academic papers'. This irritation undoubtedly contributed to the poor reception the reports had received.

Overall, viewing the affair from Cliff's point of view, his failure had been *a failure consciously to engineer a temporary system like that in Figure 12.* Our recommendations to him therefore included design of procedures which in any project he undertook would ensure that there was no disastrous breakdown in the temporary consultant/client activity system.

Mr Cliff was surprised at the outcome of our study. Although a strong advocate of the need to regard all mining operations as part of a total system, it had not occurred to him that a systems approach could apply to the activity of carrying out a study for a client in exchange for a fee. He felt that he had learnt something useful.

Now, any reader of this account of our study will find the outcome 'obvious', and may indeed feel that the examination of the project history in systems terms was an unnecessary elaboration in what was really a very straightforward problem situation. Surely progress could have been made very quickly simply by compiling the historical narrative without bothering to carry out the retrospective analysis of that history in systems terms? My response to that criticism would be twofold. The project occupied twenty working days. Of that time the whole of the systems thinking in stages 3, 4, and 5 of the methodology occupied only one morning, about $2\frac{1}{2}\%$ of the total effort. We felt that this $2\frac{1}{2}\%$ was a good investment, in terms of the shape it provided for the message to Mr Cliff. In addition, using the methodology led to real learning for us, learning which I do not believe would have emerged from an approach based on mere 'common sense'. When we were comparing the systems model of Figure 12 with the project history, it occurred to us suddenly, with some force, that in our study *we* were in role B to Mr Cliff's role A! What had we done to engineer the temporary system in which we sought to provide our client with advice he thought 'satisfactory'? The surprise with which we suddenly faced that question indicated real learning for us. Happily my colleague D. H. Brown had taken good care to maintain good communications with Mr Cliff and his partners, even when there was nothing specifically to report. But the study had the salutary effect of making us consciously aware of the client—consultant temporary system as one which needs to be engineered, not left to chance communications.

In the development of the methodology this short study was thus useful in several ways. It revealed stages 3 and 4 as a means of understanding the structure of a pattern of past events underlying the detailed happenings. It also illustrated that the methodology is time-independent; all the systems thinking was here done in half a day, where in other projects conceptual model building has itself taken several weeks. The point is that if you have only a few hours to make a systems study, the whole of the methodology can be followed in that time, but at a different *level* than will be possible in a longer period of time. Finally, this project drew attention to the existence of 'problem-owning' and 'problem-solving' systems in any situation in which a systems study is being carried out. This later led to some useful learning which is discussed at the end of this chapter.

2. *Structuring a Survey*

There are many problem situations in which the felt need is for knowledge rather than action. A survey is called for, rather than action taken. There may of course be a political content in the decision to set up the inquiry or call for the survey, and this is frequently the origin of official reports, their initiation being the minimum possible action in a situation in which, for political reasons, some public action is necessary. But this does not discount the importance of such exercises. The process of compiling and publishing reports and surveys is a part of community learning. Whether their instigators intend it or not, such reports will indicate changing concerns and evolving standards of judgement, and may well affect subsequent actions directly or indirectly. There is therefore a place for methodology relevant to the preparation of survey reports. And this will preferably be methodology which will make such reports more useful by making them more debatable. Although the 'soft' systems methodology described in Chapter 6 was evolved in situations in which action was the aim, we have also found it useful in providing a method of structuring a survey report, and one example of such a use will be described briefly.

The University-owned consultancy company (ISCOL Ltd) which was the vehicle for carrying out a number of the studies described here, was asked by the IBM Scientific Centre at Peterlee to examine the areas of waste disposal, waste recovery, and pollution control 'as they affect, or are the responsibility of local authorities.' The request was a part of the Scientific Centre's research programme, and the fact that such a body should make such a request reflects the emergence in the last decade or so of 'conservation' and 'pollution' as important matters of public concern. There was a time when, as Vickers (1970) has pointed out, to describe a project as 'uneconomic' was to end debate about it: the economic measure was assumed to be paramount. This has become less the case in recent years, in which there has arisen a much increased awareness of man's activity as one factor in the operation of the robust but sensitive natural systems we call 'ecological'. This concern, popularized as an awareness of the bad effects of 'pollution', has led in turn to the increased political power of the ecology/conservation/anti-pollution movement and to the expression of it in legislation which affects local authorities: hence the Peterlee request for this particular survey as a preliminary step in their identifying new or increased uses for computers.

The work was carried out by R. J. Burn, J. H. Collins, D. Marchant, F. Schwarz, and myself. It was a large task, and it led to three volumes of report which IBM have made generally available. It is not appropriate to attempt to summarize their content. My concern is with the way the task was approached. using such parts of the 'soft' methodology as seemed appropriate.

The problem was to find a way of organizing the structure of the detailed work, much of which consisted of finding out and bringing together a large number of relevant facts. The crux of such work is to find a way of deciding what exactly is 'relevant'. Stages 3, 4, and 5 of the methodology (Figure 6) were used to do this. The problem as posed implies a notional system which deals with waste: disposing of it, recovering some of it, and ensuring that it is not allowed to foul the environment. In a root definition due initially to F. Schwarz, all human activity (from the point of view of this problem) was taken to be a set of

200

three connected systems. In one system resources are extracted from the environment and converted into material goods and energy. This system consumes some of its own products and has outputs of goods and energy for consumption, together with material and energy wastes. A second system receives goods and energy from the first system and in consuming them converts them into waste products. A third system monitors the activity of the other two and ensures that their waste does not become an intolerable burden on the environment. The conceptual model which derives from this view is shown in Figure 13. The conversion system passes some of its waste to the 'monitoring' system to be made environmentally harmless, and also passes acceptable quantities direct to the environment. The consumption system represents man as consumer, and produces waste which is either processed or, subject to the consent of the monitoring system, passed to the environment. (Within the system there will of course be a number of recycles in which someone's waste is someone else's useful starting material.) The monitoring system both reprocesses waste from the other two systems and puts its own waste to the environment by way of harmless deposits or dispersals. It is the system responsible for defining 'harmless' for all waste flows which reach the environment, whether from itself or from the other two systems.

This model defines six waste streams:

W1 solid or liquid 'domestic' wastes arising from consumption which require reprocessing and controlled deposition;

W2 solid, liquid or gaseous 'domestic' wastes arising from consumption which the monitoring system allows to be dispersed direct into the environment;

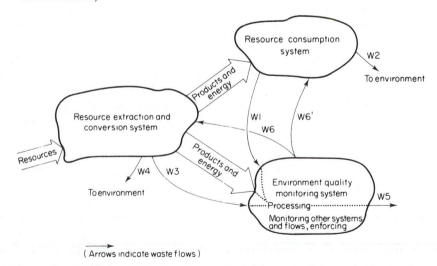

(Arrows indicate waste flows)

Figure 13. Human activity as a waste generation and environment protection system

W3 solid and liquid 'industrial' wastes which require reprocessing;

W4 solid, liquid or gaseous 'industrial' wastes similar to W2;

W5 processed outputs of the monitoring system, solid, liquid or gaseous which meet the standards of harmlessness set by the monitoring system itself;

W6 recovered waste materials or energy arising from segregation procedures applied to W1 (or W3) or from processes to which W1 and W3 have been subjected. W6 is a special case of recovered energy being used—e.g. energy from incinerators being used for district heating.

Figure 14 shows a manifestation of this system for a packaged food.

The model was extended by conceptualizing in more detail the monitoring/reprocessing system, this being the notional system whose real-world manifestations will include central government (passing legislation) and parts of local government. Putting together the waste streams and the internal functions of the monitoring system, it was now possible to define those waste streams which the survey would cover and the aspects of monitoring them which would be investigated in each case. This provided the structure of a set of 'maps' of this area of concern. Each map covered one particular waste flow (e.g. 'solid domestic waste': $W_1 \rightarrow W_5 + W_6 + W'_6$) and examined the relation to it of nine characteristics from the more detailed model of the monitoring system: inputs, required characteristics of output, processing methods, boundaries of responsibility, operational planning and control systems, monitoring systems, enforcement systems, strategic planning systems, administrative support systems. Six such maps provided the structure of the reports, and the application of the three criteria: scale of activity, degree of local authority involvement, and scope for data processing applications, enabled

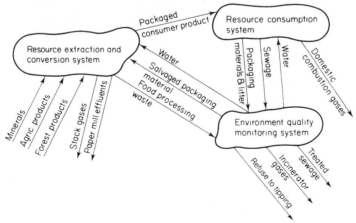

Figure 14. Major waste flows involved in manufacture and consumption of a packaged food

certain aspects from the six maps to be selected for more detailed attention.

Compiling the reports then consisted of carrying out the comparison stage of the methodology (stage 5). This entailed plotting, in the required degree of detail, the real-world manifestations of the elements in the matrix which combines the waste flows and the characteristics of the monitoring system.

In this study the choice of viewpoint as represented by the formulation of a root definition, together with the conceptual model building which derives from that definition was used not to suggest action but to bring coherence to a major task of data collection and presentation. The virtues of reports compiled in this way are, firstly, that the underlying conceptual model enables any part of the report to be related to the whole with great clarity, secondly, that the assumptions underlying the presentation are made explicit. Such reports are liable to lead to higher quality debate than those in which structure is arbitrary and tacit assumptions cannot be questioned. Many government reports are of this latter kind.

3. *Clarifying a Theoretical Concept*

In the 1970s some of the Government money spent to help improve industrial efficiency in the UK was channelled to what are known as 'the industrial technologies'. These were originally conceived as 'multi-disciplinary technologies', applicable to many different industries, whose neglect led to economic inefficiency. In the early 1970s there were four of them: corrosion technology, tribology, materials handling technology, and 'terotechnology'. We were asked to help define the latter. In 1972 there was a newly constituted Committee for Terotechnology but no agreed definition of the concept. The Committee set up a Panel under H. Darnell, Engineering Director of British Steel Corporation, to propose an argued definition, indicating exactly what was within the concept and what was not. I was a member of both the Panel and the Committee, and C. H. Pogson and I made a systems analysis of terotechnology which formed the basis of the Panel's report to the main Committee and subsequently became the basis of the official definition of the concept (Checkland, 1979d).

The problem situation was an interesting one in that it was entirely arbitrary. It was not a case of defining and describing something which existed in the real world. Rather the task was to define a concept which in the opinion of the Department of Trade and Industry and some interested industrialists ought to be taken seriously by anyone concerned with the process of generating wealth by industrial activity. The source of this belief was a 1969 report commissioned by the then Ministry of Technology from a firm of management consultants. This had indicated that a major source of inefficiency in British industry was a failure adequately to look after equipment and other physical assets (Atkinson, 1976). The consultants estimated that in 1969 some 500 million pounds could have been saved by paying modest attention to one of industry's cinderellas: the maintenance function. 'Terotechnology', from the Greek verb *térein*, 'to care for', was the name for what had to be taken seriously if the business of caring for physical assets was to make the best possible contribution towards the generation of wealth.

It is clear that systems ideas, and especially those pertaining to human activity systems, are relevant to this kind of concept. Any examination of maintenance quickly leads to the question of improved design which might reduce or eliminate it; design leads to consideration of expected life cycle, and hence to the balance of capital costs and the costs of maintenance and spares over the anticipated lifetime; the latter will bring in the market . . . and so on. Such connectivities are never-ending; can a boundary be drawn which makes terotechnology a coherent concept and not a fuzzy synonym for all industrial management?

The 'structures' in the terotechnology problem situation were those consisting of the committees, sub-committees, and panels of the industrial technologies, the 'processes' the envisaged actions needed to help British industry by means of the terotechnology concept, these actions including influencing the professional institutions, encouraging relevant education and training, the provision of case histories and manuals, and the possible setting up of a National Centre for Terotechnology. (This Centre now exists.) The relation between the structure and the processes depended primarily on the enthusiasm of some industrialists, trade unionists, academics and managers from the public sector who served, unpaid, on the various committees, formulated policies and then engaged in the activity necessary to implement them. There was something of a missionary zeal about the protagonists, and it was clear that their endorsement of the definition of the concept was crucial to the success of our study, even though, outside the Panel, there was little overt enthusiasm for the abstract business of defining the concept.

We selected as 'relevant system' a notional terotechnology system in an organization, its root definition being that it was the system encompassing all those activities concerned with ensuring that the care of physical assets made the maximum possible contribution to the 'generation of wealth'. (In a manufacturing company this might be a direct contribution to the generation of wealth. In a hospital, for example, it might be an indirect contribution. In the latter example the system would ensure that the ownership of adequate physical assets for carrying out the hospital's task entailed the minimum costs which did not impair the hospital's medical and social efficiency.) In order to avoid an unbridgeable gap at the stage 5 'comparison' (which would here consist of a matching of our systems-based definition with the notions owned by the terotechnology enthusiasts) stage 2 of the study included personal interviews with as many as possible of the

204

people who had been involved in the early discussions in the Ministry and in setting up the Committee for Terotechnology.

The main part of our study consisted of making the conceptual model of the relevant system, paying special attention to its boundary and to the links to other necessary functions within the organization. The modelling technique described above was used (using verbs as the modelling language) and the formal systems model was used to 'validate' the model obtained. The model of the terotechnology system is summarized in Figure 15. It is a system which acquires physical assets, cares for them in use, disposes of them optimally, and learns from its experiences so that it, and other functions, can improve in the future. The most significant discussions on the Panel concerned the exclusion of the activity of *using* the physical assets, and the system's measure of performance. Our argument, and later the Panel's argument to the main Committee, was that the core of terotechnology lay in the cost of ownership, and that including the activity of *using* physical plant or equipment made terotechnology a hazy version of total management. Thus, with a chemical plant, there are *costs of ownership* which stem from acquiring the plant and maintaining it throughout its useful life. There are also costs which arise from using it, which will include costs stemming from the chosen operating conditions as well as the costs dependent upon the state of management/labour relations. These costs due to *use*, though certainly important to management, are not logically a part of the set stemming from ownership. The distinction between 'owning' and 'using' was therefore used as a demarcation between

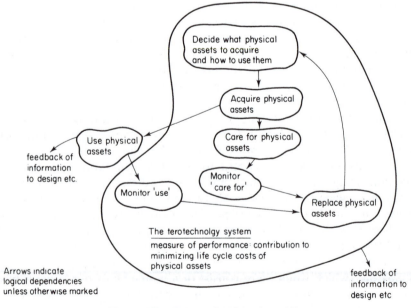

Figure 15.   Terotechnology in outline

what was part of terotechnology and what was not. This debate overlapped the other major discussion, that concerning the system's measure of performance. Because the terotechnology system is the location of the management decision to scrap a physical asset (hopefully a rational decision based on monitoring the full costs—financial and social—of owning and using it) its measure of performance is the contribution it can make to minimizing total life cycle costs. These latter will include all the costs involved in feasibility studies, research, development, design and production and all the support, training and operating costs arising from the ownership and use of physical assets. Terotechnology is concerned with that sub-set of them concerned with selecting, acquiring, caring for, and replacing physical assets, and this provides an explicit measure of performance for the terotechnology system.

Terotechnology clearly comprises a set of activities which managers will have to carry out somehow, whether or not they are conscious of their links and interactions. The terotechnology idea is that they will be better carried out if these connections are recognized and taken into account. But there does not have to be any institutionalized version of the terotechnology system, nor a new kind of professional 'the terotechnologist'. The activities of the notional terotechnology system are the concern of many people: managers, designers, engineers, and accountants, as well as purchasing managers and information system designers. These points have all been well grasped in the terotechnology activities since the definition was adopted, and this recognition is a small success for systems thinking as a way of thinking about *'what* has to be done' rather than in terms of the arbitrary organizational groupings, departments, sections, etc., which represent only one possible *how*. This point is further discussed below. Equally, it is recognized that terotechnology is not size-limited. It is in principle applicable to any physical asset, 'to a pen as well as a printing press', as Pogson put it.

The definition of terotechnology proposed by the Panel was accepted by the main Committee, and official Ministry publications now describe the concept in the following terms:

> Terotechnology is a combination of management financial, engineering and other practices applied to physical assets in pursuit of economic life-cycle costs. Its practice is concerned with the specification and design for reliability and maintainability of plant, machinery, equipment, buildings and structures, with their installation, commissioning, maintenance, modification and replacement, and with the feedback of information on design, performance and costs.

Methodologically the project provided an example of the use of 'soft' systems concepts in a purely theoretical context, and this helped to make clear the status of the methodology as a means of examining with rigour a kind of system—the human activity system—characterized by the fact that an account of a particular example will always embody one particular view of it, among other possibilities.

4. *Tackling Your Own Problems*

Given that the work described here was based in a university and that it sought practical results in a relatively short time, it is inevitable that most studies follow the pattern in which outsiders move temporarily into an organization and work in a problem situation within it. But clearly this pattern arises only because of the particular circumstances in which this work was done. There is nothing in the methodology developed which in principle requires it to be used by outsiders on someone else's problem. There is no reason in principle why the physician should not try to heal himself. We tested this by carrying out a formal systems study on a problem situation of our own, the difficulties of running a university-owned consultancy company alongside and overlapping with a university academic department. The work was carried out by Su Lau and myself.

We decided that in order to test the methodology we had to try hard not to come to conclusions in an *ad hoc* manner and then dress them up as derived from the application of systems methodology. We also had to convince knowledgeable colleagues that we were not using the study as a cover for bringing about changes already decided on other grounds. Our approach to these problems was to use the methodology with flat-footed rigour, as mentioned in the previous chapter in the discussion of the comparison stage, making each step of the thinking publicly explicit.

Over several weeks some time was spent on stages 1 and 2, taking as cool a look as possible at the structures, processes, and relations between the two which characterized our problem situation. This was not sharply defined; there was nothing obviously wrong which called for immediate action. But all were aware of difficulties in combining consultancy with academic work, and the different drives which underly the two obviously require different organizational arrangements which can lead to tensions both within and between individuals who have roles in both the department and the consultancy company. In most universities individuals are left to resolve these difficulties as best they can, and this is easy enough in situations in which the long vacation is available as a period in which the individual can apportion his time between reading, research, preparing teaching material, and doing consultancy work, in a way which suits his personal situation. But on a one-year Master's Course which starts on 1 October and ends on 30 September there is no long vacation, and the difficulties of multi-roles are much greater. In the Department of Systems we had all agreed that we would do no consultancy other than as part-time consultants through the University-owned ISCOL Ltd, and there was general agreement that organizing such work through a university company was the best way to make sure that consultancy work was in fact also academically relevant. Nevertheless ISCOL was an organization which had to survive in the market place, at least breaking even on its operations each year (its annual surplus being covenanted to the University) and this implies different modes of operation reflecting different values from those appropriate to a university department, the latter being an organization of a kind that the community as a whole is prepared to subsidize.

We decided that the 'relevant system' in our study was the Department/ISCOL entity as a whole and we formulated a root definition which was presented to a seminar of Department and ISCOL full-time staff. After modification during that discussion we had an agreed root definition on which to base the subsequent work. Later I regretted that we had not pursued the implications of several, but at the time I was pleased to have a root definition which the concerned people in the problem situation agreed was likely to be insightful. It described the relevant system as a university department (with the contractual obligations that entails for individuals) which wishes both to teach and to develop its subject by 'outside projects' and to disseminate the knowledge so gained. This embodied the *Weltanschauung* that the consultancy company was a means by which the Department did some of the things it wanted to do; it implied that the relationship between the Department and ISCOL was of the what–how variety, and that the Department could exist without ISCOL but not vice versa.

A model was made of the system described in the root definition and in order to make sure that the stage 5 comparison was done publicly, the method mentioned in the previous chapter was devised. Given the systems model, a model of the equivalent parts of the real world was made in as near the same form as possible. Overlay of one model on the other then revealed the differences between the two in pellucid clarity. These differences formed the substance of debate about possible changes, and a number were implemented in the months which followed: project load was halved by pairing postgraduates; a research programme beyond the 'action research' programme was initiated by increasing the number of Doctoral students: it was agreed to provide the full-time ISCOL consultants with 'thinking time' between projects; and it was agreed to examine the implications of ISCOL's becoming a research institute rather than a limited-liability company. These were significant management changes of the middle range, neither day-to-day details nor major structural changes. I personally was agreeably surprised at the way in which implemented changes came from the study—in a more positive way than would have happened in the ordinary day-to-day flow of events. And certainly the intellectual experience of doing the project as an 'insider' was identical with previous experience as an 'outsider'.

Methodologically there was useful learning in the finding that people who cheerfully accepted the root definition as a useful one were later much less happy with the consequences which flowed from it! It was made clear that at stage 5 it is difficult to avoid seeing the root definition and its model as normative, if there is only one definition and one model. The way to avoid this is to entertain several root definitions, best of all including incompatible ones, and to make models based on more than one of them.

5. *An Organizational Study*

The study just described had been preceded, the year before it was done, by another study which aimed to design the information system required to manage the organizational entity which consists of the Department of Systems at Lancaster University overlapping with the University-owned ISCOL Ltd. This seemed a satisfactory study at the time, and certainly produced an information system design. But we had done nothing more about it. This seemed something of a paradox: we had a systems design for the necessary regular information flows within the Department and the Company, and we believed this to be an adequate design, and yet we had done nothing to implement it. The creation of the design had failed to generate the *determination* to introduce the changes necessary to implement it. My industrial experience had taught me that corporate activity by an organization is a complex, not a simple phenomenon, but even so this seemed strange behaviour on our part! I concluded that our lack of enthusiasm for making the required changes was that the study had taken as given just what we felt was problematical, namely the Department—plus—ISCOL as an organization. The study had concentrated attention only on the information flows required by it, whereas our feeling of unease, ill-expressed, was that our 'unstructured problem' was the organization itself. This drew attention to the nature of information systems, namely that *they serve activity systems*. Systems analysis aiming at information system design, if it is to make much impact, must first concentrate on the activity system which the information system is to serve. Once the root definition(s) and conceptual model(s) of the activity system are established then the necessary information flows are easily defined by asking for each activity: What information is needed to do it? In what form? From what source? How frequently? (Checkland and Griffin, 1970). Work on information systems which *begins* by considering information flows will be inadvertently taking as given the present organizational arrangements, which ought to be questioned. But even if the structure is constrained to its present form for some reason, an activity model should still be built first as a route to an information system design which is both economical and elegant.

Rather similar considerations apply to corporate planning systems. Such systems are by definition concerned with issues relevant to an organization as a whole, asking questions of the kind: Shall we enter this new market?

What market share shall we aim for? Shall we invest in this new plant? Should we get a licence to operate the Dutch process? etc. Systems whose business it is to answer questions of this kind necessarily imply that the corporate entity is a particular kind of entity and hence that it will want to take actions of a particular kind. An idea of what the organization *is* will underly what the organization *does*, and any model of what an organization, as an entity, might do will take for granted some vision of what that organization is. In a systems study of corporate planning these assumptions ought to be exposed and debated rather than taken as given. Hence work on corporate planning systems must start by examining root definitions which embody views of what the organization as a whole now is, or might be in the future. From these views of the corporate entity will follow what it might appropriately do.

Both of these 'methodological laws'—that information systems design must stem from a model of the activity system served, and that work on corporate planning systems must start from consideration of the organization served by the planning system (Checkland, 1977b)—were illustrated in studies carried out by T. R. Barnett, C. H. Pogson, and myself for a large company having a number of operating divisions and a central headquarters staff. At headquarters were a number of senior planners of various kinds, and the defined concern of the initial study was 'the interaction between the centre and the operating divisions in the planning process'. This obviously had implications for the role of the headquarters' groups, and a second study tried to define specifically what that role should be for one of the groups of planners, that concerned with research and development.

During the methodology's stage 2, when we were assembling our 'richest possible picture' of the problem situation, we asked that the central staff should help us obtain the collaboration of two of the operating divisions, those regarded by headquarters as 'most' and 'least' corporate-planning-conscious. It was significant to our picture-building that in the opinion of the central planners collaboration would be possible only with the most planning-conscious division! The full significance of this became apparent when we began to consider the necessary information flows between the centre and a division as corporate plans were made. It was only possible to make a conceptual model of these information flows by first considering the planning *activities* in which the centre and the operating divisions were linked, and it was only possible to model those activities with reference to some image of what kind of organization this company is. Even if we had ignored this, any model built would have *implied* some company image which gave the model meaning. The 'relevant system' at the core of all this work had to be the company itself, and the cogent question was: What root definitions embody relevant views of the nature of this company as a corporate entity? The inability, or unwillingness, of the centre to arrange collaboration with a division they found 'unhelpful' as well as with the planning-conscious 'helpful' division, illustrated the very different images of the company as a whole which obtained in operating divisions and at headquarters. The operating divisions were in fact very distinct (provincial) cultures and were conscious of, and anxious not to lose, any of their own autonomy. The centre was naturally most conscious of the whole company as an enterprise with limited risk capital, one among a number of similar large companies having world markets. Root definitions which embodied 'centralized' and 'de-centralized' views of the company led to models which could not generate any recognizable version of centre-division relations as they had changed over the years. (In methodological terms, these models led at stage 5 to unbridgeable gaps between conceptual models and reality.) A more complex model, however, was sufficiently close to (but also different from) reality to generate useful debate; eventually the most useful root definition in these studies was one which embodied the image of a company which was neither managed from the centre nor consisted of an aggregate of autonomous divisions. Rather, in this model, the company occupied a position between the two which changed over the years as internal factors and the external environment changed. What endured throughout was the *intention* always to allow the operating divisions to be as autonomous as possible within the principle that the final approval for major capital investment plans had to be taken at the centre in the light of the current situation of the company as a whole.

The activity model which followed from this definition led directly both to the information flows needed in the planning process and to a definition of the various roles of centre-based planning groups.

The telling part of the project was this debate about what kind of organization the company sought to be, and this debate was the essential precursor both to definition of information flows for planning and to postulating an appropriate role for groups based at headquarters.

6. *Using the Methodology Out of Sequence*

The methodology has been described in the previous chapter as a sequence of steps each following logically from the one before. This is also in general the sequence followed in the accounts of systems studies given above. The sequence from stage 1 to stage 7 is certainly logical, and in many studies the work does unfold in approximately this order. But the most important thing about the stages taken together is *the relationship between them*, rather than their order, and as long as that relationship is remembered the work done does not have to start at stage 1 and proceed to stage 7. The point is not simply that there will be a retracing of steps between stages, although this will happen, but that work can in fact start at any intermediate stage. This was illustrated in the early days of a major study which is still underway, a study being carried out by S. Cornock and myself.

The research problem is to ascertain the extent to which systems concepts related to purposeful systems can be used to gain insight into the workings of a human activity system very different from those relevant either to industrial firms or to organizations in the public sector. The subject of the research is the art world, Stroud Cornock being a sculptor and a teacher of art. Our concern is not the so-called op art, nor 'computer art' but the whole area of the activity of artists and the other activities, cultural and commercial, which happen as a result of artistic activity. Alloway (1972) provides a preliminary and partial example of the use of systems ideas to interpret the observed activity of this particular world.

Early in the project my colleague set out to assemble a picture of the art world of the kind required by stage 2, including the work of artists, the activity of dealers and the workings of the art market, the museum and gallery system and public subsidy for art of various kinds. He had himself been an active protagonist in the art world for a number of years and naturally had certain attitudes and views which had been moulded by his experiences. Now, I had believed that the questions: What are the elements of structure? What are the elements of process? What is the relationship between the two? could be used to assemble what was certainly an explicit, but which was also hopefully a relatively objective account of a problem situation, even of a situation in which the analyst was also an actor. (I still believe this to be possible; but it is more difficult to do than I had imagined.) In the present study, when the account was presented to a seminar audience it appeared to them to be far from a dispassionate picture of art world activity, more a polemic embodying a particular point of view.

It now became clear that what was intended as a 'stage 2 picture' was in fact a stage 4 conceptual model of the art world. Working back to stage 3 by asking: If we take this to be a conceptual model, what root definition is implied by it? suggested that the model as a whole was meaningful as an expression of the view that artists constituted a tribe and that inward-looking 'tribal debate' characterized their behaviour. This is a defensible view of the art world, but clearly only one among many other possibilities.

Realization that this study had started at stage 4 and progressed to stage 3 now suggested the formulation of a wide range of other possible *Weltanschauungen* which might be found within the art world, and the building of other conceptual models to set alongside the 'tribal debate' model. This was done using a number of very different perspectives (art as profession, art as decor, art as 'cultural leavening', art as revelation, among others) and the models were tested against the perceptions of a number of people in different roles in the art world. The work has been used in analysis of an organizational problem in a subsidized organization which tries to help aspiring artists, and also used as the basis of an intervention in the art world of one city. Figure 16 shows how the methodology was used in this instance: it may provide a general methodology for problems not

212

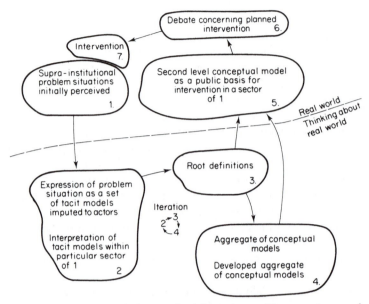

Figure 16. A use of the methodology in problems not owned within organizations ('supra-institutional problem situations'). After Cornock, 1977, 1980

owned within organizations (Cornock, 1977 gives an earlier version; see also 1980). This work continues and is not here our main concern.

What was useful in this experience and provides a lesson to bring out here, was its demonstration that use of the methodology could start at one of the middle stages, just as well as at stages 1 and 2. This is true not only when the methodology is being used as a way of deciding on, and carrying out action, but also when it is used as an analytical tool. For example, it is interesting to regard organizational changes proposed by practitioners who, as 'change agents' seek to bring about 'organizational development' (OD) (Beckhard, 1969; Bennis 1969; Schein 1969) *as if* they were 'stage 6' changes derived by this kind of systems analysis. The changes proposed, viewed through the framework of this methodology, will imply a particular kind of debate based on models embodying particular root definitions. (The general result of doing this for OD is to suggest that its essence is to identify forces for change; it does not seem to embody a theory of organization.)

A final practical example of the 'start anywhere' principle is provided by a corporate planning study for an industry Training Board. The Board had a concise public statement of its corporate objectives. By treating this as if it were a conscious root definition, and comparing the activity model implied by the definition with the Board's actual structure and activities, differences were revealed which gave insight into the Board's problems. The study later concentrated on the building up of the Board's corporate planning system; this 'starting at stage 3' was here used as a means of discussing the question which, as discussed above, precedes the design of such systems, the question What is the nature of the Board itself, as an organization?

General Outcomes from the Action Research

In developing and refining the systems methodology it was always the intention to use the sequence of undertaken studies as a means of teasing out general lessons. The link with the Master's Course helps in this, in that it provides a set of varied five-month studies each year, while the ISCOL work adds other studies ranging in length from four weeks to more than a year. The fact that it is part of the group of projects hopefully makes each individual study more valuable than it would be on its own. An individual study can be compared and contrasted with others; maverick studies can be recognized as such; and the group as a whole can gradually yield general lessons.

The first part of this chapter has described a number of studies which either help in the exposition of methodology or which provided significant learning. The remaining sections of this chapter describe some of the major generalizations which the ten years of research have produced.

The major impression which the work has made on those doing it is that 'soft' systems studies are markedly different from the 'hard' systems engineering which was the intellectual starting point. 'Soft' methodology can on occasion sharpen up into 'hard' methodology, which is a special case of it, but there can be no direct import of the 'hard' concepts into fuzzy ill-structured problems. The differences are not only of detailed method; they concern both the analyst's fundamental aim and the attitude which he brings to a systems study. It has been argued above (especially in Chapter 5) that the systems thinking which has emerged within the engineering tradition is inevitably single-valued. Engineering proper (as opposed to social or political 'engineering') aims to provide a solution to a defined problem which is optimally *efficient* in the terms in which the problem is posed. The best chemical engineering, for example, is that which defines the most efficient process, the one yielding the most product at least cost, and all of the work can be measured against the achievement of this single value, efficiency. This kind of procedure, however, turns out not to be a useful model for studying soft systems. The special nature of human activity systems means that systems studies concerned with them are always multi-valued, with many relevant and often conflicting values to be explored. The outcome is never an optimal solution to a problem, it is rather *a learning which leads to a decision to take certain actions*, in the knowledge that this will in general lead not to 'the problem' being now 'solved' but to a new situation in which the whole process can begin again. Where the general outcome of a sequence of 'hard' systems projects would be expected to be a contribution to 'the science of efficient action', what Kotarbinski (1965) calls 'praxiology', a sequence of 'soft' systems studies leads, at a meta-level to praxiology, to a learning system which decides what action to take, action which may or may not be 'efficient' with respect to various criteria.

Looked at as a learning system, the mosaic of Figure 6 embodies a sequence of basic mental acts of four kinds. Stages 1 and 2, carried out in the real world, entail *perceiving*. Stages 3 and 4, which embody a formal use of systems ideas, entail *predicating*. Systems concepts are used to build a picture, in the conceptual models, of what we can say about the perceptions if we use systems ideas as the language to elaborate them. (A similar methodology might use, say, structuralist concepts (Ehrmann, 1970; Lane, 1970; Gardner 1972) at stages 3 and 4, but I am at present particularly interested in systems ideas.) Stage 5 then entails *comparing* the systems models with what is in the real world, and stage 6 entails *deciding* what to do in the light of the comparison.

Perceiving, predicating, comparing, and deciding are all everyday mental acts, so the methodology is not outlandish. It simply provides a coherent combination of these common mental processes in a form which can incorporate the formal use of systems concepts. If the hypothesis that it is useful to take the world to consist of a complex of interacting systems is a good one, then this will be a good way of getting insight into the nature of the real world and its problems.

In different situations different parts of the methodology have seemed more important. Sometimes studies have concentrated on stage 3, as for example when R. H. Anderton and P. Thomas sought to help the professionals running a community centre to form a perception of the centre's role which would guide their actions; sometimes, as we saw in the study for Mr Cliff and Mining Support Ltd, stages 3 and 4 constitute only a few per cent of the total work. Different situations highlight different aspects; the methodology is the learning framework common to all the studies whatever their core concern, and regardless of whether they take four weeks or forty months to complete.

The fact that the methodology is a learning system, and, in tackling unstructured problems *could only be* a learning system, rather than a prescriptive tool, is due to the special nature of human activity systems. Such systems can be described, rather carefully, in what I call root definitions. But no root definition can ever provide a unique description of any actual manifestation of a human activity system. It will always be only one possibility out of a large number, and this is true even if people in the problem situation are not disposed to challenge a proffered root definition (though that would be unusual). A root definition will be a meaningful description of the relevant system according to a particular view of the world, or *Weltanschauung*. There will be other feasible *Weltanschauungen*, however, because human beings can always attach different meanings to the same social acts. The realization of the importance of this concept is a major outcome of the research, and is counter to the main concept of hard systems engineering.

General Outcomes 1: The Importance of *Weltanschauung*

According to Kant, the world outside ourselves causes only the matter of sensation. Our brains order this matter and supply the concepts by means of which we understand experience. This is a familiar process to us in everyday life. The subject of this book, 'a systems approach', provides a good example. A systems approach tries explicitly to avoid reductionism by viewing the world in systems terms. It uses systems concepts in order to see the raw data of the outer world in a particular way, namely as a set of 'systems'. It converts the raw data into a particular kind of information, and this is the process occurring in virtually all human thinking. Whether we realize it or not we view raw data via a particular mental framework, or world-view. We observe people voting and see, not 'marks being made on pieces of paper' but 'human beings taking part in the democratic process'. We attribute *meaning* to the observed activity by relating it to a larger image we supply from our minds. The observed activity is only meaningful to us, in fact, in terms of a particular image of the world or *Weltanschauung*, which in general we take for granted.

(In arguing the importance of the concept of *Weltanschauung* for systems thinking applied to 'soft' problems, I shall have to use this useful, accurate but ungainly word so often that I shall for the rest of this chapter write W instead of the whole word.)

A dramatic example of the way our thinking is dominated by unquestioned Ws, is provided by Ptolemy's cosmology. Ptolemy made accurate astronomical observations during the period of Alexandrian science when science emerged as a professional activity. His raw data—planetary movements, etc.—were seen by him through the framework of his W, in this case literally his picture of the solar system. Ptolemy took for granted, as we all do, a W which made his raw data meaningful. In his particular case he knew that the earth was at the centre of the planetary system and that both the sun and the other planets circled round it. When Mars was observed to reverse its direction of motion every so often, Ptolemy's explanation, which left his W intact, was that Mars was executing an epicyclic motion, making a small circular motion around a point which itself circled round the earth. It is no disparagement of Ptolemy to draw attention to the fact that the alternative explanation, that his W was wrong, did not occur to him. For many hundreds of years, as further aberrations in the motion of planets were discovered, epicycles were built upon epicycles to explain the observations. Doing this preserved intact the prevailing W. The change of W, what Kuhn (1962) calls a 'paradigm shift', was not initiated until the 16th century, with Copernicus's simpler explanation of observed planetary motion based on a heliocentric model of the solar system. Even then, as recounted in Chapter 2, the change was slow in coming: after all, why accept a less complicated set of motions than Ptolemy's if accepting it meant adopting a revolutionary view of the universe which the Church opposed? It is characteristic of us that we cling tenaciously to the models which make what we observe meaningful. We celebrate Newton and Einstein as the very greatest scientists precisely because they forced the establishment of new W's. Both were able to establish hypotheses which survived severe tests, and hence became public knowledge, and were based on revolutionary frameworks, on W's different from the prevailing ones of their time.

What the history of science illustrates happening on a grand scale also happens to all of us, all the time, in everyday life. Because it might be thought that the grand examples are not typical of the detailed happenings in small-scale situations, and because the concept of W is so important in 'soft' systems analysis—which must be carried out at a level at which prevailing W's are teased out and challenged—it is worth giving a few more examples.

After the Second World War the British coal mining industry introduced technically revolutionary methods of coal cutting. The industry was re-equipped with expensive machinery. Being now a much more capital-intensive industry its economic performance depended very much upon high utilization of the coal cutters, conveyers, and powered supports. In the 1950s Lord Robens, then Chairman of the National Coal Board, was concerned that so many miners were consistently working four shifts instead of the regulation five, and absenteeism was a public issue. A newspaper recorded the following exchange between the Chairman of the Board and one working miner:

> Lord Robens—Tell me, why do you regularly work four shifts instead of the regulation five?
>
> Miner—I'll tell you why I regularly work four shifts; its because I can't quite manage on the money I earn in three.

Here the same purposeful human activity, that concerned with extracting coal, is viewed according to W's which make sense to two very different roles in it.

Another interesting example comes from the American decision to make the effort to send men to the moon and return them safely to earth. This was a very remarkable decision because, as President Kennedy described it in his 'Message to Congress on Urgent National Needs' (1961) it was an open-ended commitment by America to provide whatever resources were required for this enterprise regardless of the cost and effort involved. It is interesting to speculate on the W which, for the leaders of a powerful nation, makes this a meaningful thing to do. There are a number of obvious possibilities: man as a questing animal with an insatiable desire to voyage to the unknown; high technology as an ultimate good; new technology as a worthwhile goal because of inevitable 'spin-offs' in other areas; scientific progress as an ultimate good, etc. In practice the 'true' W or mixture of Ws is probably impossible to ascertain. But an interesting feature of the President's message to Congress is that he chose to describe explicitly the W which made the moonshot meaningful in his eyes, or, perhaps, the one which he thought would be meaningful to Congress! Apparently, according to the President, unsophisticated nations committed to neither the Western nor the Soviet bloc had been so impressed by the Russian achievements in space—the sputnik and the first manned flight—that they were in danger of casting in their lot with the Russians. Hence a spectacular American achievement in space was urgently required! Using the rhetoric of international politics in the 1960s Kennedy declared

> ... if we are to win the battle that is now going on around the world between freedom and tyranny, the dramatic achievements in space which occurred in recent weeks should have made clear to us all, as did the Sputnik in 1957, the impact of this adventure on the minds of men ... who are attempting to make a determination of which road they should take ...

Hence the conclusion:

> I believe that this nation should commit itself to achieving the goal, before this decade is out, of landing a man on the moon and returning him safely to earth.

There are few examples as clear as this of such a major (and dubiously useful) enterprise being justified in terms of an image of the world which makes it meaningful.†

That the in-built images of the world by means of which we make sense of the impressions the world thrusts upon us are profoundly important for human beings is shown by our very remarkable behaviour when we find ourselves in a situation in which a W is suddenly confronted by a different W in sharp conflict with it. This behaviour, believe it or not, consists of expelling meaningless sounds and creasing up our faces! We call it 'laughing'. Consider some typical jokes:

A man entering a brothel in Marseilles finds there a girl he knew at Oxford.
—Good heavens, Cynthia, he says, how is it that a girl like you ends up in a place like this?

† ... 'dubiously useful' ... that is the author's W showing through!

—Oh, I don't know, really, replies Cynthia, I suppose some girls are just born lucky.

Or consider the wife reading a newspaper article who looks up and says to her husband
 — It says here, Andy, that men who don't drink live longer.
There is a pause for reflection before Andy replies carefully
 — Serves 'em right.

Finally we have Gloria saying to her friend Godfrey
 — I say, Godfrey, I learned an interesting thing about my family today. My great-grandfather was killed at Waterloo!
 — Oh! says Godfrey, which platform?
But Gloria is no fool, and immediately responds
 — You silly boy! What does it matter which platform?

Much wit is of the same pattern as these jokes: when a Labour minister at the time of the General Strike complained at the end of a long day 'I've got an 'orrible 'eadache', Lord Birkenhead replied at once 'Try a couple of aspirates'. Now, this list could be continued indefinitely, but the same point would be made over and over again. All of these jokes are in fact the same joke, in the sense that they have the same structure. An image of the world is established, only to be shattered by a counter-image which turns the first upside down. Koestler (1964) discussing 'the logic of laughter' in his book on creativity, describes the pattern as 'the perceiving of a situation or idea in two self-consistent but habitually incompatible frames of reference'. The central event is momentarily linked not to a single context of association but, says Koestler, 'is *bisociated* with two'. In the language I am using, a W is established by implication and then suddenly is destroyed by a counter-W, the skill of the comedian lying in a careful pacing of the initial image-building, making it rich in association, so that the demolition, when it comes, comes as a sudden shock. Tension is released by the physical action of laughing, a remarkable response when we consider that it is triggered merely by the juxtaposition of two abstract images. It would seem from this that in-built Ws are very important to us.

It was suggested in Chapter 4 that man was distinguished from the other animals by his special kind of consciousness, and this led to the idea that a set of purposeful activities could be taken to be a special kind of system: the human activity system. The present discussion suggests that his consciousness makes man, via his Ws, a *meaning-endowing* animal. And given the irreducible freedom which derives from man's self-consciousness, *he will always be able to select from a range of possible meanings*. This may be taken to be the distinguishing characteristic which makes the human animal different. Judging from their behaviour, all beavers, all cuckoos, all bees, have the same W, whereas man has available a range of Ws. The Ws of an individual man will in fact change through time as a result of his experiences. And the Ws of a group of men perceiving *the same thing* will also be different. It is because of these two facts that there will be no single description of a 'real' human activity system, only a set of descriptions which embody different Ws. In a certain sense, human activity systems do not exist; only *perceptions* of them exist, perceptions which are associated with specific Ws. In an extreme sense it is of course true also that even a description of a physical object embodies the observer's W. But it is in that case not difficult to achieve consensus on an appropriate W and hence on a generally accepted description. In these terms, the triumph of science is that it has been able to create a means (the repeatable experiment) of achieving a consensus on an appropriate W to describe particular sets of observables in a way which creates public knowledge. In the case of human activity systems, however, this kind of public knowledge—consensus on what they are, or on how we are to take them, or on what makes them meaningful—will be rare. Even if achieved at one moment in time, the consensus will break as new experiences cause perceptions to change. And in many cases Ws will remain different for people in different roles: Lord Robens' W with respect to coal mining, and the miner's W, will remain different.

The need to import the concept of W into 'hard' systems engineering methodology in order to cope with human activity systems is a main outcome of the action research. 'Hard' systems methodology is concerned only with a single W: a need is defined or an objective is stated, and an efficient means of meeting the need or reaching the objective is needed. In 'soft' systems methodology we are forced to work at the level at which Ws are questioned and debated, 'soft' problems being concerned with different perceptions deriving from different Ws. The formulation of root definitions provides a means of doing this, and in no study in the whole of the action research programme has it ever been possible to take as *given* a single root definition of a relevant human activity system. Hence the methodology emerges not as praxiology but as a learning system in which underlying Ws are exposed and debated alongside alternatives.

This is always a tense and exciting process, not least when you are using the methodology as a means of making a radical attack on your own problems! The flavour of the experience derives no doubt from the tenacity with which we all hold on to Ws which, prior to the systems analysis, were taken for granted. That it is hard to question previously unquestioned Ws has often been noted. The history of science is of course littered with major and minor examples of ingenious and/or far-fetched explanations which enable a W to remain intact. In the 16th century the explanation of the workings of the blood stream required venous blood to pass through the apparently solid septum, into the left ventricle, there to be mixed with 'vital spirits' (Butterfield, 1949). Faced with the physical impossibility of this, even Vesalius, inaugurator of modern anatomy, wrote that it shows the mighty power of God that he can make blood pass through the solid septum! But we would be ill-advised to be too critical of Vesalius for writing this. So tenaciously do we hold on to our meaning-conveying Ws that it requires a major intellectual effort to analyse and dispute them. Butterfield points out that scientific principles that are now 'obvious' to school-boys—obvious, that is, according to prevailing Ws—defeated the greatest intellects for centuries. He points out:

> . . . of all forms of mental activity the most difficult to induce, even in the minds of the young who may be presumed not to have lost their flexibility, is the art of handling the same bundle of data as before, but placing them in a new system of relations with one another by giving them a different framework, all of which virtually means putting on a different kind of thinking-cap for the moment.

The development of 'soft' systems methodology showed us that a means of putting on different kinds of thinking caps was crucially needed. The formulation of root definitions, the building of systems models implied by them, and the comparison of these models with the real world constitute an attempt to do this. The methodological guidelines make it possible to study problem situations at the level of the frameworks (Ws) involved. And the experience of the studies carried out convinces me that work at this level, though difficult, is perfectly possible. A hundred or more 'soft' systems studies using the methodology leave me agreeing with the remarks Popper makes in an essay on 'Normal Science and its Dangers' (Lakatos and Musgrave, 1970):

> . . . at any moment we are prisoners caught in the framework of our theories; our expectations; our past experiences; our language. But we are prisoners in a Pickwickian sense: if we try we can break out of our framework at any time. Admittedly, we shall find ourselves again in a framework, but it will be a better and roomier one; and we can at any moment break out of it again.

> The central point is that a critical discussion and a comparison of various frameworks is always possible. It is just a dogma—a dangerous dogma—that the different frameworks are like mutually untranslatable languages . . . The Myth of the Framework is, in our time, the central bulwark of irrationalism. My counter thesis is that it simply exaggerates a difficulty into an impossibility. The difficulty of discussion between people brought up in different frameworks is to be admitted. But nothing is more fruitful than such a discussion . . .

I see 'soft' systems methodology as a way of bringing about such discussion.

Following the emergence of the importance of the idea that every statement about a human activity system must be a statement about the system *plus* a particular W associated with it, I have noted other recognitions of the importance of the *Weltanschauung* concept; for example, Stimson and Thompson (1975) have pointed out that a problem like school bussing in America (transporting children to achieve a racial mix in schools) is approached very differently by operational researchers, who see it as a problem of *desegregation* (physically mixing the races), compared with others whose W is that the problem is one of *integration*, 'helping to bring about a viable multi-racial society'. Berresford and Dando (1978) and Mason (1969) provide further examples. But the importance of the concept has been most noticeably argued in Churchman's examination of different classes of information systems in which data are examined via different Ws whose antithesis provides the basic data for a higher level observer with a higher level, expanded, W, one which enables a new synthesis to emerge. This concept, and the rest of Churchman's analysis, is very relevant to the methodology as a whole, and is discussed in Chapter 8.

General Outcomes 2: 'Primary Task' and 'Issue-based' Root Definitions

Having completed an initial expression of the problem situation, the analyst selects some 'relevant systems' which will be named in root definitions and subsequently modelled. Each choice will express a particular way of looking at the problem situation, a particular *Weltanschauung*. What then, can be said about the process of selecting these hopefully relevant systems? The most important point is that there is in principle no limit to the analyst's freedom to make whatever choice he thinks or feels might lead to insight. He may wish to name systems which carry out basic operations which go on in the problem situation; or systems which manage—or subvert—such operations; or he may choose an *ad hoc* system of limited life, for example a system to take a particular decision. The analyst may wish to make any or all of these choices in a particular instance, and it is important that he be free to do so, if the methodology is to remain a guiding framework allowing for individual style and temperament in problem-solving rather than become a uniformly applied technique. Nevertheless experience in recent years has revealed one distinction which it is helpful to bear in mind in selecting the systems to model. It is a distinction between 'primary task' and 'issue-based' root definitions (Checkland and Wilson, 1980).

In recent years many studies have been concerned with the creation of systems which provide information. Much codified knowledge and purchasable skills are available to engineer such systems, whether computerized or manual, but much more problematical is the prior question: How can we decide which of many possible information systems to engineer?

In such studies attention is naturally paid to the activity systems which the information will serve and a version of the systems methodology has been used in which the team decides to model a notional system which can be related very directly either to an organization as a whole or to a well-established task carried out by a section, department or division (Wilson, 1978, 1979). In such studies a chosen relevant system is likely to be one which expresses the public or 'official' explicit task which is embodied in the organization, section or department. Such a choice will require a root definition which expresses this 'primary task' in a neutral manner. In work still underway, for example, a primary task root definition describes a notional system to carry out the task assigned to the engineering division of an international airline, namely to carry out planned maintenance on a fleet of aircraft effectively and efficiently under various constraints.

But not all relevant systems entail would-be neutral accounts of primary tasks of this kind. In a recent study set in a Management Services Department of a large manufacturing company, root definitions of a management services system were formulated. A 'primary task' version described a system whose function was to be very knowledgeable about the managing of the business activity in a turbulent environment in order that it could provide appropriate intellectual services which supported the management process. A different root definition for such a system, however, assumed that any 'management services' system, not being a direct wealth generator, will continuously have to justify its existence by demonstrating that the cost of providing its efforts is less than the benefits which flow from them. This definition assumes that the very existence of the system will be a permanent *issue*. It is an issue-based root definition.

Once the primary task/issue-based distinction had been recognized, we could reappraise earlier experiences and conclude that the distinction was a useful one. Bearing it in mind enables some cogent questions to be posed: Which type of root definition shall we formulate and why? If we select primary task definitions, what issues are we taking as non-problematical? Is the problem situation one requiring improved operations or resolution of issues? Are relevant systems accounts of tasks which in their real-world manifestations are on-going, or are they more usefully defined as *ad hoc* systems related to conflict over values? etc. etc.

This distinction between primary task and issue-based root definitions arose experientially in the course of action research, but the ideas are similar to ones which have arisen in other work elsewhere. The phrase 'primary task' was used initially by Rice (1958) and later by other Tavistock Institute researchers (Miller, 1976) to denote the task which an organization is 'created to perform' or 'the task which ... an enterprise must perform if it is to survive'. Also the primary task/issue-based distinction is similar to Merton's (1957) distinction between 'manifest' and 'latent' functions in his paradigm for functionalist analysis: a primary task is a manifest function; issues are likely to surround differing interpretations of latent function. There is one important distinction, however, between these usages and that in soft systems methodology. In the latter there is no commitment to taking an organization to *be* the system named in the root definition. Root definitions and conceptual models explore perceptions of the real world rather than give a would-be neutral account of the world itself or of organizations within it. Given that caveat, the distinction has been found to be useful in clear thinking about the choice of relevant systems.

It cannot be overemphasized, however, that the most important attribute of the good systems thinker is his ability to entertain a wide range of possibly relevant systems, to take his choices seriously and to model them meticulously, but to do so *without owning them psychologically*. As the debate initiated by the model/real-world comparison unfolds, as 'the real issues' emerge, the systems thinker must be able cheerfully to abandon his earlier choices of relevant system and start again. And again . . .

General Outcomes 3: The Structure of Root Definitions

It is obvious that stage 3 is a crux in the methodology. Both in the accounts of systems studies in Chapter 6, and in the descriptions of the six studies given earlier in this chapter, it is clear, I hope, that much depends upon the decision to examine particular (notional) systems as relevant to the problem, and to *formulate definitions of them with care*. It is the start of formal systems thinking to define, for example, 'a (temporary) transfer-of-advice system' as relevant to the dispute between Mr Cliff and Mining Support Ltd, or to examine the implications of taking terotechnology to be 'a system to optimise the contribution which "owning" (i.e. acquiring, caring for, disposing of) physical assets makes towards minimising their life-cycle costs'. The formulation of root definitions of this kind sees the hopeful would-be problem solver stepping back from the picture gathering of stages 1 and 2, which has been done in the real world, and starting the *systems* analysis which will lead to 'feasible and desirable' changes in the problem situation.

Given what has been said above about 'starting anywhere' in the methodology mosaic (Figure 6), and given that later stages, or further work on stage 2, may cause a return to stage 3, entering that stage is not ultimately committing; nevertheless it is the point at which 'professional' systems thinking starts, and a poor shot at selecting relevant systems or unimaginative, sloppily worded root definitions may cause unnecessary extra work later on. Because of this, many people find themselves daunted at the idea of starting stage 3, and find the formulation of root definitions especially difficult. This is a proper circumspection, in that at this stage we need to use words with a care and precision far from the carelessness of everyday speech. But the question remains: Are there positive guidelines for making up root definitions which help the hurdle to be cleared without inhibiting imagination?

It seemed to D. S. Smyth, a Doctoral student in the Department of Systems, that because conceptual models of systems described in root definitions are checked against the characteristics of the formal model of any human activity system, there ought to be characteristics in any root definition which is 'well-formed' which relate to the formal system model and make that checking process possible.

In using the methodology the sequence within stages 3 and 4 is as follows:

Formulate root definition	→	Assemble minimum necessary activities	→	Structure activities into a conceptual model	→	Check conceptual model against formal systems model

Given this sequence there ought to be a logical connection between the characteristics of a well-formed root definition and the characteristics of the formal system model which checks the outcome. The task of finding guidelines for constructing root definitions was therefore tackled in the following way. Based on intuition and the experience of using the methodology, Smyth suggested some factors which ought to be considered in thinking out a root definition. His hypothesis was that these factors would be traceable in any root definition which was well formed. First these factors were compared with the formal system model in order to establish that there was indeed a logical connection between the two; then the hypothesis was tested by examining a range of root definitions which had been used in actual studies now finished. We could examine the root definitions to find out whether or not they contained the factors in the hypothesis, or additional factors, and if they did not we could examine the now unchangeable history of those studies to find out whether the absence had mattered. The work has been described in detail elsewhere (Smyth and Checkland, 1976) and only the results will be summarized here.

The hypothesis based on intuition and experience was that an adequate root definition should contain five elements explicitly; if it does not, then omission of any of these elements should be conscious and for good reason. The five elements are as follows. The core of a root definition of a system will be a *transformation* process (T), the means by which defined inputs are transformed into defined outputs. The transformation will include the direct object of the main activity verbs subsequently required to describe the system. There will be *ownership* (O) of the system, some agency having a prime concern for the system and the ultimate power to cause the system to cease to exist. The owners can discourse *about* the system. Within the system itself will be *actors* (A), the agents who carry out or cause to be carried out the main activities of the system, especially its main transformation. Within and/or without the system will be what we call *customers* (C) of the system, beneficiaries or victims affected by the system's activities. 'Customers' will be indirect objects of the main verbs used to describe the system. Fifthly, there will be *environmental constraints* (E) on the system, features of the system's environments and/or wider systems which it has to take as 'given'.

To these five elements we add a sixth item which by its nature is seldom explicit in a root definition but which cannot ever be excluded: there will be a *Weltanschauung* (W), an outlook, framework or image which makes this particular root definition meaningful. There will by definition be more than one possible W, of course; that has been argued to be the nature of human activity systems. But, for the sake of coherent systems thinking, a separate root definition should be formulated for each W considered relevant, whether it is supplied by the analyst himself or expressed by people in the problem situation.

These six elements covered in a well-formed root definition may be remembered by the mnemonic CATWOE.

It was not difficult to trace logical connections between CATWOE and the formal systems model (see Smyth and Checkland, 1976) and having done this we then assembled for inspection a dozen root definitions from studies which were now history. Two will be examined here.

In the study of consumer magazine publishing described in Chapter 6, a root definition of a notional system to publish commercially a single consumer magazine title described it as a system concerned to assess a potential audience, provide that audience with a succession of images by means of a magazine, and ensure that the total operation earns a surplus. The transformation is clearly embodied in the three main verbs 'assess', 'provide', and 'ensure'. The customers, those who benefit or suffer as a result of the system's activities are included as 'that audience'. The environmental constraints are those commercial constraints resulting from the fact that the environment is a market: our system is not subsidized and must earn a surplus to survive. And the *Weltanschauung* is that the commercial publishing of consumer magazines is a perfectly feasible and proper thing to do. These elements are all explicit in the definition, but 'ownership' and 'actors' are covered only by implication. With hindsight it is obvious that we were simply assuming that ownership lay with a company and that 'actors' were employees. The latter was never an issue in the study and the omission did not matter. But in the case of ownership it was clear that the study would have been enriched by including this more clearly in the root definition. In the actual study, conceptual modelling eventually included a wider system which decides whether or not to give resources to the single-title system (part of that systems model is given in Figure 11). Inclusion of 'ownership' in the root definition would have enabled this point to be reached quicker; also the fact that 'ownership' here entails owning both printing and publishing expertise provides a second reason why inclusion of this element in the root definition would have speeded analysis. A crucial aspect of the Company's problems was that, as a profit centre, publishing was profitable while printing was not. But the large printing division could not be sold off or scrapped, and somehow, publishing and printing had to operate together for the benefit of both. A root definition meeting CATWOE requirements would have driven us more quickly towards aspects which with hindsight we know were finally crucial; we got there in the end, but with CATWOE we would have been quicker!

As a second example it is instructive to consider a study tackled by a postgraduate student, C. Y. Yuen. He was one of a group of students who made regular visits to a local mental hospital to chat to the patients. A root definition of a system of this kind was taken to be 'a regular, volunteer-student-manned, medically approved, mental hospital patient "comforting" system'. The basic transformation process, the customers, and the actors are all fairly clear, as are some environmental constraints: the students have to volunteer, and doctors have to approve. But 'ownership' is not clear, and this makes the W of the definition ambiguous. Is the system created by (and for) medical practitioners glad of unpaid helpers? Is it self-generated by the student actors, a system which makes them feel good? Or is it the result of self-help by the patients? Asking these questions, based directly on the CATWOE elements, itself opens up the analysis, and incidentally leads to iteration to stages 1 and 2 of the methodology. It is clear with hindsight that it would have been better to set up *three* root definitions, expressing the three Ws which stem from taking doctors, students, and patients respectively to be system owners. The real-life manifestation of such a system will almost certainly reveal itself as a blurred ill-thought-out compromise of these three models, and radical debate about improvements will follow from comparing the three competing versions with reality.

These two examples are typical of the dozen which were examined. The highly significant outcome was that 'ownership' and 'actors' were common omissions. There was real learning in this discovery. The reason for it, I believe, lies in the history of our attempt to develop 'soft' systems thinking. We have found that the most difficult step to take in freeing our thinking has been to avoid the unthinking assumption that organizations, departments, divisions, and sections are *ipso facto* systems. Such groupings, an administrative convenience, always represent only one possible choice out of many possibilities. They will usually be reasonably coherently defined ('Production Department', 'Design Section', etc.) but the number of functions to be coped with in an organization will always be greater, usually very much greater, than the number which can be institutionalized as departments, sections, etc. In making up root definitions it *may* be useful to take such sub-groups as systems, but this should be done consciously, and only after questioning whether or not other definitions might provide more insight. If we identify system boundaries with organization boundaries then it is easy also to take for granted the 'ownership' and 'actors' concerned, and this I believe is the reason for our finding that these elements had frequently been missing from real-life root definitions.

This discovery made further sense of a number of completed studies, and helps to convince me that it is worth using the CATWOE elements as a base for the formulation of root definitions. It is true that on the basis of a strict Popperian logic our hypothesis did not survive its test unscathed, in that some of the historical root definitions in *successful* studies were found to be not 'well-formed' according to CATWOE. However knowledge of the history of those particular applications of the methodology convinces me that well-formed root definitions would have enriched the debates which took place during those studies. This, together with the learning associated with the discovery of the frequent omission of 'ownership' and 'actors' in the past, generates some faith in the use of CATWOE at stage 3.

There remains the criticism that even the most painstaking compilation of a root definition incorporating all CATWOE elements does not guarantee a good or insightful definition. This is true but it does not matter. Even if the systems thinker is so dull-witted that he cannot think of any transformation other than the one he perceives in the real world, no matter! Modelling the system incorporating it, and comparing that with the real-world manifestation may well reveal deficiencies in the latter. If it does not, and stage 5 is a comparison of X with X, then this signals that a more radical root definition is required. At the very least, even if we are bereft of imagination, since every definition will imply a *Weltanschauung*, we may at least entertain an alternative definition embodying a counter-W which opposes the first one. Or we can examine definitions embodying Ws appropriate to different roles: actors, owners, victims, beneficiaries, or the analyst himself. On the other hand, if the root definition is *too* radical, then at stage 5 an unbridgeable gap will be revealed between conceptual models and reality, and this signals that more constrained root definitions are required. This further illustrates that no stage of the methodology is complete or autonomous on its own: the role and meaning of each stage derive from its place in the dynamics of the methodology as a whole.

General Outcomes 4: 'Whats' and 'Hows'

The previous chapter discussed the several different ways in which, at stage 5 of the methodology, a comparison is made between the conceptual models and the real-world problem as expressed in stage 2. It was pointed out that this stage is not precisely a comparison of like with like, and this point, an important one, will now be elucidated.

At stage 5 we have available some systems models which themselves derive from the careful naming, in root definitions, of human activity systems which we hope are relevant to the problem situation and to its improvement. In stage 5 (and we are now back in the real world) we examine the models alongside the expression of the problem situation assembled in stage 2. The comparison between the two is the formal structure of a discussion about possible changes, a discussion held with concerned people in the problem situation. In order that the discussion shall be rich and wide-ranging, we wish to question *whether* various activities in the models are discernible in the real world, as well as—if they are present—*how well* they are being done. We also wish to discuss possible alternatives to the real-world activities, alternatives suggested by the models. The real-world activities will always represent one way of doing things, one particular *how* related to a *what* which is usually implicit rather than explicit. The conceptual models, on the other hand, ought to represent *whats* rather than *hows*, since inclusion of constraints in the root definition which reduce the models from accounts of 'what' to accounts of a particular 'how' will limit the range of the debate about change.

It is this distinction between 'whats' and 'hows' which makes the word 'comparison' a somewhat crude description of what is happening in stage 5.

The relation between 'what' and 'how' is a hierarchical one, a 'what' being at a logically higher level than a set of possible 'hows' related to it. Thus, if *what* you must do is 'avoid a mental breakdown' there is, at a lower level, a set of possible *hows* by which you might achieve this. They might include: take a holiday; change your job; fall ill physically; etc. Each of these possibilities, a 'how' to the previously defined 'what', is itself a 'what' with respect to lower level considerations. At this level, 'take a holiday' is a 'what' to a set of 'hows' which might include: canoe down the Thames; walk through the Black Forest; spend a week in Bognor Regis; etc. The what/how relation is the same as that between system and sub-system: the latter is 'sub' with respect to the former, but is in its own right a system, and may itself have sub-systems.

We can now see why stage 5 is not a straightforward comparison. A conceptual model defines a particular system at a level of 'what'. As long as the root definition does not contain specific constraints, it defines a *class* of human activity systems. The real-world situation being 'compared' with the conceptual model will contain specific activities which are a particular 'how'. In bringing the two together we wish to question whether a version of the conceptual does (or should) exist *and*, if a version does exist, whether a different version might not be superior. The outcome might be to change the way things are now done (better 'hows') or to introduce new activities or a new version of the whole system (a new 'what').

The what/how distinction at the comparison stage has been important in most of the studies done as part of this research. It may be illustrated by a study carried out by my colleague T. R. Barnett and postgraduate student D. C. Nevin for a large science-based manufacturing company which competes with other large companies in world markets. The problem situation centred on various aspects of work done within the Research and

Development Department. Given the nature of this Company's business it seemed to us that to maintain a position among its competitors the Company had to be an innovator, using the word to include the total sequence of bringing an invention through to a successfully marketed product. 'An innovation system' was therefore selected as relevant, and a conceptual model of such a system appropriate to this particular Company was built. At the comparison stage this model (a 'what') was set alongside the Company's activities. There were considerable differences between the two. The Company contained no innovation system institutionalized as such. Their particular 'how' was that the various essential activities of a notional innovation system were spread among several departments. Research and Development did many things which were a part of the innovation system model, but also did many things which were not, such as improving existing processes to make existing products. Some innovation activities were the responsibility of Marketing Department, who also did many other things unconnected with innovation. Similarly Engineering Department, Planning Department and a Management Services Group were in part concerned with innovation. Thus a fragmented version of an innovation system was discernible among the various departmental responsibilities; but as a system it was far from perfect. Some activities were missing and some were not done well. Some information links were poor. The study concentrated on improving the present 'hows' so that they made up a more coherent version of an innovation system, this being taken to be a desirable 'what'. There was a slight disappointment at implementation, when the sponsors of the study (in R and D Department) felt politically constrained to make only changes which came within their purview; but the clarity of the issues at the comparison stage largely derived from the recognition of the what/how distinction.

General Outcomes 5: Gathering and Enriching the Initial Impression

The initial stages of any systems study are difficult because of the need to gather as rich an impression of the problem situation as possible without at that stage imposing a structure upon it. The well-tested guidelines for this procedure are that it is helpful to look for elements of structure and elements of process, and to examine the relation between the two. These guidelines are at the same time positively helpful and virtually neutral with respect to any subjective interpretation of the problem situation. However, many people have found them too sparse, and there is a call for more help with stages 1 and 2. Although in general I would not wish to elaborate in greater and greater detail the specifications of the methodology's stages, since this will encourage the false view that the methodology is technique, nevertheless some enriching of the structure/process guidelines is supported by the experiences in the whole set of studies carried out in this research.

Before describing two particular kinds of structures and processes which need to be examined, it is worth reiterating a point made in Chapter 6, namely that stages 1 and 2 should not be pressed in systems terms. In 'hard' systems thinking it is entirely appropriate that the initial analysis should describe the problem as being located in a hierarchy of manifestly obvious systems, since 'hard' methodology takes such systems as given. But if in a soft unstructured situation we express the problem, at stage 2, in terms of a systems hierarchy, then we have begged all the questions which should be posed at stage 3 by the selection of relevant systems and the formulation of various root definitions to describe them: we have taken as given the very features which in 'soft' problems are problematical. What must be avoided in enlarging upon stage 2 is any introduction of the idea that the problem situation *is* some obvious hierarchy of human activity systems. It is necessary to re-emphasize this because the additional help available at stages 1 and 2 comes from the recognition that every soft problem situation will contain what *in everyday language* are usually described as 'systems': social systems and political systems.

If social science had provided some generally accepted and well-tested models of social and political systems, then it might be appropriate to use those models in gaining the initial picture of the problem situation. But no such public knowledge has been produced by the social sciences, and the additional guidelines are necessarily based only upon some elements of social and political systems. It is of course the case that any individual analyst who thinks highly of, say, Parsons's (1951) conceptualization of social systems, Homans's (1961) psychology-based model of social interaction as 'exchange', Kuhn's (1974) model of social transactions, or Deutsch's (1963) cybernetic (information feedback) model of political systems, is free to use them in his analysis. My position is that what has been found to be relevant in every instance in which the methodology has been used is something less than any of these models. What has been useful is the idea that, whatever else it contains, the problem situation will contain the special kind of structures, processes, and relations between the two which we recognize as constituting, in everyday language 'social systems' and 'political systems'. It has been found useful explicitly to look for some of the elements of such systems rather than to try and use any complete models of such systems.

In the problem situation, among the elements of structure, will be discernible a set of *roles* which are recognized by the participants. In learning the characteristics of the problem situation it is important to ascertain what roles seem relevant to the problem, and also which roles participants regard as significant.

The concept 'role' has a long history and a large literature in social science (see Jackson, 1972) and I do not propose to discuss it in detail here. Happily the theatrical analogy and the dictionary definition—a role as 'what one is appointed, or expected or has undertaken to do'—convey essentially the same concept as the social science literature. There, roles and the taking *and making* of roles are seen as fundamental to the process by which social groupings become more than mere aggregates of people. Schutz (1943) describes the familiar process in everyday life in which we assume roles:

> By isolating one of our activities from its interrelations with all the other manifestations of our personality, we disguise ourselves as consumers or tax-payers, citizens, members of a church or of a club, clients, smokers, bystanders, etc.

and goes on to emphasize that closely allied to any role are the expectations which others have of the behaviour of anyone occupying a role:

> The traveller, for instance, has to behave in the specific way he believes the type 'railway agent' to expect from a typical passenger . . .

while emphasizing that we remain free to drop (or alter) a role:

> And we keep—even in the role—the liberty of choice. . . . This liberty embraces the possibility of taking off our disguise, of dropping the role. . . . We continue to be subjects, centres of spontaneous activities, actors.

These expectations of behaviour in a role are the social *norms*. Cohen (1968) describes norms as 'specific prescriptions and proscriptions of standardised practice' and points out that the word is used in two senses: that which regularly occurs and more important, 'what members of a society have a right to expect'. In fact neither roles nor norms are fixed, but change steadily through time, sometimes slowly sometimes remarkably quickly. Every individual who occupies the role 'prime minister', for example, or 'shop steward', is both changed by his occupancy of the role and himself changes the norms associated with the role as a result of his behaviour in it. Ascertaining what roles are significant, and discovering the norms associated with them are important parts of the systems methodology, as is the discovery of the third element always associated with roles and norms: the *values* according to which behaviour in a role is judged. 'Values', says Cohen 'express preferences, priorities or desirable states of affairs', and points out that 'men . . . evolve values which limit the range of norms which they are willing to adopt or reject'. Values, like roles and norms, continually change, and again this is familiar to all of us in everyday life: in the community in which I was brought up thirty years ago one of the worst things which could happen to a young girl was that she found herself in the role 'unmarried mother'. Attitudes, and the values they embody, are very different today.

These three elements of roles, norms, and values are important aspects of the structures and processes of any 'soft' problem situation and should not be neglected in the initial stages of a systems study. The studies which have been described above provide many illustrations of their importance. In the Airedale Textile Company, for example, the new Marketing Director, coming from a large corporation, brought with him a concept of his role which was virtually meaningless in the unsophisticated small firm he had joined, and the norms and values he associated with the role struck no chord in Airedale. In Index Publishing and Printing Company it was extremely important to appreciate the nature of the role 'consumer magazine editor', as well as to ascertain the norms and values associated with it. Later it was important to tease out the different values accepted by printers and publishers, the former being technology-oriented, the latter media/business-oriented. Perceiving that different sets of values were present within the same situation was also important in the study we made of our own problems. Some of the difficulties of operating together a university department and a consultancy company must arise from the fact that, with prevailing norms, the prime value of the former is something like 'the acquisition and dissemination of knowledge' while the latter values most highly the successful transmission of practical advice to a client. In fact, gaining an appreciation of roles, norms, and values has been important in every systems study we have done.

The other particular kind of structures and processes important in stages 1 and 2 are those associated with the exercise of *power* in the problem situation. I take it that power is *the* concern of what we call 'political systems': obtaining it, legitimizing it, exercising it, preserving it, passing it on. Again the social sciences do not yet provide agreed models of political systems, and even if they did it would probably be essential to take every unstructured problem situation to be unique and hence requiring a specific analysis. What can be said from the research experiences is that the structures and processes associated with power (which are of course also related to roles, norms, and values) are always an important characteristic of situations perceived as problems.

As in the case of 'role', there is an extensive social science literature on the concept of power and social control (see Buckley, 1967, for a useful introductory discussion in a systems framework) but again the everyday notion corresponds closely with professional definitions, from Parson's 'control over the action of others' (1951)[†] and MacIver's 'capacity to control the behaviour of others either directly by fiat or indirectly ... ' (1947)[†] to Lynd's conception of power as a resource which can be democratically organized (1959)[†] and Buckley's (1967) 'consensual

[†]Discussed in Buckley (1967).

authority'. (Lynd and Buckley are here emphasizing that coercion is not the only 'how' associated with power as a 'what'.)

In stages 1 and 2, having defined the important roles and their associated norms and values it is important to find out how power is gained, legitimized, held, exercised, and passed on in the situation. Even if it is not itself part of 'the problem' as that is finally perceived, it will be highly relevant to defining in stage 6, changes which are both desirable and feasible as well as to taking action to implement changes in stage 7. A good example of this has been mentioned above, in the study in which at a late stage a Managing Director decreed that any changes suggested in the study report must be procedural rather than structural. He regarded defining the organization's structure as an important part of his power. And my colleague Brian Wilson had an experience of power exercised at lower levels in a study in which middle and senior managers readily agreed on rational structural reorganization of their engineering company. The new structure involved the merging of what had previously been two separate drawing offices. The new arrangements meant that one of the drawing office superintendents would become head of the new single office, and this arrangement was acceptable to all the middle and senior managers involved. When implementation of this change was announced, a group of fitters on the works, who maintained useful informal contacts with design engineers announced that if the proposed changes were implemented they would cease to co-operate with management. Their 'working to rule' would end their valuable liaison with designers. This response astonished all concerned with the study, who had not imagined that the fitters would be particularly exercised over a managerial rearrangement having no effect at all on the fitters' day-to-day work. However the power of the fitters was such that the Company was not prepared to make an issue of this, and a different reorganization of the drawing offices was agreed. The response of the fitters was due to their perception of recent history in the organization. They felt that they had been losing status. The drawing office issue was a vehicle by which they could exert themselves corporately rather than a specific comment on the drawing office reorganization. This episode was not crucial to this particular study, and it is perhaps asking a lot of the systems analyst to be prepared for every possible turn of events of this kind. But this story does illustrate that an examination of power, and who wields it, and how, is very relevant to the initial assembly of a rich picture of the problem situation. It *may* be crucial to the choice of insightful root definitions; it will certainly be important at the comparison stage and when feasible and desirable changes are defined. The rich picture is frequently further enriched at that stage when the realities of power emerge.

General Outcomes 6: Methodological Laws

If the systems *Weltanschauung* is eventually established as a useful framework within which to describe the world—and it has had some modest successes on that road—then no doubt there will gradually accumulate a body of systems laws which summarize the properties of systems as such, regardless of the specific phenomena to which they are applied. Ashby's Law of Requisite Variety, 1956 (that the variety of a regulator must equal that of the disturbances whose effects it is to negate) and Bertalanffy's Law of Equifinality, 1950 (that open, but *not* closed systems can reach the same final state from different initial conditions) are early examples of systems laws of this kind. Given the generality of systems concepts the content of such laws will necessarily be very general and these two examples illustrate that fact. They are summaries of the *logic* of systems models rather than empirical laws: Ashby himself says of his law that it 'owes nothing to experiment'.

As the use of systems concepts in problem-solving is developed we may expect a parallel emergence of laws deriving from the logic of the *use* of systems ideas. Two manifestations of one such law have been described above in the fifth of the six studies described at the start of this chapter; there we established the need to examine the systems *served* by information and by planning systems before going on to the information or planning systems themselves. Like the general systems laws quoted above the content of this law is logical rather than empirical, and it seems obvious enough once stated. But it was derived from much hard-won experience, and there are plenty of examples of problems arising from its neglect. (It is of course perfectly possible to disobey a logical law in practice. Regulators which do not obey Ashby's law can be built; but they will not be very good ones.)

The law has been illustrated in its applications to information and planning systems but more generally states that if a system A serves the purpose of another one B, then it is not possible to form a root definition and conceptual model of A without first doing so for B. The logic of this derives from the *Weltanschauung* concept. If A exists to serve B then it cannot be considered on its own. If the law is ignored, any root definition and conceptual model of A will take for granted some unquestioned version of B which makes the account of A meaningful. Logic requires that the 'version of B' should itself be made explicit, in root definition and conceptual model. This applies wherever one system serves another, and thus applies to all information systems, all planning systems, and all control systems. The former two cases have been illustrated above; as an example of the latter, consider a quality control system: it serves a system which produces something. In order to define and model the quality control system it is necessary first to define and model the system which produces, since its characteristics will dictate the necessary characteristics of the sub-system which monitors its performance and takes action to control the quality of what is produced. Applied to the situation implied in Ashby's law, a regulator cannot be designed without first defining (either by examination if it exists physically, or by a design) the system to be regulated.

An outcome of this law of conceptualization is a logical procedure for model building at stage 4. Starting from the root definition, the minimum set of necessary activities needed in a system which is to be that described in the definition is assembled. Logic dictates their sequence and their mutual dependences (a plan cannot be implemented before it has been formulated! etc.) Each major activity in this first model is now notionally a sub-system which exists to serve the purpose of the system as a whole. Model building may now proceed by taking each major activity together with its inputs and outputs as itself a root definition, and assembling the minimum necessary sub-activities in this (sub) system. And so the process goes on, working at lower and lower levels of detail until, in the model-builder's judgement, there exists a model worth bringing to the real world for a comparison. Our experience is that the difficulty of conceptual model building lies in maintaining consistency of resolution level. The first level model ought not to contain more than, say, five to ten major activities, and should itself be a complete model at that level of resolution. It may then be expanded by the method described to yield another complete model, now more detailed than the first one, and so on to higher resolution levels. An illustration of the kind of thinking involved in conceptual model building is given in Appendix 1.

A second methodological law, which, like the first, is obvious enough without being easy to follow in practice, stems directly from various experiences of model building and comparing the models with real-world manifestations. It concerns the kind of activities it is appropriate to put in conceptual models at stage 4.

There are *direct* activities, verbs describing something an actor can proceed to do, and there are other activities, also verbs, which are only the hoped-for consequences of direct activities. The distinction is clear if we imagine the verbs to be in the imperative. Taking examples of direct activities from the conceptual model of Figure 11, we could take each of the main verbs to be an order, and proceed to obey it: 'appraise resources', 'collect information', 'make plans', 'appraise future technology-based possibilities', 'monitor performance', etc. These are all direct activities; actors could do them. On the other hand, activities of the kind 'raise morale', or 'reduce costs' are only the hoped-for outcomes of various direct activities. Only direct activities should be included in the models built at stage 4; inclusion of 'consequent' activities leads to much poorer clarification of issues at the comparison stage. A model containing a consequent activity like 'improve industrial relations' is liable to be greeted by actors with cries of 'We know we want to do that; tell us something we didn't know'.

The difference between direct and consequent activities is very similar to the distinction between *manifest* and *latent* functions in a sociological analysis by the structural–functional school, mentioned above in discussing primary task and issue-based root definitions. In clarifying a paradigm for functional analysis (in which the social scientist, taking a holistic view of social systems, seeks to identify the contribution which social customs, acts, and institutions make to the system as a whole) Merton (1957) drew attention to the important conceptual difference between 'manifest functions', referring to 'those objective consequences contributing to the adjustment or adaption of the system which are intended and recognised by participants in the system' and 'latent functions' which are 'neither intended nor recognised'. This distinction is implicit in the work of many previous writers, as Merton himself points out; his point is that the distinction is invaluable in achieving a clear functional analysis.

One of Merton's examples is the purchasing of a consumer item such as a car; while having the manifest function of use of the car, it has a latent function of maintaining social status. Arguing that in 'soft' systems methodology the conceptual models must contain only direct activities, and not consequent activities, is equivalent to saying in Merton's language that the models must be in terms only of manifest, not latent, functions. If his example arose as part of conceptual model building, then the model should contain the activity 'purchase car', not 'maintain social status'.

Merton, of course, is saying that a good functional analysis of a social system must be in terms of both manifest and latent functions. If it seems odd that I am arguing for the exclusion of all but manifest functions in the conceptual models built as part of the methodology, then this may serve as a useful reminder of the nature of the methodology itself. Merton is interested in the richest possible account of the reality of a social system. Conceptual models built at stage 4 of the methodology are neither descriptions of actual human activity systems nor accounts of such systems which ought to exist. They are a working out of the (direct, intended) consequences of taking particular views—those summarized in root definitions. Their purpose is *only* to generate a high quality discussion with concerned participants in the problem situation. That discussion may well go on to cover possible unintended consequences of making the models into reality, but the models themselves, being only a means to an informed debate, should remain a clear exposition of intended consequences. Later on, when action is decided and changes are implemented, it will certainly be the case that the outcome is not quite that intended: nobody will be able to predict all the consequences of introducing changes in real human activity systems, because nobody can predict the *Weltanschauungen* which will then be operative. 'World is suddener than we fancy it; World is crazier and more of it than we think', as MacNeice's poem has it. That is why the outcome of any use of the methodology will be a new problem situation, why the methodology itself is an on-going learning system whose task is never done because learning can never be complete.

It has been argued by Prévost (1976) that the methodology is in fact part of the structural—functional tradition in sociology. There are similarities, certainly, as there are bound to be given the holistic framework of structural functionalism, but the aims of the social scientist and the systems-oriented problem solver are different, and this makes an application of the methodology different from a functional analysis. The social scientist wants the most accurate possible, testable account of what a social system *is*. The systems man using the methodology wants improvements in what is taken to be a problem situation. Given these aims, the functionalist sociologist wants the richest possible model he can get, including manifest and latent functions; the systems analyst wants his systems thinking to be as clear and coherent as possible, leading to clear-cut *debate*, and hence he makes his systems models models of possibilities.

Summarizing the two laws of procedure which have emerged most clearly from this research:

The law of conceptualization states that a system which serves another cannot be defined and modelled until a definition and a model of the system served are available;

The law of model building states that models of human activity systems must consist of structured sets of verbs specifying activities which actors could directly carry out.

General Outcomes 7: Problem-Content and Problem-Solving Systems

Most of the work in the action research programme has been directed either to establishing tentative methodology and then testing, modifying, and re-testing it, or to drawing general lessons from a number of systems studies considered together. This section, however, describes some general outcomes from the work which concern not the content of the methodology itself but what the availability of a systems methodology teaches us about trying to bring about improvements in real-world situations. Every application of the methodology can be seen as a specific illustration of a general model of circumstances in which a problem is perceived or improvement is thought desirable, and action is taken (by means of the methodology) to do something about it. That there must be such a general model is of course implicit in the existence of 'a systems methodology for problem-solving', and the model would not be worth describing for its own sake. However, it is worth including because it has led to a practical outcome: a workbook which provides help in the very difficult preliminary stages of a systems study, when the desire to get started is restrained by the fear that a false step at this point might take the whole study in a direction which will later be seen to be mistaken.

Real-world problems, whether they are explicit disasters or no more than a vague ill-defined feeling of unease, will be associated with human activity. Given the nature of human beings it will be the case that we are concerned not with 'problems' but with 'perceptions of problems'—and of course a wide variety of perceptions will be possible; the development of supersonic passenger aircraft for example, may be taken to be a technical problem by some, an economic problem by others, an environmental problem by a third group. But no matter what particular perceptions or mix of perceptions seem obvious to particular individuals or groups, a fixed element in every problem situation will be the existence of the role *problem owner*, occupied by those who perceive the problem. A second fixed element will be the role, the would-be *problem solver*, the occupants of which wish to tackle the perceived problem(s). It is important to emphasize that these are roles, not individuals. The problem owner may be one person, a group of people, an organization, or society as a whole. And the role problem solver may be occupied by one person or by a group of people prepared to try and work together. Also, one person, or a group, might be the would-be solvers of problems they themselves own.

Every systems study we have carried out can be seen as taking place in circumstances in which there is a perceived problem and a readiness to take action to solve it. In such a situation there is a 'problem-content' system, containing the role of problem owner, and there is a 'problem-solving' system containing the role problem solver. The problem solver uses the systems methodology to take action to improve aspects of the problem-content system. There will be several potential occupiers of the roles, and there will be several possible ways of describing problem-content and problem-solving systems. *Asking questions about them, deciding how to describe them, is the best way to start a systems study.* If it is clear what the 'problem-content system' is being taken to be, and who is 'problem solver' with what resources, then organized work according to the stages of the methodology can begin. And if subsequent experience causes the initial perceptions to be modified, as is quite likely, that can be done without the study drifting into incoherence.

Figure 17 pictures this system to use the methodology.

The problem solver defines one or more problem owners and the problem content(s) associated with that ownership. He uses the methodology to recommend or take action in the problem-content system, or to redefine it and/or its owner(s). And he uses the experience as a means of further developing both the methodology and the concept of a human activity system. Note that the problem-content and the problem-solving systems are not separable in the way that they are when the problems are those of natural science: they are linked parts of a single

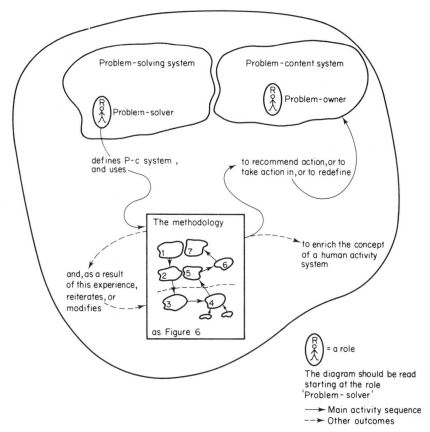

Figure 17.   An outline of a system to use the methodology

system. In physics, chemistry, or biology there will frequently be agreement on a definition of problem content, and various would-be problem solvers in different laboratories may actually engage in a race for the solution, as happened recently, for example, in the celebrated chase after a structure for the DNA molecule which would yield the patterns of Rosalind Franklin's X-ray photographs (Olby, 1976; Watson, 1968). In such cases not only is the problem content public, but what will constitute a solution is also generally agreed. Neither characteristic is typical of 'soft' problems. There, the problem solver's perceptions as well as his attitudes towards real or notional problem owners are an integral part of the problem situation and of *learning one's way to a solution*.

The important distinction between the problem-content system and the problem-solving system first became clear to us in the study for Mr Cliff of his failed project for Mining Support Ltd, presumably because there the problem-content system happened to be his problem-solving system. From that point in the research experience this idea has been found to be useful both in structuring new studies and in understanding studies in the past. Asking questions about the ownership of the problem and the resources available to tackle it, and *writing down the answers* is important because it establishes the *Weltanschauung* of the study itself, that which makes it a meaningful undertaking for the problem solver. A work book which asks some appropriate questions is included in Appendix 2; filling it in gets a study underway, and, if the answers are recorded, then a retracing of steps and an exploration of other lines is always possible as learning proceeds.

It might be thought that this discussion is similar to Churchman's (1971) discussion of the roles 'client, decision maker, designer' in the chapter on 'the anatomy of teleology' in *The Design of Inquiring Systems*, but there are two important differences. Churchman does not distinguish between an account of the real world in systems terms and a manipulation of systems models as a means of exploring the implications of various viewpoints (such as happens in stages 3 and 4 of the methodology when root definitions are formulated and models built). The clear distinction between the 'above the line' activity in Figure 6, carried out in the real world, and the 'below the line' systems thinking, was found to be very important in the action research programme. Based on this research experience, Figure 17 represents as far as I can go in justifying description of the real world in systems terms. As has been argued earlier, this is something which *is* justified in 'hard' problems, in which the problem is to find an efficient means of meeting a defined objective, taken as given. This brings us to the second difference between this discussion and Churchman's. His 'client' is a man whose problem is to choose between alternative ways of approaching a future he already knows to be desirable:

> The client can be described in terms of his value structure. For him there are a set of possible futures, i.e. states of nature, and *he has a real preference for one state over others.*† More specifically, we can describe the possible futures in terms of a set of properties which we call 'objectives' or 'goals'. . . . Since his resources are limited, he cannot create his ideal future, but instead must create a future which comes as near to what he wishes as his resources will allow.

In other words the client's problem is a 'hard' one. My 'problem owner' *may* be lucky enough to know a desirable future state, but more likely he will not; this may indeed be his ('soft') problem, that he does not know what future state is desirable, or how to define 'desirable'. The methodology is intended to help him in this situation, and the model in Figure 17 indicates a way of starting to use it without having to make any prior judgement as to what attainable end is also desirable.

---

†Present author's italics.

Conclusion

This and the previous chapter have described the systems methodology for real-world problem-solving which has emerged from a programme of action research. The methodology makes careful distinction between action in the real world (stages 1, 2, 5, 6, 7) and the use of systems ideas to explore via systems models the implications of taking particular views of the problem situation (stages 3, 4). The seven stages, which do not have to be followed in sequence, themselves constitute a model of a system of a particular kind, a *learning* system which aims to increase knowledge in and understanding of a real-world situation which is regarded by at least one person as a problem. Use of the methodology occurs within a loosely-defined wider system in which the roles 'problem-owner' and 'problem solver' are identified. Problem-solving is seen as a never-ending process in which changes are made, problems are redefined and situations, or views of them, change. If a particular problem is regarded as 'solved', that is equivalent to saying that the problem owner feels it to be so.

At this stage two outstanding questions concern the efficiency of the methodology: Does it work? and To what extent can it be *proved* to work? These questions were raised in the introduction to this chapter, where it was pointed out that the practical proving of methodology, as such, is not possible.

In the present case it is true to say that those who have been involved in the action research programme, who have tried to use the methodology and have been involved in the struggles to develop it in a broad range of problems for 'problem owners' of many different kinds, have felt that the methodology does provide a means of bringing systems thinking to bear on problem-solving. Confidence in the use of it among that group is now high, and there seems to be no problem large or small arising in real manifestations of 'human activity systems' which is in principle excluded from its scope. But the question of *proof* that it works is of course a different matter. If the methodology had the status of a scientific finding, then I could merely await the confirmation of repeatability by other workers. Unfortunately no methodology can claim scientific status. Firstly, no two real-world problems are ever exactly the same, not in the sense in which two reactions between copper and nitric acid are exactly the same when carried out under similar conditions. Secondly, even if complex human situations could be exactly duplicated, methodology will always fall short of the scientific criterion of testability. If someone says to me: 'I have tried the methodology and it works', I have to reply on the lines: 'How do you know that better results might not have been obtained by an *ad hoc* approach?' This is an undecidable question. If, on the other hand, the assertion is: 'Your methodology does not work,' I may reply, ungraciously but with logic, 'How do you know the poor results were not due simply to your incompetence in using it?'

The problem is analogous to that of the irrefutability of philosophical theories (Popper, 1963) and the way of avoiding (or 'redefining' or 'solving') it is the same: methodology can be tested only in conjunction with a problem to which it is applied. We must take, not methodology, but methodology-plus-problem and ask, not about the methodology, but about the problem—was the problem solved? In dealing with human activities as perceived problems, the best we can hope for is that in the eyes of concerned people former problems are now rated as 'solved' or that problem situations are rated as 'improved'. On these criteria the ten-year development of the methodology has been 'successful'. But I shall never be able to prove it in the sense of obtaining scientific proof.

Apart from the methodology itself the main outcome of the work must be its implications for systems thinking, for the fundamental proposition of the systems movement that it is useful to take the world to consist of a set of overlapping and interconnected wholes called 'systems'.

In the first part of the book an examination of systems thinking as a response to certain difficulties within the intellectual tradition of science led to the idea of 'a systems map of the universe' in which sets of purposeful human activities could themselves be regarded as systems. Such systems I have termed 'human activity systems', and, within the limitations discussed above, have tried to test the hypothesis that real-world problems, being problems of such systems, can be tackled by making use of systems concepts. The attempted testing has led to the methodology. Whether or not the methodology is useful (and it could have utility in practice even if unsoundly based theoretically!) it is important to ask: What are the implications of this systems *practice* for systems *thinking*? This is the topic which must now be discussed.

Part 3

CONCLUSION

Chapter 8

Implications of Systems Practice for Systems Thinking

The last forty-odd years have seen the emergence of the systems movement within the broad sweep of the scientific way of trying to understand the world. I have tried to show that systems thinking arises historically as an attempt to cope with the kind of complexity—whether in natural or in human and social phenomena—which defeats the reductionism of the classic scientific method. The programme of the systems movement as a whole has been mapped in Chapter 3 (Figures 2 and 3) and subsequent chapters have described an attempt by means of an action research programme to use systems ideas to tackle 'real-world' problems (as opposed to those which the scientist can define for himself in the laboratory). The results of this work are hopefully useful (and usable) in their own right. But whether or not the outcome is useful in an 'engineering' sense, and independently of that, the work also represents an approach to the question of whether or not systems thinking can give an insight into the nature of the complexity of social phenomena.

Part 1 described some systems thinking. This led to the concept of the 'human activity system' and to the use of that concept in the action research of Part 2. Now we can ask: what are the implications of successful systems practice for systems thinking?

The way of working in the research was to start with 'hard' systems engineering methodology and modify it as it failed in situations in which the problems were 'soft' and ill-structured. To the extent that we could make it so, the modifications were dictated by the problem situations and were not theory-based. The outcome is the soft systems methodology, developed in practice. Writing of the methodology itself, Jones (1978) says of it, quite correctly: 'General theory is nowhere in evidence'. But since the arena in which the methodology is used is that of human interactions, we can ask, after the event: What model of social reality is implied by the (successful) methodology? What is the methodology's theory? This is the subject matter of this final chapter.

The questions being discussed here cannot be posed unequivocally as scientific questions, at least, not in a way which would satisfy the strict Popperian position outlined in Chapter 2. Social scientists in general, and especially action researchers who study human beings interacting with each other, will feel a rueful sympathy for the position expressed by Chomsky (quoted by Lyons, 1970) namely, that in this kind of work empiricists have to regard as meaningful any hypotheses which meet only the condition that 'they not be entirely neutral with respect to all conceivable evidence'! In fact the major outcome of this discussion of the implications of the systems practice is that the social and natural sciences cannot be regarded as similar enterprises using, or seeking to use a common method. Rather the work lends support to the view that the investigation of social reality is *fundamentally* different from the investigation of the natural world.

Reflections on the Action Research

The main outlines of the methodology emerged in about a dozen studies. The lessons recorded in Chapters 6 and 7 were clear by the time the methodology had been used about fifty times. Now that it has been used more than a hundred times it is useful to look back over the whole learning experience and to distil the main impressions gathered. Three aspects of the learning stand out most clearly.

The first is that although at the start it seemed obvious that studies started by finding out about a situation and proceeded by formulating root definitions, building conceptual models and then following a sequence of activities which eventually led to action in the problem situation, it gradually became clear that the sequence from stage 1 to stage 7 is only one rational way in which the methodology can be used, rather than the way in which it has to be used. This is discussed in Chapter 7 in the section 'Using the Methodology out of Sequence', but its importance is worth emphasizing. The relationships between the stages are what is important, and it is frequently possible, not only to start at any stage, but to do some work in the problem situation which simultaneously contributes to more than one stage. In discussing OR projects (and incidentally deploring the fact that in 'the dreary 1960s' OR was 'presented to managers as "solving specific problems by models" ') Churchman (1979) writes: 'I'm often inclined to put the implementation questions first, i.e. "can anything be changed"?' This strange proposition is perfectly sensible within the experience of developing the methodology. If there is a proposed and well-defined course of action about to be undertaken in the situation studied, it is perfectly possible to work *back* to the changes involved and from those to clashing root definitions and conceptual models which might lead to these particular changes being considered relevant. Another example is provided by Singh's (1979) suggestion that in research in problems of industrial relations it is important to distinguish 'operational' models (of real-world arrangements) and 'conceptual' (systems) models, and to proceed by comparing the latter with the former—a procedure closely akin to the methodology under discussion. Here again this accords with the experience of developing and using the methodology. If the problem situation studied is a strike, for example, this may be taken as an extreme manifestation of stages 5 and 6, in which the strikers are proposing some changes which derive from their view of reality and the employers are opposing the changes as unreasonable or unrealistic according to their different view of reality. By starting at stages 5 and 6 it is possible to work back to the root definitions implicit in the stances of the two opposing sides. This offers the prospect of changing the terms of the argument, enabling the appreciative systems of strikers and employers to be modified. Over a number of investigations, the relation between the systemic interactions investigated 'below the line' (stages 3 and 4) and the happenings observed in the real world, will provide the theory of industrial relations which Singh seeks.

The second aspect of the learning accompanying the action research which seems in retrospect to be most significant, is the gradual realization of the significance of the difference between the stages carried out in the real world (stages 1, 2, 5, 6, 7) and the systems thinking *about* the real world in stages 3 and 4. It is obvious that the systems thinking involved in formulating root definitions and building conceptual models is a different kind of activity from finding out about a perceived problem and using conceptual models in a real-world debate which eventually leads to action to alleviate the problem. No doubt this always was obvious, at least in a shallow way. But over a long period of using the methodology a deeper appreciation of this difference has grown up. Put succinctly, the deeper appreciation is of the fact that systems is an epistemology (making a statement of the kind: 'A certain type of knowledge may be expressed in systems language') before it is an ontology (which would make a statement of the kind: 'The world *is* systemic'); and in the case of work whose concern is social reality it may never be possible to make ontological statements in systems terms. This summary no doubt needs some expansion to be comprehensible.

I am saying that the work described in the previous chapters leads to, or lends support to, a particular intellectual position. It is, I suppose, a philosophical position. It may be described as follows.

1. There is a real world of great complexity in which we find ourselves. Our species has a curiosity about complex reality, and has developed ways of finding out about it. We are able—and we do this all the time—to make intellectual constructs pertaining to complex reality. The constructs (which, though themselves abstract, may of course be expressed in words on paper or as physical artefacts) are themselves simpler than reality, but may be checked against it.

   (This initial position has been elegantly expressed by Michael Frayn (1974) in the opening of his book *Constructions*):

   The complexity of the universe is beyond expression in any possible notation.

248

Lift up your eyes. Not even what you see before you can ever be fully expressed.
Close your eyes. Not even what you see now.

Our notations are by their very nature simpler than what we denote. This is the point of them: to reduce the multiformity of the world to common forms, so that things can be brought into a logical and conceptual relationship with each other. . . .

Our reading of the world and our mastery of notations are intimately linked. We read the world in the way that we read a notation—we make sense of it, we place constructions upon it.

2. The activity we call 'science' happens to be the most powerful means we have of making valid some of our intellectual constructions, our notations, by checking them against the real world itself. The power of science arises from the fact that its constructs, or their consequences, are *publicly* testable, as discussed in Chapter 2.
3. When the intellectual constructs survive severe tests we tend to slip into describing the world as *being* what the simplified constructs say it is. It would be pedantic to do otherwise. Nevertheless, we ought occasionally to remind ourselves that our descriptions and models of the world, even when well tested, *are not the world itself*. The most spectacular example of this discrepancy between models and reality is in physics (again, see Chapter 2). Newton's intellectual model of the world as consisting of space in which there are objects survives some severe tests. Not unnaturally we talk about the world as if it actually consisted of space and objects-in-space. In fact, the model which survives the most severe tests so far is Einstein's, in which space and objects-in-space are not separable as in Newton's model. The discrepancy between Newton's model and reality does not matter for everyday purposes, or for many scientific ones. And even Einstein's model, though more powerful than Newtons, is not reality itself, only the best *model* of reality that we have so far. Occasionally, we do need to remind ourselves that there is a distinction between complex reality and our notations of it.
4. Systems ideas constitute one particular set of intellectual constructs, one particular notation, which may be used to make descriptions which can be tested against reality itself. The system idea, in summary, is one of hierarchical wholes showing emergent properties and characterized by mechanisms of communication and control (as discussed in Chapter 3). If this paradigm is used to investigate natural phenomena, then the intellectual constructs, here systems models, can be tested against reality in a straightforward manner—can they reproduce the repeatable observables? If they can do so successfully, we slip into saying that the natural world actually contains or consists

of systems. We forget that we are talking about our notations of reality, rather than about reality itself. This probably does not matter very much, especially if the systems models are empirically well tested.

5. Now suppose, however, that the systems thinking is applied to human beings and their interactions, and involves the concept of a 'human activity system' as discussed in Chapters 3 and 4. Because a model of a human activity system will express only one particular perception of a connected set of activities out of a range of possibilities, we cannot expect the kind of match between reality and model which natural science seeks, and which it is possible to achieve in the case of natural systems. If our model of an English pub as a human activity system, for example, sees it as 'a system to provide a particular social *milieu*' this could not possibly match the actual complexity of our perceptions of the English pub. And even a set of models from a number of pure well-defined viewpoints (a notional landlord's view, or those of barmaids, licensing magistrates, police, brewers, Salvation Army officers, under-age would-be drinkers, etc. etc.) could not cope with our freedom to change our perception of a pub—or of any other bit of social reality—*arbitrarily*. Suppose that a systems model of a chemical system, say a model of the kinetics of a chemical reaction, does not match the (repeatable) measured kinetics: then the fault must lie with the model builder. But when a model of a human activity system does not match observed human activity the fault might be the model builder's but it might also be due to the autonomous real-world behaviour of human beings. We cannot expect a match between model and reality in the latter case both because of the multitude of autonomous perceptions and because those perceptions will continually change, perhaps erratically.

6. This means that in the case of human activity systems we need to be particularly aware that they are mental constructs, not would-be accounts of reality. Our purpose in building them cannot be to grope towards a systemic ontology. They are tools of an epistemological kind which can be used in a *process of exploration* within social reality.

7. If, in trying to find out about the world around us we use that part of the systems notation which deals with 'human activity systems' we must be especially careful not to talk glibly as if human activity systems exist in the world. Methodologically, we need to be especially aware of the line separating stages 1, 2, 5, 6, and 7 from stages 3 and 4. When using the methodology—in whatever sequence—we need to be aware of crossing that line, and of the different modes of working appropriate when we are above it and below it. Above it, we are operating in and on the world of reality. Below the line, we are constructing models for epistemological use above it.

The third most significant lesson learned from the whole experience of developing and using the methodology concerns the systems ideas underlying the formulation of root definitions and the building of conceptual models in stages 3 and 4. Although the systems literature, and especially that of General Systems Theory, offers a number of ideas relevant to thinking holistically, including the notions of requisite variety, equifinality, negentropy, homeostasis, etc., this work emphasized that, at least when using the concept 'human activity system', the core systems ideas are the two pairs of ideas discussed in Chapter 3: emergence and hierarchy, communication and control. If, in stages 3 and 4, on the principle of Ockham's Razor, we pare down the ideas used to the minimum, then the irreducible residue includes: the notion of whole entities which have properties *as* entities (emergent properties, described in the root definition); the idea that the entities are themselves parts of larger similar entities, while possibly containing smaller similar entities within themselves (hierarchy: systems, sub-systems and wider systems); the idea that such entities are characterized by processes which maintain the entity and its activity in being (control: a key idea in the formal systems model); and the idea that whatever other processes are necessary in the entity, there will certainly be processes in which information is communicated from one part to another, at the very minimum this being entailed in the idea 'control'.

These are the four ideas at the core of systems thinking about the real world. They are neutral ideas which do not themselves require any particular disposition of power within a model (although of course a particular root definition may be chosen deliberately to embody a specific power structure); nor do they, in themselves, imply any particular ideological view of the real world. This ought to be apparent, but it is a further piece of learning from the action research that many people automatically (and wrongly) associate 'hierarchy' with authoritarian management and 'control' with coercion. The user of the soft systems methodology will frequently find himself reminding others that he uses 'hierarchy' in the sense of levels of reality and meta-levels above those levels, and 'control' in the control engineers' sense of deciding which variables govern the behaviour of the whole. In any case, whatever is embodied in the systems thinking, stages 3 and 4 are not intended to convey normative statements about the real world, which ideologically will be in a state describable only as a complex changing mix of ideologies from totalitarian through liberal to anarchic.

External Comment on the Action Research

In approaching the question of what model of social reality is implicit in the methodology, it is useful both to place the work in relation to other similar work reported in the literature and to learn from published comments on the work itself.

Taking the latter first, Jones (1978) suggests that using the methodology is akin to constructing 'an *ad hoc* theory' about the problem situation which is derived 'neither from general theory nor from scientific testing, but from a pre-formed set of concepts developed in experience'. Assembling a theory from the conceptual language is thus a creative act and the methodological process is one of catalysing what he calls 'conditioned inspiration', in a process similar to that in which a scientific hypothesis emerges. Although I would argue that the methodology provides not only 'pre-formed concepts' but also some well-tested procedures for using them, this view of the methodology usefully underlines that the use of it, unlike the application of a technique, does not lead automatically to a solution. Selecting a number of 'relevant systems' in stage 3 gives an opportunity for personal insights to be brought to bear on the problem, and the best systems thinker is the person whose relevant systems turn out to be most relevant!

Jones remarks also that it is not the purpose of the methodology to construct models of real-world 'systems'. This is correct, but it is a point which many people find difficult to grasp. Prévost (1976) for example, has argued that the soft methodology is located in the structural–functional tradition of sociology, as represented by Parsons and Merton. The difference, he claims, is only one of scale, Parsons working towards a comprehensive theory of all social action, Merton developing 'theories of middle-range', the methodology constituting a tool to tackle particular real-world problems comprehensively. Having asserted this connection, he then levels against the methodology the usual charge made against structural functionalism, namely that it has a static and conservative bias, being unable to deal satisfactorily with conflict and change. Prévost's argument rests I believe, too much on the evidence of early systems studies in which a single root definition and conceptual model were used to furnish the debate about change. When this happens (and it frequently did happen in the earlier years of the action research programme) it is easy to imagine that both the root definition and the conceptual model which flows from it refer to some real-world system and are some kind of normative representation of it. Once it is grasped that the root definitions and conceptual models express pure perceptions having a single-mindedness rarely found in the real world (except among fanatics), it becomes clear that the methodology sets up a debate in which various alternative perceptions can be compared and contrasted with the real world. The root definitions most usefully express several conflicting *Weltanschauungen,* and the debate, far from embodying a static bias, is then intrinsically concerned with conflict and change. (This is in fact the case even when a single conceptual model is compared with 'what is'.) As Naughton (1979b) points out in answering Prévost's argument, a conceptual model in the methodology is 'a device for helping actors to explore aspects of the problem situation' rather than a description of part of the real world. 'Soft systems research', says Naughton 'is more accurately viewed as a kind of social "technology", interpreted as "the systematic application of scientific or other organised knowledge to practical tasks" . . . consequently Prévost is really not comparing like with like.' The proper comparison would be between the methodology and *applied* sociological research conducted within a functionalist framework.

Naughton does however conclude by conceding to Prévost that 'there is . . . a "whiff of functionalism" about the methodology even if it is difficult to pin down precisely'. What this useful exchange highlights, I think, is the fact that during the period in which the methodology was developed in use the appreciation grew steadily that the outcome in a problem situation was a process of learning rather than a specification for a system. Hard systems engineering methodology, from which the soft methodology was derived, makes a positivist assumption that there is a system in the real world whose objectives can be defined. This links it broadly with some aspects of Parsonian structural functionalism, in particular with the goal-seeking model of human behaviour. As soon as the earlier experiences in the action research revealed that the definition of 'the system' was never obvious, and that definition of system objectives was usually impossible or unhelpful, it was implicit in the work that the outcome would be exploration rather than engineering. But realization of the significance of this grew only gradually, and the "whiff of functionalism" derives from the early years of the action research programme, especially from those studies in which much effort went into a single root definition and conceptual model.

In addition to rehearsing the case against the methodology's being considered functionalist, Naughton has also provided an exegesis and commentary, written originally because of difficulties in teaching the methodology to systems students of the Open University (Naughton, 1977). In it a most useful distinction is made between 'Constitutive Rules' which 'must be obeyed if one is to be said to be carrying out a particular kind of inquiry at all', and 'Strategic Rules' which are more personal, which 'help one to select from among the basic moves . . . those which are "good" or "better" or "best" '. The rules of the two types, slightly modified from Naughton (1977) are summarized in Table 6. The distinction made is useful not only because it emphasizes that the methodology is a set of firm guidelines within which individuals may operate personal strategies, rather than a technique guaranteeing an answer, but also because it makes the user aware that if he chooses to break a constitutive rule he may lose some of the help which the methodology can provide when it is used *as a linked whole*, as an enquiring system.

Finally Bryer (1977), correctly noting the emphasis on the process of enquiry rather than on the substantive properties of the ultimate object of enquiry (i.e. social reality), suggests that this version of a systems approach is inspired by the hope that it will be possible 'to find objectivity in the properties of the process itself'. In fact the action research programme did not consciously seek for 'objectivity' in the process of the methodology.

Table 6   Constitutive and strategic rules of the 'soft' systems methodology (after Naughton, 1977)

Constitutive rules

o The complete methodology is a 7-stage process.
o Each stage from 2 to 6 has a defined output:
    Stage 2: rich picture; relevant systems
    Stage 3: root definitions evaluated by CATWOE criteria (Smyth and Checkland, 1976)†
    Stage 4: conceptual models of the systems described in the RDs built by assembling and structuring verbs†
    Stage 5: agenda of possible changes (derived from comparison of CMs with 'rich picture' expression of problem situation)
    Stage 6: changes judged with actors in the situation to be (systemically) desirable and (culturally) feasible
o Conceptual models should be checked against RDs and 'formal system' model
o Conceptual models should be derived logically from RDs *and from nothing else*
o Conceptual models are *not* descriptions of systems to be engineered (although stage 6 *may* yield a decision to engineer a system)

Strategic rules (some examples among many possibilities)

o Preliminary expression conducted by searching for elements of *structure* and *process* and examining the relation between the two
o Expression not conducted as a search for 'systems' in the problem situation
o Expression may be facilitated by asking 'resource allocation' questions:
    What resources are deployed in what operational processes . . . , etc.
    How is this monitored and controlled? (See Checkland, 1972)
o Problem themes—i.e. one- or two-sentence blunt statements—used to focus attention on interesting and/or problematic aspects of the situation
o Iterate, especially: relevant system → RD → CM → comparison → relevant system
o Set up stage 5 as a *debate* with important actors in the situation
                    . . . etc. . . . . etc.

†It is of course possible in principle that these ways of tackling stages 3 and 4 will later become strategic rules; they are constitutive at present, in the absence of alternatives which experience has shown to be valid.

The criterion by which the research was judged internally was its practical success as measured by the readiness of actors to acknowledge that learning had occurred, either explicitly or through implementation of changes. The development of the methodology was a practical,

254

self-constructing, closed system of the kind:

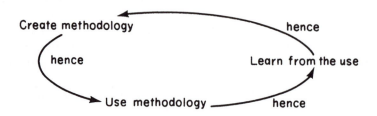

entry being forcibly effected by initially taking hard systems methodology as given. After many operations of this cycle, the methodology being 'successful' on the criterion named, we are now in a position to extract, if not totally objective, at least well-tested substantive knowledge about the area of application of the methodology—namely social reality. To do this it is necessary to adopt the weak sufficiency relationship that the success of the methodology implies at least a reasonable match between the model of social reality implicit in the methodology and the actual nature of social reality itself. Such a relationship cannot of course be proved (hence the adjective 'weak') but it does seem unlikely that a problem-solving methodology embodying a completely false model of social reality would be successful. Thus the final outcome of the research is substantive knowledge concerning the nature of what in everyday language are called 'social systems', this knowledge being guaranteed, to the extent that such knowledge can be guaranteed, by the methodology's success on the criterion defined above.

Related Work Elsewhere

Given completion of the work on developing the methodology, at least in its fundamentals, it is possible to compare and contrast it with other work which tackles the problem of bringing a scientific approach (in the broad sense) to the problems of the real world. Doing so has helped understanding of the work itself. It has been useful both to appreciate similarities between this and other work elsewhere, and to note differences between it and other work which initially might be taken to be similar in nature. The two contributions to which the action research programme can be most closely related are both essentially theoretical, although both embody reflections on real-world problem-solving. They are Churchman's (1971) work on 'inquiring systems' and Vickers's (1965 *et seq.*) work on 'appreciative systems'. These are discussed later in this section; but first: what the work is *not*, despite superficial resemblances.

Clearly, given the work's initial discovery that hard systems engineering methodology is not applicable to soft problems, it is not systems engineering. Neither is it RAND Corporation systems analysis; indeed, it is fundamentally distinct from any version of systematic thinking which deals in real-world systems and their creation, or with models of parts of the real world and their optimization (i.e. with areas 4.1 and 4.2 of Figure 2). This is not to say that it bears no resemblance at all to these activities. Because the methodology orchestrates a process of learning it could in fact be applied *to* these other activities: a systemic RAND-style 'systems analysis' could in principle be built upon the present systematic foundations—a possibility which helps to illustrate the differences between the present work and existing 'systems analysis'.

Another contribution which resembles soft systems methodology, but also turns out to be distinctly different from it in important respects is the development at the Wharton School of the University of Pennsylvannia of what is described as 'complex planning' or 'participative planning'. This is a planning methodology which appreciates that an analytical reductionist approach to social problems is inappropriate in an age in which problems are 'systems of problems', or 'messes' (Ackoff, 1974). Much of the basis of that work matches what emerged during the development of soft systems methodology: for example the idea that teleological modes of thought are useful; the idea that problem situations rather than problems must be the object of concern; and that 'problems' and 'solutions' are in constant flux, so that 'problems do not stay solved'. But in spite of these similarities the 'complex planning' paradigm differs from that of soft systems research in making central use of the goal-seeking model of human behaviour (just as does hard systems methodology) something my colleagues and I found distinctly unhelpful in our work.

Ozbekhan (1976, 1977) has provided an example of 'complex planning' in two presentations made in London in late 1976. They concern a study of the future of Paris, its possible roles and functions during the next thirty years, carried out for the French Government. The planning methodology which the study illustrates requires the activity of planning to be participative, co-ordinated (as between different aspects of the *problématique*), integrated (between different organizational levels) and continuous (Ackoff, 1974). Four phases of activity are identified by Ozbekhan (1977). In the first the problem situation is structured into 'some kind of model capable of suggesting causal linkages'. In the second or normative phase, 'it becomes necessary to conern ourself with the *design* of those future states of the system which are deemed to be good, hence "desirable" '. In the third, the means needed to pursue the objectives are identified, and in the fourth an 'organisational and implementation plan' is defined. More bluntly, Ackoff describes five interacting phases as:

Ends planning (determining what is wanted . . . short-run, intermediate, and ultimate ends).

Means planning (determining how to get there . . .).
Resource planning (determining what types of resources and how much . . .).
Organizational planning (determining organizational requirements and designing organizational arrangements and the management system . . .).
Implementation and control planning (determining how to implement decisions and control them . . .).

Now this is very much the language of hard systems thinking as discussed in Chapter 5: the implication is that there are systems to be engineered and the way to do this is by defining system objectives. But the context in which 'complex planning' is advocated is explicitly one of soft ill-structured problem situations in which the planning process is more important than any plan and in which 'problems do not stay solved'. 'Complex planning' argues that the needs of such a situation can be met by designing 'an idealised future for the system being planned for'. Ackoff adds:

> Once an idealised design has been prepared on which consensus has been obtained it is possible to begin planning the approach to that ideal. The output of such planning† should be treated as tentative, subject to continuous revision in the light of experience with it. The system for making such revisions . . . must itself be planned.

This is somewhat different from our experience in a hundred admittedly rather small projects. In our modest studies we have done nothing as grand as concluding that Paris should cease to be capital of France, becoming instead a 'global city' (for that was the ideal future defined in Ozbekhan's study). We have rarely found it possible either to define ideal futures or to obtain consensus on them. Although always ready to see a soft study sharpen into the simpler special case of an agreed need for an efficient means of reaching a defined end, we rarely found ourselves with that option. We were driven to conclude that there is a large class of soft problems for which the language of 'ends' and 'means' is inappropriate, problems concerned not with achieving goals but with managing on-going relationships through time.

Closer to soft systems methodology than is 'complex planning' but also importantly different in some respects is the socio-technical systems approach associated with researchers of the Tavistock Institute (Emery and Trist, 1960; Trist et al., 1963). This rests on two ideas: that organizations may be regarded as open systems, and that in such a system the technology, the sentiments of members as a social group, and the organization's environment are all interdependent; none is of prime

---

†That is, as I read it, steps in the approach to the ideal, not the ideal end itself.

importance over the others. The concept is that in an organization groups of people organize to carry out a *primary task* (i.e. a task whose absence would indicate that the organization was not functioning at all, such as, for the Post Office, 'collecting and delivering mail') but that in working to improve organizational performance the object of concern has to be not simply the primary task but the system as a whole, including interactions with the environment and social factors. Clearly there are some overlaps here with soft systems methodology, especially for the sub-set of problem situations in which the concern is overall organizational performance. Expressing it in the language of this book, the open socio-technical systems approach makes a statement of the following kind: 'If the problem is one of organisation performance then the relevant system will be the organisation itself; what is more there is one particular root definition and conceptual model which will be appropriate, one linking technology and social relationships in an entity involved in exchanges with an environment; comparison of that model with existing arrangements will enable useful changes to be defined.' This interpretation is confirmed in a recent admirably clear account of three projects carried out within this framework (Warmington *et al.*, 1977). It is accepted there that 'a socio-technical model of a particular situation usually deals with a large number of specific variables (manning levels, rates of absence . . . output per man . . . market trends . . .)' but the classes of variables are said to provide a general model of any socio-technical system. The nine groups of variables are those concerned with: product, resource, and labour markets; technical design; mediating mechanisms; attitudes; unofficial manipulatory devices; behaviour affecting performance; and cost and technical performance. As long as a problem may be sharply defined as one of improving organizational performance such a check-list, together with the interactions between the classes of variables, will obviously be helpful, always as long as there is a readiness to allow experience to change the model. My colleague Dr. B. Wilson has done somewhat similar work using the soft methodology in the problem of making a rational choice of organization structure. In a sequence of projects he has taken the 'relevant system' to be a notional system to carry out the organization's primary task. A root definition of that system then leads to a conceptual model which forms the rational basis of an organization structure (Wilson, 1979). But here as in the rest of our work it is the methodology which is generalized rather than the content of models. There is nothing in principle wrong with general models, at least in certain situations of a well-defined kind, but part of the strength of systems thinking lies in the analyst's power to select a viewpoint he considers relevant and to denote systems whose boundaries do not coincide with organizational boundaries, the latter being in the final analysis arbitrary. This is a strength not to be given up lightly, this ability to make models specific to an individual situation not only in terms of the specific values of the variables but also in terms of the classes of variables included. In our programme of action research the value of the open socio-technical systems model is greatest in the (weak) validation of conceptual models in stage 4b of the methodology, as discussed in Chapter 6

In spite of these differences, the open socio-technical systems approach as used by Warmington, Lupton and Gribbin is closer to soft systems methodology than are hard systems engineering or 'complex planning'. Warmington *et al.* regard their approach as seeking 'description, analysis and understanding' rather than the efficient achievement of a goal. They say specifically that

> we have neglected the concept of organisational goals, considering that its usefulness for us and its relevance to any criteria of organisational performance we need, are very limited.

Their core paradigm is thus one of *learning* rather than *optimization* and that is where lies the resemblance to soft systems methodology.

It is a core concern with learning rather than optimization which also makes the link between systems methodology and the two main theoretical contributions mentioned above: Churchman on enquiring systems and Vickers on appreciative systems.

For some years now Churchman (1968a, 1968b, 1971, 1974) has been assiduously exploring the philosophical foundations of a systems approach. The work itself contains certain tensions which could, I believe, be resolved by making a clear distinction between hard systems thinking (the optimization paradigm) and soft systems thinking (the learning paradigm). Thus the overall conclusion of one discussion of the systems approach (1968a) is that '*the* systems approach really consists of a continuing debate between various attitudes of mind with respect to society'; and yet most of Churchman's writings are based implicitly on hard systems thinking, as when he writes, for example: '[the systems approach] views a problem in terms of objectives (goals, ends, purposes). But, more important, the systems approach unifies all the variety of goals into one "measure of performance" ' (Churchman *et al.*, 1975). Another tension arises in a rich discussion of the problem that whatever you take to be 'the system' will itself be a part of wider systems which, since they will affect 'the system', ought also to be taken into account: the attempt to be holistic appears to be never-ending (Churchman, 1968b). This dilemma of idealism probably reflects Churchman's background in management science rather than in the mainstream of the systems movement. Systems thinking in the tradition stemming from organismic biology (see Chapter 3) regards its holism as manifest in thinking in terms of 'wholes' rather than '*the* whole'. But these internal tensions in Churchman's work are always fruitful tensions, and his writings provide a perspicacious discussion of systems thinking.

Most relevant to soft systems methodology is his discussion of the different ways in which we may design systems for finding things out, *The Design of Inquiring Systems* (Churchman, 1971), a difficult but rewarding book.

Churchman considers that 'the most important intellectual activity' is 'the formulation of social problems'. He takes as his problem the question as to whether man can be bettered by his own designs, and examines the history of man as a designer of systems which produce knowledge—i.e. enquiring systems—over the period from the renaissance of epistemology in the 17th century, 'when everything was so open to speculation and imagination' until the present day. The book's method is to examine the work of five historical figures—Leibniz, Locke, Kant, Hegel, and the American philosopher E. A. Singer, taking them to be designers of systems to produce sure knowledge. The examination of the work of the last four philosophers is done using a basic model of any enquiring system which itself stems from Leibniz. The Leibnizian enquirer examines sentences and determines which ones are neither tautologous nor self-contradictory but are what Leibniz calls 'contingent', that is to say, which imply or are implied by earlier sentences now in the system's memory. Such contingent sentences are candidates for 'likely truths', and such truths belong to fact nets: they imply other contingent truths and are implied by them. As the system continues to enquire, the fact nets grow; in modern language the enquirer is a model builder, embryonic models evolving to full-fledged formal models.

This is Churchman's basic account of any enquiring system. In a Lockean version of the system, the nets of contingent truths will grow according to the extent to which there is consensus among the community of enquirers: a selection among alternative fact nets will be made according to the extent that the community can agree with each other about their sensory impressions. The natural sciences provide many examples of such Lockean enquiring systems, which can be highly successful while operating under the umbrella of a general theory which the whole community accepts. But Kant raises a subtle problem. Does not the very existence of a system capable of receiving data imply that the enquiring system has built into it certain *a priori* sciences? Kant himself believed that geometry, arithmetic, and kinematics were 'givens' in this sense. The problem arises as to how the enquiring system could validate by its experience the assertions of these *a priori* sciences themselves. If, for example, the Lockean consensus is based upon an unquestioned geometrical image of the world as consisting of objects in space, how could the common-sense model (incidentally shown by Einsteinian physics in this century to be inadequate) itself ever be validated? It looks as though a Leibnizian enquirer could create fact nets of almost any kind, each telling a particular 'story of the world', or *Weltanschauung*. And here Hegel in the 19th century, building on Kant's philosophy, offers a strategy for the operation of enquiring systems. If we accept that a complete picture of an enquiring system has to include its designer as well as the operation of the system itself, then the appropriate strategy becomes clear. We can examine the content of the enquiring system—its fact nets—from the point of view of asking: What must the designer's *Weltanschauung* be to make that particular fact net meaningful? And as soon as that image of the world is available, its counter-image, what Churchman calls 'its deadliest enemy', can also be constructed. At this stage an observer of the conflict between these *Weltanschauungen* can 'build a new world-view in which the nature of the conflict is understandable . . . The conflict is devoured in the higher level *Weltanschauung*' and 'the very act of creating such a world-view in which the observer can observe the conflict also creates a strong conviction about its truth'. This is Churchman's exposition of Hegel's dialectic. For Hegel the dialectic of thesis-antithesis-high level synthesis was part of the progressive creation by this process of an Absolute Mind. Churchman tackles the question of whether or not the operation of an enquiring system is in some sense progressive, rather than simply being the operation of a process, through the work of E. A. Singer. A Singerian enquiring system is interested both in defending and destroying the consensus of a Lockean community; it accepts that enquiry is a continuing process without end and emphasizes the need for a particular *mood* on the part of the enquirer, an 'heroic mood' in which two opposing processes are at work:

> One is the process of defending the status quo, the existing 'paradigm' of inquiry, with its established methods, data and theory. The other is the process of attacking the status quo, proposing radical but forceful paradigms, questioning the quality of the status quo.

This is a sparse account of a long discursive argument, but sufficient to indicate Churchman's categories of enquiring or 'information' systems. The complexity of his argument derives to an extent from its being based on the idea of design. This is essentially a 'hard' concept: any design is necessarily based upon a prejudgement that some particular purpose is accepted as desirable; the notion of design is inseparable from the notion of a taken-as-given purpose which the design will fulfil. This concept is not a good model for the constantly changing ambiguities of soft problem situations, and it is inevitable that it should be a rather tortuous path which leads to Singerian enquiring systems. These are far from being systems which design a means to an end; they are concerned with the choice of ends. The line of argument Churchman follows can in fact be regarded as an equivalent in the world of theory to the path followed in practice in the action research programme I have described. Both start from the simple certainties of hard situations and find their way to a method of trying to deal with the complexities of soft problem situations.

The work on the development of the soft systems methodology was not, during its course, based on any explicit theory. As described in previous chapters, it took hard systems engineering methodology as given, applied it to unsuitably soft problems and introduced modifications dictated by the problem situations themselves. Subsequent to the work described in Chapters 6 and 7, examining it to ascertain what theory was implied by it, I noticed that to a remarkable degree it could be seen as an operational version of Churchman's theory of enquiring systems. Within the methodology, the debate about feasible and desirable changes in stages 5 and 6 seeks a Lockean consensus of concerned actors, those who own the problem situation. This debate is fed and structured by systems models which represent constructed images and counter-images; the structure of this debate derives from rival conceptual models based on opposed root definitions. It is a Kantian/Hegelian structure. And overall the methodology's acceptance that no soft systems study is ever finally complete, stage 7 being a redefinition of the problem situation, shows the methodology as a whole to be a Singerian enquirer. Certainly those who have used the methodology have no doubt that Singer's 'heroic mood' is the appropriate emotional state for the investigator.

The other body of work which, after the event, was seen to relate closely to the experience of developing soft systems methodology was Sir Geoffrey Vickers' work on 'appreciative systems' and the social process he calls 'appreciation' (Vickers 1965, 1968, 1970, 1973). Vickers has had several professional careers which he describes as 'practising the law and helping to administer public and private affairs'. He was a solicitor in the City of London, Director of Economic Intelligence at the Ministry of Economic Affairs during the Second World War and then legal adviser to, and a member of, the National Coal Board in charge of manpower, training, education, welfare and health; in addition he has been a member of such bodies as the Medical Research Council and the London Passenger Transport Board. Subsequent to these experiences he has reflected upon them and sought to understand the familiar but mysterious processes by which policy is continuously decided, executed and changed, seeking to understand it 'both as a mental activity and as a social process'.

The words 'policy making' are used in different ways by different authors; sometimes they describe the setting of objectives, sometimes the defining of priorities, sometimes the act of making a plan or the plan itself. For Vickers (1965) the words fundamentally entail the setting of 'governing relations or norms, rather than ... the setting of goals, objectives or ends'. This quotation summarizes the main thrust of the theory Vickers has developed to describe and explain the process which characterizes social systems. Through all his books he has argued steadily that the goal-seeking paradigm, while adequate to explain the behaviour of rats in mazes, is totally inadequate to explain what goes on in the Cabinet, in board rooms, in Trade Union branch meetings, on committees, and in our everyday life. There, the bulk of our activity is concerned with establishing and modifying relationships through time, rather than seeking an endless series of 'goals', each of which disappears on attainment. Vickers's writings against goal-seeking as an adequate model of human behaviour culminate in a bravura passage in *Freedom in a Rocking Boat*:

> The meaning of stability is likely to remain obscured in Western cultures until they rediscover the fact that life consists in experiencing relations, rather than in seeking goals or 'ends'. The intrinsic confusion about means and ends arises from the fact that no end can ever be more than a means, if an end is equated with a goal. To get the job or marry the girl is indifferently an end, a means and a goal; it is an opportunity for a new relationship. But the object of the exercise is to do the job and live with the girl; to sustain through time a relationship which needs no further justification but is, or is expected to be satisfying in itself. The barren self-contradiction of life, where this truth is overlooked seems to me to be well, though unconsciously, expressed in lines by Louis Untermeyer which have, significantly, become a favourite quotation in North America:
>
> > From compromise and things half done
> > keep me with stern and stubborn pride,
> > and when at last the fight is won,
> > God keep me still dissatisfied.
>
> Whilst there is some conceivable context in which almost any utterance makes sense, this seems to me to epitomise one of the most barren and self-defeating creeds ever conceived by man. In identifying compromise with things half done, in equating doing with fighting and in making a virtue of chasing what cannot satisfy, it expresses in twenty-six words nearly everything that is wrong with Western culture.

This remarkable passage brings in some general cultural judgements; ignoring those and returning to the bones of Vickers's theory: having rejected the goal-seeking model of human behaviour Vickers (1974) also rejects as inadequate the cybernetic paradigm. There, the course to be steered is available from outside the system, whereas systems of human activity themselves generate and regulate 'multiple and mutually inconsistent courses'. Vickers suggests replacing the goal-seeking and goal-seeking-with-feedback (cybernetic) models by one in which personal, institutional or cultural activity consists in maintaining desired relationships and eluding undesired ones. The process is a cyclic one which operates like this: our previous experiences have created for us certain 'standards' or 'norms', usually tacit (and also, at a more general level, 'values', more general concepts of what is humanly good and bad); the standards, norms and/or values lead to readinesses to notice only certain features of our situations, they determine what 'facts' are relevant; the facts noticed are evaluated against the norms, a process which both leads to our taking regulatory action and modifies the norms or standards, so that future experiences will be evaluated differently. Thus Vickers (1970) argues that our human experience develops within us 'readinesses to notice particular aspects of our situation, to discriminate them in particular ways and to measure them against particular standards of comparison, which have been built up in similar ways'. These readinesses are organized into 'an appreciative system' which creates for all of us, individually and socially, our appreciated world. The appreciative settings condition new experience, but are modified by the new experience; such circular relations Vickers takes to be the common facts of social life, but we fail to see this clearly, he argues, because of the concentration in our science-based culture on linear causal chains and on the notion of goal-seeking.

Probably the most spectacular recent example of change at a societal level which can be expressed in the concept of shifting appreciative settings is the change in attitudes towards the environment and its conservation. The 'facts' about industrial depredation were available for many decades, but the readiness to notice them and evaluate them in a particular way seems to have been triggered by the publication in the early 1960s of Rachel Carson's book about the over-use of pesticides, *Silent Spring*. Thereafter the standard of what was acceptable changed sharply as new considerations became regarded facts and new norms emerged. The rise of the conservation movement followed, and no industry or government can now ignore in its activity or legislation the care and maintenance of our small planet; the settings of the appreciative system have changed.

That Vickers's ideas and soft systems methodology are closely related to each other is, I trust, fairly clear. Since the state of an appreciative system is a function of its own history, appreciative systems are learning systems, and the basic social process, in Vickers's perspective, is a process of learning. Now, in stages 5 and 6 of the systems methodology, conceptual models are compared with 'what is' in the real world as a means of structuring a debate about feasible and desirable changes. This debate will reveal what facts are noticed by the owners of the problem situation, what different aspects are regarded as important, and what conflicting norms and standards underly actors' interpretations of the observed happenings. In other words, the root definitions and conceptual models provide a vehicle for an explicit examination of the appreciative settings in the situation studied and how those settings are changing, and might change in the future. In the Cordia Engineering study described in Chapter 6, for example, the Company activity was examined by using a root definition and conceptual model of a system to carry out a major engineering *task*, together with a definition and model of a system consisting of an aggregate of specialist *functions*; both models were compared and contrasted with the Company's actual organization and procedures. The debate which ensued revealed an unwillingness to take seriously the project organization. In the

Company's appreciative system the prime value was technical excellence in a specialist function. Promotion to the most senior ranks was normally achieved as a result of demonstrated competence in a specialist area—electrical engineering, product design, etc.—and as a result there was a considerable undervaluing of such project arrangements as did exist. The senior managers' previous experience had given them no readiness to notice, or value highly, skill in something as nebulous (to them) as project management. The appreciative settings in this instance were so fixed, in fact, that no implemented change was achieved during the course of the five-month study. Change came only subsequently with the appointment of a new Managing Director with significant experience outside the Company. In another very recent study the concern was the possible organizational role of a professional information service in a large company. Half a dozen root definitions and models embodied theoretically possible conceptions, from a service system whose skill lay in rapid response to requests, to a proactive system which decided for itself what were business information needs and supplied them to appropriate people. Comparison between the range of concepts and the traditional role of the group, together with discussion of resource and training needs implied by the six root definitions and conceptual models, changed the appreciative system of the information professionals. They then made a presentation to managers whose departments used their library and information service. The discussion following the presentation was the epitome of the social process of 'appreciation' with managers and information scientists exploring a modified relationship with consequentially changed norms; one Research Manager said that he had come along expecting to hear about a service he already knew well, but instead had acquired a new perception of the information group in relation to his own department.

These are simply two examples illustrating the general point that stages 5 and 6 represent a formal articulation of the dynamics of appreciative systems. In a sentence, the work of Churchman and Vickers can be related to soft systems methodology in the following way: the methodology is a Singerian enquiring system whose mode of operation provides a formal means of initiating and consciously reflecting upon the social process of 'appreciation'.

The Nature of Social Reality

These reflections on the action research and on the relation between it and other work, especially that of Vickers and Churchman, serve the essentially practical purpose of enabling us to use the systems methodology with greater finesse and skill. The assumption is that the more we understand it the better shall we be able to use it.

In principle, however, we can also get from the work an outcome of a theoretical kind. We can ask: What model of social reality is implied by the methodology? This is a question worth trying to answer. The methodology's success, as measured not only by various practical outcomes but also by the readiness of people in social situations to agree that insights have been gained by its use, does at least suggest that the model of social reality implicit in the methodology is a well-tested one. To argue this is to contribute to the perennial debate about the nature of social science and its object of concern: social reality. This section, then, is a brief attempt to relate the experience of the systems research to the context of social science, something the systems movement as a whole has been markedly reluctant to do. Bryer and Kistruck (1976) claim not to have found a single attempt by a systems theorist to justify his approach in sociological terms. What follows is a preliminary sketch for such a justification, an attempt *after the event* to relate the research experience to the main strands of thinking in the social sciences.

Were the social sciences able to point to a significant body of empirically-derived publicly-repeatable results, then the social science literature would be very different from its actual state. It would contain substantive accounts of the laws governing social interactions, both individual and institutional, rather than offering, as it does, a plethora of discussions of the nature of social theory and of the relation between it and philosophy. But the fact that the literature is as it is does not necessarily establish the waywardness of social scientists. Rather it reflects the peculiar difficulties faced by a science which cannot assume that repeatable happenings characteristic of an external reality can be discovered by disciplined observation. The difficulty was discussed briefly in Chapter 3, where the problems faced by the method of (natural) science when it is faced with great complexity were reviewed. There it was argued: (*a*) that social phenomena are extremely complex in terms of both the number of relevant variables and the variegated nature of data; and (*b*) that special complexity is introduced into scientific investigation which tries to examine the behaviour of a part of the natural world—the human being—which can attribute meaning to his own and others' actions and can, if so inclined, act wilfully to falsify predictions made on the basis of previous observations. A would-be sociologist has first to discuss, as a budding natural scientist does not, whether or not there exists an observable entity which is to be the focus of his attention. And because the object of concern, social reality, is problematic, so too is the method by which it might be investigated scientifically. These problems occupy a significant place in the literature of sociology in particular.

The boldest possible position on the relation between social and natural science would be to assert that they are exactly similar enterprises. This is the position of Popper (1972, 1976), who has been so influential in formulating the logic of the method of natural science. Unfortunately he has devoted little attention to the particular problems of sociology, or any other social science, and he has not been a main protaganist in the debate about social theory; nevertheless his extreme position helps to indicate the general shape of that debate.

Popper assumes that facts can be gathered in the social sciences in much the same way that they are gathered in natural science (Ackermann, 1976). He does not deal convincingly with the problem that human beings can react to external predictions of their behaviour, and the examples he offers of sociological laws (for example: 'You cannot make a revolution without causing a reaction') lack empirical significance. In a paper given at a meeting of the German Sociological Association at Tübingen in 1960, a famous occasion which led to the exchanges known as 'the positivist dispute in German sociology' (Adorno, 1976), Popper begins by summarizing his view on the proper method of science and asserts in the sixth thesis of his paper:

> The method of the social sciences, like that of the natural sciences, consists in trying out tentative solutions to certain problems: the problems from which our investigations start . . .

He concedes (seventh thesis) that 'objectivity in the social sciences is much more difficult to achieve (if it can be achieved at all) than in the natural sciences' but does not concede that the task of getting agreement on what constitutes the problem, and hence on what constitutes a refutation of a hypothesis is different in principle in the social sciences. His concession to the complexity of social science is to develop the concept of *the logic of a situation*. In a note in the English edition of Adorno's book Popper suggests that in social science an explanation will usually consist of a model of a situation and a 'rationality principle' which defines action rational in that situation. This is a considerable concession on Popper's part, since such explanations will be operational only if human beings act rationally. Thesis twenty-five spells out what Popper terms this 'purely objective method in the social sciences':

> [the] method consists in analysing the social *situation* of acting men sufficiently to explain the action with the help of the situation. . . . Objective understanding consists in realising that the action was objectively *appropriate to the situation*.

Later he speaks of the method as providing 'rational, theoretical reconstructions' which can only in the strictly logical sense be 'good approximations to the truth'.

It is interesting to find Popper starting from the position that natural and social science are similar, and then arguing himself to a position in which sociology emerges as an account of what action would be rational in a given situation, rather than an account of the actual acts of real men. For Popper, the situation itself will be the source of a version of what is rational, this being a kind of social fact independent of any actual human agent. This is at the other pole from the position which argues that the social sciences are in fact intrinsically different from the natural sciences precisely because what is rational for one human being in a situation may be irrational for another who attributes different meanings to what he observes in the situation. Winch (1974), for example, argues against Popper that the method of conjecture and refutation cannot simply be transferred from the natural to the social sciences because refutation requires both repeatable experiments *and* agreement on what hypotheses are in question and what constitutes falsification; this agreement may be permanently unavailable in the case of social phenomena since actors and observers are both free to attribute their own meaning to what is observed. The trial-and-error concept of the method of science, highly satisfactory where physical phenomena are concerned, is problematic when one man recognizes 'error' where another sees 'success'—as is frequently the case, for example, in questions of social policy.

This brief discussion of Popper's views on social science illustrates the general nature of the debate about social theory. Is it the study of objective social facts which transcend the individuals who make up a society; or is it the study of the individual subjective understandings which men acquire of their social situations? Both strands of thinking are heavily represented in the literature of social science, either in the form expressed above or as a dichotomy between two views of social reality: the tradition of *functionalism* sees social reality as consisting of social structures which transcend the individuals who constitute them, while the so-called *action approach* gives primacy to the individual actors who pursue their own ends and in so doing create social reality as a process (Cohen, 1968). The literature itself may be seen as a debate conducted from these two stances; at the level of philosophy they are the stances of *positivism* and *phenomenology*.

It is not necessary or possible to give here an account which does justice to the richness of this debate; neither am I competent to do so. The intention here is to provide an account just sufficient to enable the theory of social reality embodied in soft systems methodology to be expressed in sociological terms.

Social theory is as old as man, in the sense that all societies have myths which provide an account of their origins and present state, but modern social theory emerges in the 18th and 19th centuries. Auguste Comte (born 1798) envisaged the new science he christened *sociology* as the apex of the hierarchy of sciences; but the idea that there could be such a science of man similar to the sciences of the natural world stems from 18th century writers such as Montesquieu. In his writings on law as 'the necessary relations which derive from the nature of things' Montesquieu postulated that some simple principles underlay the diversity of forms of government and social customs. He makes social facts a proper object of scientific study and lays down that the correct method in such study involves comparison and classification, leading to the formulation of generalized 'ideal types' which can be used as an analytical tool, a method later developed by both Durkheim and Weber (Bronowski and Mazlish, 1960).

The first major figure within the new science, important both for empirical studies and for methodological developments, is Emile Durkheim, whom we may regard as the central founding figure in the tradition which takes sociology to be concerned with the objective study of social facts. (For a brief introduction to Durkheim's thought see: Durkheim, 1895; Aron, 1968; Fletcher, 1971b, Giddens, 1978; Tiryakian, 1979). Durkheim sought to establish the independence of sociology from psychology and to eschew explanations of social phenomena in terms of the psychological states of individuals. His claim for the autonomy of sociology as a science rested upon the existence of social facts as emergent properties of social groups as a whole, properties which transcend the individuals who constitute the group—a famous Durkheimian example of a social fact being the suicide rate characteristic of a society. Durkheim argued that social facts were the subject matter of sociology, set out rules for their scientific study and illustrated the method in his own empirical work. His book *The Rules of Sociological Method* is a short bold manifesto for such work. After a discussion of 'What is a social fact?' the second chapter opens with the assertion: 'The first and most fundamental rule is: *Consider social facts as things*'. Following the controversy which greeted the book, Durkheim added a preface to the second edition in which he explains that by 'thing' he means 'all objects of knowledge that cannot be conceived by purely mental activity, those that require for their conception data from outside the mind . . .'. Social facts were manifestations of society as a whole which were external to individuals and exercised constraints on their behaviour. The sociologist should discover the law-like patterns in which such constraints operate and analyse at a level beneath that of the surface manifestations by 'substituting a limited number of types for the multiplicity of individuals'. Sociological *explanation* is then either causal (what is the succession of social phenomena?) or functional, the latter being an account of how a particular social fact or characteristic meets a societal need, a need of the social collectivity rather than those of individuals. In the *Rules* Durkheim writes:

> When, then, the explanation of a social phenomenon is undertaken, we must seek separately the efficient cause which produces it and the function it fulfils. We use the word 'function' in preference to 'end' or 'purpose' precisely because social phenomena do not generally exist for the useful results they produce. We must determine whether there is a correspondence between the fact under consideration and the general needs of the social organism . . .

This is virtually a defining statement of the position now described as 'structural functionalism' or, more generally, 'functionalism'. This stance, with its view of a social system as a set of.relationships persisting through time as a result of functional sub-systems which contribute to the equilibrium-maintaining processes of the system as a whole, is frequently taken to be the paradigm example of the application of systems thinking within social science. This is a limited and false view, I believe, and it will be discussed further below. But first it is useful to contrast the Durkheimian tradition of sociology with the counter tradition which sees the subject as being concerned not with the objective study of social facts but with the actor's subjective understanding of social action. This is the tradition associated with the name of Max Weber.

Weber (1864–1920) was a contemporary of Durkheim (1858–1917) but although his study contained a complete set of the influential review *L'Année Sociologique* which Durkheim helped to found (Aron, 1968) his writings make no mention of Durkheim or his work. This is surprising not only because Durkheim represents the approach which is counter to Weber's own (and hence, one would think, worth attacking) but also because they share some methodological principles, notably a belief in the need for 'ideal types' for analytical purposes. The differences, however, outweigh the similarities. Durkheim's search for 'social facts' at the level of the group rather than the individual inevitably leads him to regard as material things concepts (such as that of 'the group') which are in fact abstractions created by the observer. Weber is much opposed to such reification. For him the basic concept in sociology is the single deliberate action by an individual directed to affecting the behaviour of others. Sociology's concern is the scientific understanding of the subjective meaning associated with such action. (For a brief introduction to Weber's thought see the selections from Weber's voluminous works edited by Runciman, 1978; and by Shils and Finch, 1949; also Fletcher, 1971b; MacRae, 1974; Runciman, 1972; Freund, 1979).

Max Weber was a jurist and historian at a time when sociology was only beginning to be institutionalized as an academic discipline. He taught law and then held a chair of political economy; after the First World War he held a specially created chair of sociology at Vienna. He published extensively, his work including major historical studies as well as scattered writings on the methodology appropriate for the new science. His written style has the obscurity typical of 19th century German scholarship, being characterized by what MacRae calls 'a liquid and evasive richness'; Weber himself is said to have remarked that he didn't see why it should be easier for a reader to read his books than it was for him to write them! But in spite of presenting such difficulties to his readers, he is accepted as a (perhaps the) major figure in the emergence of sociology, the person whose success and range

> qualify him better than any sociologist, anthropologist, or historian of this century to select and define the controversies which require to be

resolved if we are ever to have an account of the social sciences which is at the same time philosophically defensible and also consistent with the actual achievements (few though these may yet be) of those engaged in them (Runciman 1972).

For Weber the fact that human interactions were affected by something missing from the non-human world, namely *meaning*, meant that the social sciences were fundamentally different from natural science. The core concept for sociology was the intentional action, the action which has meaning for the individual who acts. Society was the result of the totality of such actions, and the task of sociology was to analyse and explain social action by the study of subjective meanings by which individuals direct their conduct.

It might be thought that on this view a true social science, as objective as natural science, would not be possible. This however was not Weber's view. His aim was not to abandon positivist science, causal explanation or objectivity. His aim was to create an interpretive social science based upon analysis of meaningful action, action interpreted by the observer in terms of a means–end scheme of rationality. This would be possible because the actions of individuals are not random but fall into structures or patterns. They would be accessible by the method of *Verstehen*, placing oneself in the role of the individuals observed, and the interpretation would be made using generalizations about typical pure processes of action.

This line of thought led Weber to the creation of a number of methodological tools, notably a typology of rational action and the 'ideal type'. Rational action is the model for all meaningful action, and Weber distinguishes four types of social action. Affective action stems from emotion; traditional social action is 'determined by ingrained habituation'; a third kind of action is determined by an unquestioned acceptance of some value, ethical, religious or aesthetic, while *zweckrational* action is rational throughout, as regards both ends and means. These four types of social action are themselves pure *ideal types* rather than accounts of real-world action. Weber discusses the ideal type in an essay on '*Objectivity' in Social Science and Social Policy*, written in 1904 and included in the Shils and Finch (1949) selection. An ideal type, whether it is a type of action or an institutionalized pattern of action (as in Weber's account of bureaucracy, for example) is not a description of reality but is arrived at by the analytical accentuation of certain elements of reality:

An ideal type is formed by the one-sided accentuation of one or more points of view and by the synthesis of a great many diffuse, discrete more or less present and occasionally absent concrete individual phenomena, which are arranged according to those one-sidedly emphasized viewpoints into a unified analytical construct.

Such a construct is *ideal* only in the strictly logical sense of the term, and the points of view from which they can become significant 'are very diverse'. Weber makes the point many times that such constructs are not normative:

> An ideal type in our sense, to repeat once more, has no connection at all with value-judgements, and it has nothing to do with any type of perfection other than a purely *logical* one. There are ideal types of brothels as well as of religions.

The purpose of this methodological framework was to establish a procedure for social science which would ensure that propositions could be formulated which were capable of empirical test by observation of the actions of individuals.

Weber's substantive studies of the economic order and the sociology of religion are not relevant to my concern here. But it is obvious that the sociological methodology he developed for such studies has interesting parallels with soft systems methodology: root definitions and conceptual models precisely meet Weber's definition of the ideal type, although the way they are used is not quite a Weberian analysis. This point will be taken up later.

The distinction established above between social science either as the objective study of social facts (the 'Durkheimian' tradition) or as the subjective understanding of social action (the 'Weberian' tradition) provides a basic pattern which is helpful in placing and understanding more recent developments. For my purpose, which is to give an account in sociological terms of the social theory implicit in soft systems methodology, it is only necessary briefly to trace further developments within the two traditions. Doing this, we find that the application of systems thinking in sociology is always described (under the name 'functionalism') as a contribution within the Durkheim tradition. I shall go on to argue, however, that the work described in this book is not another contribution of this kind; instead, and perhaps surprisingly, it establishes sytems ideas within the processes evolved within the other tradition.

Functionalism, sometimes 'structural functionalism', has been a major concern in social theory in the last fifty years: 'the dominant framework for academic sociology in the 20th century' (Burrell and Morgan, 1979); it is the concept which prompts Lazarsfeld (1970) in a good short survey, to write: 'Sociologists can neither live with nor without functionalism'. The basic idea is very simple, and is systemic. It is the idea that social entities contain units which are differentiated but also interdependent; in other words the differentiated parts contribute functionally to the social entity as a whole, which is a *system*. Functional analysis then requires an account of the parts and the way they interdependently contribute to the system's self-regulating mechanisms and so help the system to survive. This style of thinking can be traced back to Durkheim (1895) one of whose dictums is that the explanation of a social phenomenon must seek separately 'the efficient cause which produces it *and the function it fulfils*'. The ideas and the consequent method of analysis were taken up by the social anthropologists Malinowski and Radcliffe-Brown (though the latter disavowed the term), their concern being to establish how a particular pattern of behaviour, for example a puberty ceremony or co-operation in building a canoe, functioned within a particular culture to reinforce social structures and to help maintain their equilibrium. Later the ideas entered sociology in a major way in the global theory constructed by Parsons (1951). Starting from a 'voluntaristic theory of action', at the core of which is the individual actor interpreting his situation and acting in it according to his goals, interests, and normative standards, Parsons moved to a highly elaborated positivist theory of all social systems. According to this theory any social system is subject to four 'functional imperatives' which must be met adequately if the system is to survive. The system must adapt to environmental changes, define system (rather than individual) goals and manage resources to achieve them, establish control procedures which regulate internal relations within the system, and maintain the motivating patterns whereby 'cultural values' are institutionalized (Parsons, 1977). With Parsons we reach an explicit systemic theory of social systems as real-world entities existing independently of individual actors. Parsons himself has been much influenced both by the systems thinking deriving from biology and by more recent developments, remarking in his 1977 volume, for example:

Clarification of the problem of control, however, was immensely promoted by the emergence, at a most strategic time for me, of a new development in general science—namely, cybernetics in its close relation to information theory.

Since the 1960s, writers in social science have seen in systems theory a new version of functionalism. Buckley (1967, 1968) claims that general systems theory is not functionalism in another form because the former focuses on interdependencies where functionalism concentrates on 'system needs'. In fact his point has not been taken, and the contribution of systems theory is seen by virtually all other commentators as providing one of the many new versions of functionalism which emerge as soon as apparently damaging criticisms have been mounted against it. (It is this Phoenix-like behaviour which causes Lazarsfeld to write that sociologists can neither live with functionalism nor live without it.) Although systems theory has been seen by some as a potential source of intellectual weapons to counter the normal criticisms of functionalism (namely that its emphasis on equilibrium gives it a conservative bias, that it cannot treat conflict as endemic in social systems and that it cannot propose mechanisms which relate parts to the whole) the normal view in the literature is that systems theory is simply modern functionalism.

Given that in everyday language we all glibly refer to the mysterious social groupings we see and feel ourselves to be a part of as 'social *systems*', it is not surprising that the emergence of systems thinking has led to attempts within the Durkheimian tradition to describe and explain social reality in systems terms. The assumption is that social systems exist as social facts and can be described using the language developed to describe systems of any kind. Indeed, unless we stop and think very carefully, we might simply take it as utterly obvious that systems thinking in the social sciences is bound to offer a version of functionalism. If so, we can stop this brief survey of social theory at this point. However, placing the experience of developing soft systems methodology in the context of social theory requires that some further attention be paid to developments within the Weberian tradition: in particular to phenomenology and hermeneutics.

Philosophically, the Durkheimian tradition of 'social facts' is underpinned by the philosophy of positivism, according to which all true knowledge is based upon empirical data (Kolakowski, 1972, provides an excellent short survey of the positivist position). But there is in this tradition a fairly clear distinction between two kinds of work: developing the philosophical base and the working out of practical methods. In the alternative tradition of phenomenology, which links very directly to Weber's interpretive social science, the distinction is much less clear. Perhaps because it is a younger tradition all accounts hark back to the founding philosphers—Brentano, Husserl, Dilthey—and practical methodology and substantive findings are harder to find. There is no phenomenological equivalent of Merton's paradigm for functionalism, which he presents as 'a codified guide for adequate and fruitful functional analyses' (1957). Although much discussion of phenomenology concentrates on the business of making a phenomenological analysis of a social situation, this is usually at a highly abstract level.

The most important founder of this tradition is Edmund Husserl (1859–1938), a mathematician by initial training before he became a professor of philosophy in a German university. From his teacher Brentano, Husserl took the idea of intentionality, the concept that all conscious mental activity is thinking *about* something. This is one of the main ideas in phenomenology, whose central concern is the nature and content of our thinking about the world rather than the world itself as something independent of all observers of it, which is the world of positivism. (For some useful accounts of phenomenology see: Schutz, 1962, 1964, 1966; Pettit, 1969; Roche, 1973; Gorman, 1977; Luckman, 1978). Husserl wishes to reform philosophical thinking by drawing attention to the fact that much of it takes too much for granted; it starts not from the basic data of consciousness but from concepts which already presuppose various theories. Husserl wanted a fundamental reappraisal, and himself returned to Descartes's initial position of extreme doubt in which he doubts everything except that he is himself thinking his doubt. For Husserl, Descartes's doubt was neither radical enough nor sufficiently sustained, the 'I am' of Descartes being for Husserl only 'a pure possibility generated by the meaning-constituting activity of transcendental subjectivity' (Roche, 1973). Distinguishing between the 'natural attitude' in which, in order to live our everyday lives, we make common-sense judgements about the reality of the world and its events, and the phenomenological attitude, in which common-sense belief is suspended, Husserl tried to develop a new method for philosophical thinking based on the latter. According to this the philosopher makes a resolute attempt to suspend his naive beliefs about the nature of the world and to think his way to the data of pure consciousness. Borrowing an expression from mathematics, Husserl speaks of putting the real world 'between brackets'. It is not that the philosopher denies that there is a real outer world, rather that he signals that he is not going to take it for granted in his quest for the universal types among the data of consciousness.

From this argument that the basic reality lies in our thinking about the world, rather than in the world itself, Husserl proceeds to build a philosophy which elucidates each type of cognition and the ways in which cognition correlates with an object. Philosophy becomes a study of meaning: according to Husserl the whole philosophical process is one of seeing, clarifying, determining, and distinguishing meanings. In the first lectures in which the ideas were developed, given in 1907 and published in 1950 under the title *The Idea of Phenomenology*, Husserl states it this way:

> The task is just this: within the framework of pure evidence or self-givenness to trace all forms of givenness and all correlations and to conduct an elucidatory analysis.

We see this process at work in, for example, Husserl's (1936) essay on the origins of geometry in which he argues that our acceptance of the deductions of Euclidean geometry is a matter of cultural learning, not a matter of direct self-evidence. He uses this example to illustrate the extent to which the everyday world we take as given is in fact 'constructed through human activity'.

Husserl wrote mainly about the methodology of his particular philosophy. The central figure in developing a phenomenological orientation within social science is Alfred Schutz, but here too the emphasis is on methodological issues rather than substantive studies.

Schutz (1899–1959) like Weber, takes as a problem the need to reconcile the notion of the individual free to attribute meaning to what he observes with the requirements of a rigorous scientific method. Husserl, he feels, provides a more useful theory of subjectivity than does Weber himself. Schutz's writings are based upon Husserl's distinction between the natural attitude of common-sense belief about the world and the phenomenological attitude in which that belief is suspended. But where Husserl considers the lived-in everyday world of experience only as a preliminary to making the 'phenomenological reduction' to the pure data of consciousness, the lived-in world, or *Lebenswelt*, is Schutz's main concern. His programme is to discover the structure of that world: to investigate the types of everyday taken-for-granted knowledge and to find out how they are socially structured and distributed. Roche (1973) sees the sociologist in Schutz overcoming the phenomenologist! In an important paper 'Some Structures of the Life-World' (Collected papers Vol. III, reprinted in Luckmann, 1978) Schutz argues that in the natural attitude

> We ... experience culture and society, take a stand with regard to their objects, are influenced by them and act upon them. In this attitude the existence of the life-world and the typicality of its contents are accepted as unquestionably given ...

He proceeds to analyse the nature of structures which are taken as given, using a general framework of: the ego's existence; the intentional/rational nature of social action; and (hence) its ability to be understood by other people. For the individual his stock of knowledge, a sedimentation of previous experiences and his definitions of them, will be relevant in three senses. It may be thematically relevant ('a theme for our knowing consciousness'), motivationally relevant as a result of current interests plans or problems, and interpretationally relevant according to how typical or atypical we judge our current experiences to be. In the light of this formulation Schutz describes the task of social science to be

> to investigate to what extent the different forms of systems of relevancy in the life-world—motivational, thematical and, most of all interpretational systems, are socially and culturally conditioned.

When he writes of the individual that

> The interest prevailing at the moment determines the elements which the individual singles out of the surrounding objective world ... so as to define his situation (Schutz, 1966)

we are reading a precise account of what Vickers, reflecting upon his own forty years of experiences in the world of affairs, has put into his theory of 'appreciative systems'.

The similarities between Vickers's formulations and the models of the life-world in phenomenological sociology bring us near to a position in which we can attempt to place soft systems methodology within the general picture of social theory. But finally, in building that picture, it is useful to mention briefly the contribution of Wilhelm Dilthey (1833–1911) philosopher of hermeneutics, the theory, art or skill of interpreting and understanding the products of human consciousness (see Makkreel, 1975, for a detailed discussion; Brown and Lyman, 1978; and also Dilthey's essay translated by Kluback and Weinbaum, 1957).

Dilthey was a historian of culture and philosophy as well as a philosopher in his own right; Husserl recognized his work as an anticipation of phenomenology. His chief concern was to establish that the subject matter of the human sciences was intrinsically different from that of the natural sciences, being concerned not with external facts but with expressions of the human mind which become cultural artefacts by a process of 'objectification'. Gaining valid knowledge of the process of objectification required the method of *Verstehen*. In an essay on 'The Rise of Hermeneutics' Dilthey's basic model uses the concepts of lived experience (an experiencing of meaning), objectification or expression, and understanding. He recognizes three categories of objectification: of ideas, of action, and of lived experience. These ideas inform his interpretive method of understanding society and history, the method comprising a circular process of discovery called 'the hermeneutic circle', a means of perceiving social wholes as both wholes and parts. A preliminary overview of subject matter is used to guide an examination of what the parts denote; this clarifies the concept of the whole, which at the end of the cycle must be perceived so that all the parts can be related to it. Thus there are no fixed or absolute starting points, only an iterative cycle which gradually leads to increased understanding of social reality. The hermeneutic circle opposes a Cartesian faith in any self-evident starting point, and there is no 'right' answer to the question of where to begin or end. This makes hermeneutic enquiry fundamentally different from that of positivist natural science. As Makkreel says: Dilthey 'all but banishes hypotheses from the human studies'. There are clear similarities here with the iterative approach essential in systems studies carried out in soft problem situations. And there are connections too between the formulation of root definitions of human activity systems embodying a specific, pure *Weltanschauung* and Dilthey's delberations on the basic structure of such world-images and the common types which recur. According to Dilthey a *Weltanschauung* is a totality compounded of three elements: our cognitive representation of the world, our evaluation of life, and our ideals concerning the conduct of life. Analysis of *Weltanschauungen* as they are expressed in philosophy and in great literature then suggests to Dilthey that there are three basic world-images. These are 'naturalism', embodying the view that man, relying on his senses to understand nature, adopts a goal of manipulating nature for his own ends; 'subjective idealism', in which the holder projects moral ideals that transcend given reality; and 'objective idealism' in which the world is neither manipulated scientifically nor dominated by the assertion of moral will, but is viewed as an object to be appreciated aesthetically—the world as a universal harmony. Dilthey sees Hume, Hobbes, and Balzac as naturalists, Kant and Corneille as subjective idealists and Hegel and Goethe as objective idealists.

In summary, we have in phenomenology and hermeneutics an attitude towards social science which takes as its prime datum not the world external to observers of it, but the observer's mental processes. This extension of the interpretive tradition of social science offers a 'human–culturalistic' approach to compare and contrast with the 'positivistic–naturalistic' approach of the Durkheimian tradition (Morris, 1977). In the humanistic–culturalistic tradition of phenomenology and hermeneutics human beings in the social process are constantly creating the social world in interaction with others. They are negotiating their interpretations of reality, those multiple interpretations at the same time constituting the reality itself. There is no 'pre-given universe of objects' but one which is 'produced by the active doing of subjects' (Giddens, 1976). Clearly in this tradition it is believed that special methods, not simply those of natural science, are required to understand such uniquely human processes.

The Social Reality Implied by Soft Systems Methodology

With the account of social science in the previous section as a framework, I can now give an account of the theoretical implications of soft systems methodology. From what has been said earlier in this chapter about the matching of this work with Churchman's philosophical account of enquiring systems and Vickers's theory of appreciative systems, it is apparent that I shall argue that soft systems methodology is not a new version of functionalism. The question to be faced is the extent to which its placement in the phenomenological tradition of social science can be justified. This issue is best approached by starting with a major finding in the action research—namely that hard systems engineering methodology could not be used in ill-structured problem situations in which the naming of desirable ends was itself problematical—and asking: if the hard methodology *had* been successful, what image or model of social reality would have been implied by that success?

All of the hard methodologies discussed in Chapter 5 have a built-in positivist ontology. They assume that inspection of the world by the observer will reveal it to contain systems — organizational systems, manufacturing systems, legal systems, transportation systems, etc. etc. Everyday language, in which we casually and constantly refer to such entities as 'systems', reinforces this view. Systems engineering, RAND systems analysis, and most operational research are all positivistic in this sense, making the assumption that the system of concern exists, can be named, and can be manipulated in the interests of efficiency. The tradition in the literature which assumes systems thinking to be a recent version of functionalism is not foolish, for the hard methodologies make the same (positivist) assumptions as functionalism. In his book on functionalism from a systems point of view, Sztompka (1974) says that he considers functionalism to be 'a particular implementation of the systemic approach', while functional analysis is 'a specific form of systemic analysis'. He finds five types of functional analysis, which use systems models of increasing complexity (systems models which are 'simple, teleological, functional, purposeful or multiple') but nowhere does he question the assumption that out beyond the analyst is a given social reality on to which one or other of these systems models will map.

This is the defining assumption of hard systems thinking and it is a well-justified assumption if the system of concern is a designed physical system or a procedural system so unthreatening that no one will bother to challenge the need for it—such as an 'order-processing system' in a manufacturing company. Had it been possible to use hard systems methodology in soft problem situations, the implication would have been that social reality is indeed systemic, and that the problem for the analyst is simply that of finding an appropriately sophisticated systems model.

In fact, the dramatic—not to say traumatic—finding in our formative studies, some of which are described in Chapter 6, was that it simply was not possible to take any system as given. In both the Airedale Textile Company and in Cordia Engineering, for example, it was apparent that there was no account of 'the system' appropriate to the study on which everyone would agree; and the disagreement extended not only to the choice of system but also to how it should be described. In Airedale, in selecting initially a system which generated and accepted orders for a defined range of products and used its expertise to meet them quickly and efficiently, we were very conscious that this was an arbitrary system selected by the analyst, not an account of part of the real world. And in Cordia Engineering, too, we were very clear that our choice of a system to carry out a task under various constraints was our arbitrary choice, and that there were potentially as many descriptions of the relevant system as there were concerned actors in the problem situation. These early experiences have been multiplied many times since then, and the general conclusion has to be that this research experience does not support either the positivist assumptions of hard systems thinking or the positivist account of the nature of social reality. The question remains: Does the experience support a phenomenological account of 'social systems'?

It is possible to give a broad answer to this question as follows: both hard (positivist) systems thinking and traditional management science assume that systems exist in the real world and can be unequivocally described. This leads logically to the idea of manipulating models of the assumed reality in order to discover a solution which is either optimum or at least 'good enough' in a particular situation. In contrast to this paradigm of *optimizing* (or satisficing), soft systems methodology embodies a paradigm of *learning* (Checkland, 1979b). The notion of 'a solution', whether it optimizes or satisfices, is inappropriate in a methodology which orchestrates a process of learning which, as a process, is never-ending. To this extent the methodology as a whole clearly articulates a phenomenological investigation into the meanings which actors in a situation attribute to the reality they perceive. And at a more detailed level, too, there are many parallels between the operations within the methodology and the philosophical/sociological tradition of interpretive social science.

Firstly, the methodology declines to accept the idea of 'the problem'. It works with the notion of *a situation* in which various actors may perceive various aspects to be problematical. It tries to provide help in getting from a position of finding out about the situation to a position of taking action in the situation. In the real world of managing, this is probably most often done by relying on finding similarities with previous experience. The methodology, in contrast to this, offers the chance of making the transition from finding out about the real world to taking action in the real world by means of some systems thinking *about* the real world. Its emphasis is thus not on any external 'reality' but on people's perceptions of reality, on their mental processes rather than on the objects of those processes.

Secondly, using the idea of 'human activity system' as the name of a particular kind of intellectual construct, the methodology accepts that any real-world purposeful human activity will be describable in many different ways within many different *Weltanschauungen*. Each root definition will embody one particular world-image, and we may note in passing that, in principle, observation of the images which continually recur in stages 5 and 6 of the methodology provides an experimental means of testing Dilthey's theory of the structure of *Weltanschauungen*, first published in 1931.

Thirdly, the conceptual models built in stage 4 are explicitly 'ideal types' in Weber's (1904) sense quoted in the previous section of this chapter. Each model of a relevant human activity system embodies a single one-sided concept of such a system, a view much purer than the complex perspectives we manage to live with in our everyday world.

Fourthly, in the stage 5 debate initiated by the comparison between conceptual models and the expression of the problem situation assembled at stage 2, the process is a formal way of elucidating, comparing, and contrasting different individuals' typifications of real-world events and structures, very much in the phenomenological manner. At this point the methodology offers a way of describing 'the universal structures of subjective orientation in the world' (which will also entail emergent properties of social wholes which transcend individuals) rather than 'the general features of the objective world', these quoted words being Luckmann's (1978) account of 'the goal of phenomenology'.

Finally, the fact that in practical use it is virtually possible to start at any point in the methodological cycle and proceed in any direction makes it resemble Dilthey's account of the hermeneutic circle. In positivist natural science the activity has a starting point in the scientist's selection of his experimental object; the acquisition of repeatable experimental results then provides an end point. In hermeneutics, and in using soft systems methodology, it is necessary to break into the hermeneutic circle rather arbitrarily, and follow a process which is characterized by gradually enriched learning as iteration follows iteration. Both natural science and phenomenological social science reveal *long-term* patterns of iterative learning; but in the case of the latter this characterizes the short term as well. It is the reason why the preferred operational mode in soft systems studies has to be that of action research (Susman and Evered, 1978).

The result of giving this sociological account of the methodology is to emphasize not only its position within phenomenology but also the highly positivistic nature of most of the applied systems thinking developed elsewhere. This point is underlined if we try to place the work described here, and hard systems thinking, on the two-dimensional typologies of social science which appear with some regularity in the literature of social science. Runciman (1963) contrasts holism–individualism with positivism–intuitionism; Robertson's (1974) axes are subjectivity–objectivity and sociality–culturality. For my purposes the most useful typology is that recently advanced by Burrell and Morgan (1979). They construct their map of social science in order the better to understand the sociological underpinning of theories of organization. Their axes of regulation–radical change on the one hand, and subjective–objective on the other, produce the framework of Figure 18.

Clearly the regulation/objective combination yields the paradigm of functionalist sociology, and one cannot object to their placing 'social system theory' within this quadrant. They specifically mention Bertalanffy (1950, 1956), Parsons (1951), and Katz and Kahn (1966) among others. This is where the implicit social theory of systems engineering, RAND systems analysis, and formal OR resides. The social theory implicit in soft

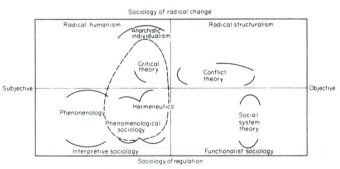

Figure 18. A partial reproduction of the Burrell–Morgan typology of social science (Burrell and Morgan, 1979)

systems methodology, however, would lie in the left-hand quadrants with hermeneutics and phenomenology, although the position would be not too far left of the centre line because the methodology will over a period of time yield a picture of the common structurings which characterize the social collectivities within which it works. Also, given the analyst's complete freedom to select relevant systems which, when compared with the expression of the problem situation, embody either incremental or radical change, the area occupied must include some of the 'subjective/radical' quadrant. Here the extension of the area towards the 'radical change' axis will be limited by the desire to achieve action in the real world; in practice, defining changes which are 'culturally feasible' has led to rather conservative use of the methodology—but this is a matter of practice so far, rather than principle.

This typology illustrates sharply the difference between hard and soft systems thinking. It is also interesting to note that the approximate area covered by the social theory implicit in soft systems methodology covers the area in which Burrell and Morgan locate the critical theory of the Frankfurt School. Since this particular conjunction had not been postulated in advance, examining whether or not the two theories do overlap provides at least a modest test of the proposition that soft systems methodology embodies a social theory in the interpretive sociology and radical humanism quadrants—assuming only the coherency of Burrell and Morgan's typology. A second reason for examining the similarities and dissimilarties between critical theory and soft systems theory is that Jürgen Habermas, a leading social theorist of the Frankfurt School, has mounted an attack on systems theory as he perceives it in the work of Niklas Luhmann (Habermas and Luhmann, 1971; Sixel, 1976).

'The Frankfurt School' is the group of scholars associated with the Institute of Social Research at the University of Frankfurt from 1930 when Max Horkheimer became its Director. The Institute moved to America in 1934, and was reconstituted in Europe when key members returned to Frankfurt in 1949. (Connerton, 1976, edits an anthology of articles by and about the group; Quinton, 1974, provides a short introduction; Jay, 1973, a comprehensive one.) The work of the School as a whole combines an attack on positivism as an adequate base for social science with a belief that adequate social theory must perceive society as a set of processes which emancipated men can change. In particular, advanced capitalist society is seen as a repressor, a cultural manipulator of men; Marx's analysis needs further distancing from positivism and needs to be extended to cover more than economic factors. The School's rejection of what they see as the science-dominated culture of Western societies can go to extreme lengths. In a television discussion of the School (Magee, 1978) Herbert Marcuse was asked why its writings were 'not just difficult to read but usually turgid and sometimes unintelligible'. Marcuse in reply quoted Adorno's view that language itself has been 'so much permeated by the

Establishment' that clear intelligible writing itself 'expresses . . . the control and manipulation of the individual by the power structure'!

In developing his critique of Western society, Habermas contrasts purposive–rational action 'governed by technical rules based on empirical knowledge' and communicative action, or symbolic interaction, 'governed by binding consensual norms which define reciprocal expectations about behaviour' (Habermas, 1971, in the essay 'Technology and Science as "Ideology" '). In science-based capitalist societies purposive–rational action has become dominant through its institutionalization; and positivist natural science, supposedly value-free, has supplied an apparent legitimation for this process. We have, then, a society in which behaving in accordance with technical recommendations is the only 'rational' way to behave; our culturally-defined self-understanding of a social life-world (as in pre-capitalist traditional societies) is replaced by 'the self-reification of men under categories of purposive–rational action and adaptive behaviour'. For Habermas systems analysis (he means of the RAND variety) supplies a sinister means by which cultural control can be exercised:

> It is possible in principle to comprehend and analyse individual enterprises and organizations, even political or economic sub-systems and social systems as a whole, according to the patterns of self-regulated systems. It makes a difference, of course, whether we use a cybernetic frame of reference for analytic purposes or *organize* a given social system in accordance with this pattern as a man-machine system. But the transferral of the analytic model to the level of social organization is implied by the very approach taken by systems analysis (Habermas, 1971).

If we are to avoid the negative Utopia implied by systems analysis it will be necessary, in Habermas's view, to remove cultural restrictions on communication, to achieve 'public unrestricted discussion, free from domination, of the suitability and desirability of action-orienting principles and norms . . .' To achieve the necessary 'communicative competence' (Habermas, 1970) requires both equal participation in discussion, undistorted by power relationships, and unlimited scope for a radical questioning of societal structures and procedures.

Such is a bare outline of Habermas's critique. It is interesting to find that in the debate with Luhmann, Habermas rejects Luhmann's theory of sense-making systems, even though Luhmann, no conventional functionalist[†], and aiming to combine Husserl's phenomenology and systems theory, argues that progress in society can take place to the extent to which its sense system can produce societal alternatives (Sixel, 1976; McCarthy, 1978). Habermas, however, fears that the theory would

---

[†]Luhmann accepts 'meaning' as a fundamental category of (social) systems analysis.

legitimate the political power of experts and lead to a sociology which took the form of a social technology—just what he fears from traditional systems analysis.

In comparing Habermas's thought with soft systems methodology Mingers (1980) finds three major points of agreement. Firstly, both take seriously the problem of human action—at the same time purposive/rational (hence capable in principle of being engineered) and natural, or unchangeable, as a result of the characteristics of the human animal. Secondly, both conclude that hard systems analysis, tied to technical rationality, cannot cope adequately with the multi-valued complexities of the real world. Finally, both deny the inevitability of the divorce between rationality and values which characterizes natural science, and both try to bring the two together in rational communicative interaction. Habermas's communicative competence would enable social actors to perceive their social condition in new ways, enabling them to decide to alter it; Checkland's methodology aims at consensual debate which explores alternative world-views and has as criteria of success 'its usefulness to the actors and not its validity for the analyst'. The *differences* between the two approaches, for Mingers, stem from critical theory's more overtly political stance. Soft systems methodology does not yet have any theory of how the structure of society—especially its stratification—might limit fundamentally the range of debate about change.

I think that Mingers has established significant compatibility between critical theory and soft systems methodology. It is perfectly possible to see the latter eventually as a vehicle for what in an interview with Frankel (1974), Habermas calls 'radical reformism', namely an attempt 'to challenge and to test the basic or kernel institutions' of present-day society. What I hear Habermas arguing is that the debate at stages 5 and 6 of the soft systems methodology will be inhibited by society's structure. I think that it is the nature of society that this will be so. The point is well made by Popper, usually regarded as the implacable enemy of the Frankfurt School (in Schilpp, 1974, page 1168)

> At any moment of time we and our values are products of existing institutions and past traditions. Admittedly this imposes some limitations on our creative freedom and on our powers of rational criticism . . .

But these 'limitations' do not mean that the use of the methodology cannot in fact be emancipatory for the actors concerned, in a way which Habermas would approve.

Finally, to end this section, it seems appropriate to try to answer explicitly the question about the nature of social reality implicit in soft systems methodology. The success of the methodology in real situations suggests the following answer: social reality is the ever-changing outcome of the social process in which human beings, the product of their genetic inheritance and previous experiences, continually negotiate and re-negotiate with others their perceptions and interpretations of the world outside themselves.

Conclusion

Looking back over the history of ideas we can see that occasionally there arises an idea so powerful that it leads to irreversible changes in the world and how we perceive it. The idea of 'the capacity of a physical system to do mechanical work' is one such idea. This is the idea to which we give the name 'energy', although in the everyday world we usually think of it in terms of its embodiments, such as electricity, rather than as an abstract idea. This abstraction can, however, be regarded, crudely, as the idea powerful enough to underpin the Industrial Revolution; it is the idea which makes it possible to think about, and to seek, ever more efficient science-based technology. An equally powerful idea, and one which is frequently thought of as the crucial idea in the Second Industrial Revolution, that of the last few decades, is the idea 'information', connoting order or something which embodies it. The rise of the systems movement has of course depended much upon this idea; it is the one idea central to the four basic systems concepts of emergence, hierarchy, communication, and control. Unfortunately, as discussed in Chapter 3, present information theory deals only with the statistics of message transmission, not at all with the meaning of the message transmitted. I suspect, however, that the idea of 'meaning' will eventually be seen to be as important as 'energy' and 'information'. It may even herald a revolution in social thinking as significant as the two Industrial Revolutions. The work on which this book is based may be seen as a small contribution towards making possible that third revolution.

The work began by trying to extend hard systems thinking to situations involving everyday human decision-making. No doubt I brought to the work the inevitable biases due to previous experience as scientist, technologist, and manager. Certainly I was consciously trying to carry out the work in the spirit of science, with some imprecise expectation of eventually deriving generalizations about systems of various kinds. In parallel with the action research, an attempt was made to think out the basis of systems thinking. This led to a view of systems thinking as a response to difficulties which confront the method of natural science when it faces phenomena of great complexity, notably those of the social world. This in turn led to consideration of the unsolved methodological problems of the social sciences. As this was happening, the experiences in the action research were convincing all those engaged in it that whether they liked it or not, carrying out systems studies in real problem situations involved wrestling with the problems of social, not natural science.

The outcome, which at least for the researchers has had the kind of unexpectedness associated with true learning, is a methodology for finding out about the social world. In Churchman's language it is a 'Singerian inquiring system' in which enquiry is never-ending; in Vickers's language it offers a way of discovering 'the appreciative settings' and establishing a debate in which they will change; in Habermas's language it provides a means of increasing 'communicative competence'.

The implication of this for social theory is that no once-and-for-all substantive account of social reality is possible because there *is* no social reality to set alongside what appear to be the well-tested physical regularities of the universe. Even natural science, so the 20th century has taught us, acquires only provisional knowledge; but it is at least reasonable to assume that there is an unchanging physical, chemical, and biological world 'out there' and that we can find out about it by using the method of the natural sciences. Our knowledge of social reality cannot hope to achieve even this kind of certainty; but the way one finds out about it may in principle be reasonably stable: hence the importance of methodology rather than findings, of process rather than content.

Finally, since this work has emphasized the importance of the unquestioned world-image by means of which we perceive the world, it is proper to ask what unquestioned assumption is built into an acceptance of soft systems methodology as an enquiring system? The theoretical and practical work described here leads to the conclusion that the unquestioned prime value embodied in 'a systems approach' is that continuous, never-ending learning is a good thing. This means that soft systems thinking will not appeal to determinists, dictators, or demagogues. It *will* appeal to all those people in any discipline who are knowledgeable enough to know that there is much they do not know, and that learning and re-learning is worth-while. For such people a systems approach is not a bad idea.

Appendix 1

Building Conceptual Models†

The technique for building conceptual models is based on very simple principles which have been tested in many systems studies carried out over several years. A model of a human activity system will contain a set of activities connected together. The basic language used for model construction is therefore all the verbs in the analyst's speaking language. The model will contain the *minimum* number of verbs *necessary* for the system to be the one named and concisely described in the *root definition*. These will need to be connected together in order to represent the system as an entity, and the most basic form this connectivity may take is a number of arrows which indicate logical dependencies. Where it seems essential to represent a flow (whether of materials, money, energy, or information) its nature must be indicated.

The aim is to build an activity model of *what* must go on in the system. Particular *hows* (including such things as roles, organizational structures, and specific ways of carrying out the activities) must be included only if they are specifically named in the root definition. They may of course be included in subsequent more detailed models obtained by the expansion of the first-level model.

A General Example

To illustrate these principles before discussing a specific example, Figure 19 is a topic-free indication of the general nature of a conceptual model of a human activity system. In this notional system six connected activities constitute a transformation process which produces periodic reports and requires information inputs. Activity 4, described by verb 4, requires the prior completion of Activities 1 and 2 (verbs 1 and 2). Activity 5 depends upon 4 and 3, and enables Activity 6 to take place, this being the origin of the system's output.

The task in conceptual model building is simply that of assembling the list of verbs describing the activities required by the root definition, connecting them according to the requirements of logic, and indicating any flows which appear essential at this first resolution level.

---

†Extracted from Checkland (1979c).

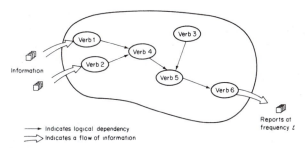

Figure 19. An illustration of the general form of a conceptual model of a human activity system

Once this version of the model has been constructed it may be used as a basis for further expanded versions. Some may show activities at more detailed levels, or record all flows in the system, material and abstract; also, noun-based versions of the model may include organizational entities which might carry out the activities in the basic model. A frequent elaboration is to ask of each activity: What information is needed to carry out this activity? From what source? With what frequency? In what form? The basic activity model then becomes the origin of an information-flow model which can be used to enquire into present information flows or to design new information systems.

A Specific Example

The specific example of building a conceptual model from a root definition concerns a system relevant to a problem in a science-based industrial firm which is active in world markets. Economic survival requires the firm both to make technical and commercial innovations and to protect its innovations by means of Patent and Copyright law. That at least is the view taken by the analyst in his selection of a 'relevant system'. The root definition is then based on the legal notion of 'intellectual property'.†

> Root Definition: A professionally-manned system concerned with the overall management of 'intellectual property' so that by this management the system makes the best possible contribution to the business success of a science-based company.

Taking this RD as given, the task now is to build the conceptual model implicit in it.

It is best to start from the CATWOE elements. The basic transformation (T) which this system brings about is evidently to take in professional skill,

---

†It is interesting to note that British Post Office Telecommunications has recently renamed its Patents Branch as the Intellectual Property Unit of the Procurement Executive.

knowledge of the company's business and potential intellectual property (which *its* professionals can recognize) and use these inputs to generate action concerning intellectual property which makes the best possible contribution to business success. The actors (A) in the system are the professionals mentioned in the RD and the system's beneficiary or victim (C) is clearly the company as a whole. The Company is also the system's owner (O) since it is the company which could decide to abolish this system and get rid of these particular 'actors'. The main environmental constraint (E) which this system has to take as given is the state of the law concerning intellectual property, and the *Weltanschauung* (W) implicit in the RD is that the company is operating in a world in which respect for the law concerning intellectual property makes this a resource which can be managed in a way which makes a contribution to business success. This suggests a system which will have a number of information inputs (covering the company's technology/markets and the state of the law on intellectual property); which will take managerial action concerning intellectual property; and which will monitor the success of all its operations in order to take control action aimed at increasing the system's contribution to business success. The actions the system could take with respect to intellectual property, in order to 'manage' it, are obviously to *acquire* it, to *protect* it (using Patent Law, for example) and to *exploit* it (for example, selling licences to other companies or exchanging rights to protected products, processes, or machines). The outputs from the system will be both informational and financial intellectual property being a saleable commodity.

These considerations require as a basic structure for the model that shown in Figure 20.

We can now consider the activities minimally necessary in the three sub-systems of Figure 20. Consider first the awareness which will have to be gained before the system can 'acquire, protect, and exploit' intellectual property. The system will have to appreciate the company's science base, its technology; it will have to appreciate the company's business, since it aims to help foster business success; it will have to appreciate the law concerning intellectual property (Patent, Copyright and Trade Mark Law); and it will have to acquire and maintain its professional skill in doing these things, since all of them apply to changing environments. These considerations lead to the structure shown in Figure 21.

Considering the 'operational' sub-system, this will have available the knowledge gained in the 'awareness' sub-system, and its main activities will be acquire, protect, and exploit. Since the latter is ambiguous (an actor told to 'exploit some intellectual property' might not know what to do) it is justifiable to break this down into more explicit activities. Given the nature of the law governing intellectual property—and here the systems analyst might have to make some enquiries about this aspect of the real world *which the system takes as given*—we can justify breaking down 'exploit' into selling licences and making exchanges of patented products, processes,

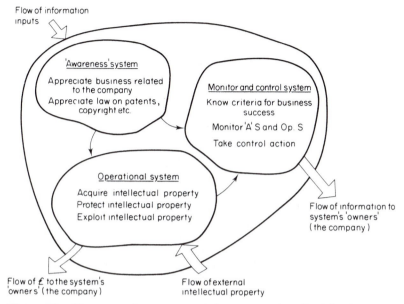

Figure 20. The general structure of the system described in the root definition

or machines. These considerations give us the model of the 'operational' sub-system in Figure 22.

In order to ensure that the system's managing of intellectual property makes a positive contribution to the owning company's business success, it is necessary for the system to monitor and control its operations in the light of the contribution it makes. This will require measures of performance for the system to be defined. There are obviously a number of possibilities, such as the net positive or negative cash flow from licence

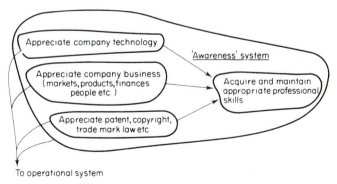

Figure 21. The 'awareness' system modelled in more detail

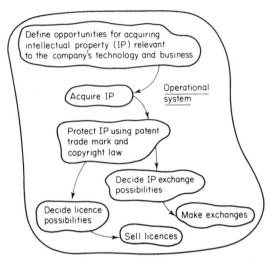

Figure 22. The 'operational' system modelled
in more detail

management, the legal strength of the company's patents, designs, and trade marks, etc. At this first level of modelling, however, it is probably legitimate to record the activity of setting measures and monitoring with respect to them, rather than including any specific measures.

Adding a 'monitoring and control' sub-system to the 'awareness' and 'operational' sub-systems gives the final model in Figure 23. Every concept in the root definition finds expression in the model, and the model hopefully reflects all aspects of the root definition but no others. The aim is to achieve a pairing of root definition (what the system *is*) and conceptual model (what the system *does*) which are mutually consistent. The model can then be used coherently in other stages of the methodology.

Discussion

The technique for building a first-level conceptual model from a root definition, illustrated above for a system of 'techno-commercial' activity in a science-based industrial company, may be generalized in the following sequence, which should, however, be used flexibly:

1. From the RD and its CATWOE elements, form an impression of the system as an autonomous entity carrying out a physical or abstract transformation process.
2. Assemble the small number of verbs which describe the most fundamental activities necessary in the system described. Try to maintain one resolution level, avoiding the mixing of activities defined at different levels of detail.

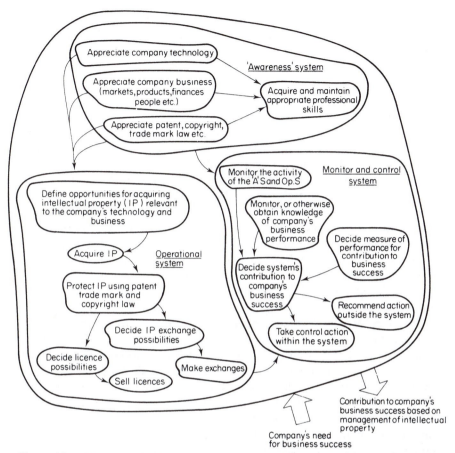

Figure 23. The conceptual model of the system described in the root definition

3. If it can be justified from the RD, structure the activities in groups which bring together similar activities (e.g. grouping those which together generate some output which passes elsewhere in the system).

4. Connect the activities and the groups of activities by arrows which indicate logical dependencies.

5. Indicate any flows (concrete or abstract) which are *essential* to expressing what the system does. Distinguish these flows from the logical dependencies of 4 above, and in any case keep the number of flows to a minimum at this stage.

6. Check that the root definition and conceptual model together constitute a mutually-informing pair of statements: what the system *is* and what the system *does*.

Once the model has been built according to this sequence, it may be used as the source of other versions, either models of the same kind at higher resolution levels or models expressing flows and/or possible structures. Each basic conceptual model which expresses *what* the system does, for example, can be expanded into a group of models which express possible *hows*. Decisions about the value of further modelling of this kind can be taken only in the context of the overall problem situation which led to a particular RD being regarded as 'relevant', not on the basis of the RD itself.

The experience of about a hundred systems studies involving conceptual modelling of the kind described suggests that certain errors recur. Anticipating them may help in avoiding them.

There is a tendency to formulate root definitions which contain several transformations and express several *Weltanschauungen*. It is virtually impossible to build a model from such definitions: a root definition of an English pub, for example, which tried to express the several major W's and T's discussed above†, would be unmanageable. Also, if several major transformations are included, the definition tends to become a prose version of a conceptual model, and we lose the richness gained by the pairing of *being* and *doing* which the root definition—conceptual model relationship provides.

Within CATWOE the most common mistake is to define as C, the system's beneficiaries or victims, some persons who are affected by the system, but at several removes. For example, customers in the market place are frequently named as representing CATWOE's 'C' in root definitions of production planning systems. For such a system the correct 'C' (the system's direct beneficiaries or victims) are the people responsible for carrying out the production process.

Failures of logic in representing logical dependencies are of course common, at least in the first attempts at building a model. More important, and equally fundamental, are inconsistencies in the use of verbs as the modelling language. The inclusion of a few systems elements as nouns is not only inconsistent, it frequently signals the incorporation of a particular 'how' not justified from the root definition.

An error which sometimes survives the effort to weed out errors is to include in the model activities which hopefully arise as a result of doing other things but are not themselves something an actor could positively carry out. Thus a conceptual model ought not to contain the element 'generate profit' or 'gain a high reputation'. The model should include the activities which will hopefully lead to these outcomes. In the example above 'exploit intellectual property' was expanded because of this kind of ambiguity.

In general, the most common error is undoubtedly to slip into modelling part of the real world, rather than building the model of the system named in the root definition. Root definitions may well contain *constraints* which derive from the real world (such as the legal framework concerning

† See Checkland (1979c)

'intellectual property', in the example above) and hence these aspects of the real world may legitimately be included in the model. But it cannot be overemphasized that conceptual models are simply models which will help to orchestrate a debate in the real world. They are not models of part of the real world, for the simple reason that there is no unique account which tells us what some real-world human activity *is*. There are only sets of different perceptions of any real-world human activity, and root definitions and conceptual models provide a means of exploring these perceptions.

Appendix 2

A Workbook for Starting Systems Studies

Definition of Terms

'Client': He who wants to know or do something and *commissions* the study. The implication is that he can cause something to happen as a result of the study. (He may also be in the 'decision taker' role.)

'Decision taker': The role player *in a human activity system* who can alter its content (its activities) and their arrangements within the system (sub-systems) and who can decide resource allocation within the system.

'Problem owner': He who has a feeling of unease about a situation, either a sense of mismatch between 'what is' and 'what might be' or a vague feeling that things could be better and who *wishes something were done about it*. The problem owner may not be able to define what he would regard as a 'solution', and may not be able to articulate the feeling of unease in any precise way. (The analyst may assign to the role 'problem owner' someone who does not himself recognize his own ownership of the problem; and the problem owner may not be 'decision taker'. However, most systems analyses have been carried out for 'decision takers' who were 'problem owners'.)

Note that technically 'client' and 'problem owner' are real-world roles, where 'decision taker' is the name of a role in a human activity system, i.e. it is a 'below the line' conceptual construction. However, most systems studies aim to map conceptual models on to reality in a way which eventually enables some action to be taken in the real world. Some *real-world* 'decision taker' has to initiate that action, hence 'decision taker' then becomes a real-world role. Often this is clearly the case from the start, and frequently there is a decision taker/problem owner who seeks guidance on what action he should take. This is indicated in the workbook in 2.2, 2.3, 2.4, and 2.5 as 'decision taker'/problem owner; but it may be that at the start of a study only a 'problem owner' can be suggested.

1. *Take the study situation to be one in which a client has commissioned the analysis; there will be a problem-solving system (containing the analyst as*

problem solver) whose efforts are brought to bear on a problem-content system (containing the roles, problem owner, and 'decision taker', which may be coincident).

1.1 Who is the client?

1.2 What are his aspirations?

2. *Consider the problem-content system*
   2.1 Who are the occupants of the roles 'problem owner' and 'decision taker'?

   2.2 The 'decision taker'/problem owner's version(s) of the nature of the problem is (are):

   2.3 The 'decision taker'/problem owner's *reasons* for regarding 'the problem' as a problem are:

2.4 The 'decision taker'/problem owner's expectations of the problem-solving system are:

2.5 The answer to 2.4 suggests that the following are highly valued by the 'decision taker'/problem owner:

2.6 Some possible names for the problem-content system are:

2.7 In describing the problem-content system initially the following elements seem likely to be relevant:

(a) Nouns—

(b) Verbs—

2.8 Environmental constraints on the problem-content system are:

3. *Consider the problem-solving system*
   3.1 The occupant(s) of the role problem solver is (are):

   3.2 The other persons (and roles) in the problem-solving system are:

   3.3 The resources of the problem-solving system are:
       (a) People—

       (b) Physical resources—

       (c) Skills—

       (d) Finance—

       (e) Time—

   3.4 Likely or known environmental constraints on the problem-solving system are:

3.5 The problem solvers will know the 'problem' is 'solved' when:

Bibliography

Abraham, J. H. (1973) *Origins and Growth of Sociology* Harmondsworth: Penguin Books

Ackermann, R. J. (1976) *The Philosophy of Karl Popper* Amherst, Mass.: University of Massachusetts Press

Ackoff, R. L. (1953) *The Design of Social Research* Chicago: University of Chicago Press

Ackoff, R. L. (1957) 'Towards a behavioral theory of communication' in Buckley (1968) q.v.

Ackoff, R. L. (1971) 'Towards a system of systems concepts' *Management Science*, **17**, No. 11

Ackoff, R. L. (1974) *Redesigning the Future* New York: Wiley

Ackoff, R. L. and Emery, F. E. (1972) *On Purposeful Systems* London: Tavistock

Adorno, T. W. (Ed.) (1976) *The Positivist Dispute in German Sociology* London: Heinemann (Published in German 1969)

Allen, G. (1975) *Life Science in the Twentieth Century* New York: Wiley

Alloway, L. (1972) 'Network: the art world described as a system' *Art Forum XI* (1) September 1972

Anderson, P. W. (1972) 'More is different' *Science*, **177**, 393

Anderton, R. H. and Checkland, P. B. (1977) 'On learning our lessons' Internal Discussion Paper 2/77, Department of Systems, University of Lancaster

Armytage, W. H. G. (1961) *A Social History of Engineering* London: Faber and Faber

Aron, R. (1968) *Main Currents in Sociological Thought*: Vol II London: Weidenfeld and Nicholson (Vol. I 1965, deals with Comte, Montesquieu, Marx, Tocqueville, Vol. II with Pareto, Weber, Durkheim)

Ashby, W. R. (1956) *An Introduction to Cybernetics* London: Chapman and Hall

Atkinson, L. (1976) 'Industrial rearmament', Lord Wakefield Gold Medal Paper to the Institute of the Motor Industry, May 1976

Ayala, F. J. and Dobzhansky, T. (1974) *Studies in the Philosophy of Biology* London: Macmillan

Ayer, A. J. (1973) *The Central Problems of Philosophy* London: Weidenfeld & Nicolson

Bar-Hillel, Y. (1955) 'An examination of information theory' *Philosophy of Science*, **22**, 86

Bar-Hillel, Y. and Carnap, R. (1953) 'Semantic information' *British Journal for the Philosophy of Science*, **4**, 14

Basalla, G. (1968) *The Rise of Modern Science: External or Internal Factors* Lexington, Mass.: D. C. Heath

Bateson, G. (1971) 'The cybernetics of "self": a theory of alcoholism' in Bateson (1973) q.v.

Bateson, G. (1973) *Steps to an Ecology of Mind* St Albans (Herts): Paladin

Bayraktar, B. A., Müller-Merbach, H., Roberts, J. E. and Simpson, M. G. (Eds.) (1979) *Education in Systems Science* London: Taylor and Francis

Beck, W. S. (1957) *Modern Science and the Nature of Life* New York: Harcourt Brace

Beckett, J. A. (1971) *Management Dynamics: The New Synthesis* New York: McGraw-Hill

Beckhard, R. (1969) *Organization Development: Strategies and Models* Reading, Mass.: Addison-Wesley

Beer, S. (1970) 'Marlow seventy' London: OR Society Ltd

Beer, S. (1972) *Brain of the Firm* London: Allen Lane

Bell, D. A. (1968) *Information Theory and its Engineering Applications* London: Pitman

Bennett, R. J. and Chorley, R. J. (1978) *Environmental Systems* London: Methuen

Bennis, W. G. (1969) *Organization Development: Its Nature, Origins and Prospects* Reading, Mass.: Addison-Wesley

Berlin, I. (Ed.) (1956) *The Age of Enlightenment: The 18th Century Philosophers* (Selections) New York: New American Library (Mentor)

Berlinski, D. (1976) *On Systems Analysis* Cambridge, Mass.: MIT Press

Bernal, J. D. (1939) *The Social Function of Science* London: Routledge

Bernal, J. D. (1967) *The Origin of Life* London: Weidenfeld & Nicolson

Bernstein, J. (1973) *Einstein* London: Fontana/Collins

Berresford, A. and Dando, M. R. (1978) 'OR for strategic decision-making: the role of world view' *Journal of Operational Research Society*, **29**

Bertalanffy, L. von (1940) 'The organism considered as a physical system' Reprinted in Bertalanffy (1968) q.v.

Bertalanffy, L. von (1950) 'The theory of open systems in physics and biology' in Emery (1969) q.v.

Bertalanffy, L. von (1956) 'General system theory' in Bertalanffy (1968) q.v.

Bertalanffy, L. von (1968) *General System Theory* New York: Braziller

Bertalanffy, L. von (1972) 'The history and status of general systems theory' in Klir (1972) q.v.

Black, M. (Ed.) (1963) *The Social Theories of Talcott Parsons* Englewood Cliffs: Prentice-Hall

Blackett, P. M. S. (1962) *Studies of War* Edinburgh: Oliver and Boyd

Blair, R. N. and Whitston, C. W. (1971) *Elements of Industrial Systems Engineering* Englewood Cliffs: Prentice-Hall

Blum, F. H. (1955) 'Action research—a scientific approach?' *Philosophy of Science*, **22** (1)

Boas, M. (1962) *The Scientific Renaissance 1450–1630* London: Collins

Boguslaw, R. (1965) *The New Utopians: A Study of System Design and Social Change* Englewood Cliffs: Prentice-Hall

Bottomore, T. and Nisbet, R. (1979) *A History of Sociological Analysis* London: Heinemann

Bottomore, T. B. and Rubel, M. (Eds.) (1956) *Karl Marx: Selected Writings in Sociology and Social Philosophy* London: C. A. Watts

Boulding, K. E. (1956) 'General systems theory—the skeleton of science' *Management Science*, **2** (3)

Bradley, F. H. (1893 and 1962) *Appearance and Reality* Oxford: University Press

Breck, A. D. and Yourgrau, W. (Eds.) (1972) *Biology, History and Natural Philosophy* New York: Plenum Press

Broad, C. D. (1923) *The Mind and its Place in Nature* London: Kegan Paul, Trench and Trubner (Reissue 1980)

Bronowski, J. (1973) *The Ascent of Man* London: British Broadcasting Corporation

Bronowski, J. and Mazlish, B. (1960) *The Western Intellectual Tradition* London: Hutchinson

Brown, R. H. and Lyman, S. M. (Eds.) (1978) *Structure Consciousness and History* London: Cambridge University Press

Bryer, R. A. (1977) 'The status of the systems approach' OR Society Conference 'The king is dead, long live the king' University of Aston, December 1977

Bryer, R. A. and Kistruck, R. (1976) 'Systems theory and social science' Working Paper 735/76 Warwick University School of Industrial and Business Studies

Buchdahl, G. (1961) *The Image of Newton and Locke in the Age of Reason* London: Sheed and Ward

Buckley, W. (1967) *Sociology and Modern Systems Theory* Englewood Cliffs: Prentice-Hall

Buckley, W. (Ed.) (1968) *Modern Systems Research for the Behavioral Scientist* Chicago: Aldine

Bugliarello, G. and Doner, D. B. (Eds.) (1979) *The History and Philosophy of Technology* Urbana: University of Illinois Press

Bunge, M. (1966) 'Technology as applied science' in Rapp (1974) q.v.

Bunge, M. (1973) *Method, Model and Matter* Dordrecht: D. Reidel

Burrell, G. and Morgan, G. (1979) *Sociological Paradigms and Organisational Analysis* London: Heinemann

Burton, J. W. (1968) *Systems, States, Diplomacy and Rules* Cambridge: University Press

Butterfield, H. (1949) *The Origins of Modern Science 1300–1800* London: Bell

Calder, N. (Ed.) (1973) *Nature in the Round: A Guide to Environmental Science* London: Weidenfeld & Nicolson

Cannon, W. B. (1932) *The Wisdom of the Body* New York: Norton

Chapman, G. P. (1977) *Human and Environmental Systems* New York: Academic Press

Chase, W. P. (1974) *Management of Systems Engineering* New York: Wiley

Checkland, P. B. (1970) 'Systems and science, industry and innovation', *Journal of Systems Engineering*, 1 (2)

Checkland, P. B. (1971) 'A systems map of the universe', *Journal of Systems Engineering*, 2 (2)

Checkland, P. B. (1972) 'Towards a systems-based methodology for real-world problem-solving' *Journal of Systems Engineering*, 3 (2)

Checkland, P. B. (1975) 'The development of systems thinking by systems practice—a methodology from an action research program' in Trappl and Hanika, q.v.

Checkland, P. B. (1976a) See Smyth and Checkland (1976)

Checkland, P. B. (1976b) 'Science and the systems paradigm', *International Journal of General Systems*, 3 (2)

Checkland, P. B. (1977a) 'The problem of problem formulation in the application of a systems approach' in Bayraktar *et al.* (1979) q.v.

Checkland, P. B. (1977b) 'Systems methodology in problem solving: some notes from experience' in Trappl, Hanika and Pichler (1977) q.v.

Checkland, P. B. (1979a) 'The shape of the systems movement' *Journal of Applied Systems Analysis*, 6

Checkland, P. B. (1979b) 'The systems movement and the "Failure" of management science' 25th Anniversary Meeting of Society for General Systems Research, London, August 1979. Also *Cybernetics and Systems*, 11 (1980)

Checkland, P. B. (1979c) 'Techniques in "soft" systems practice'. Parts 1 and 2 *Journal of Applied Systems Analysis* 6

Checkland, P. B. (1979d) 'A systems approach to terotechnology: defining the concept' *Terotechnica*, 1, 83–88

Checkland, P. B. and Griffin, R. (1970) 'Management information systems: a systems view' *Journal of Systems Engineering*, **1** (2)

Checkland, P. B. and Jenkins, G. M. (1974) 'Learning by doing: systems education at Lancaster University' *Journal of Systems Engineering*, **4** (1)

Checkland, P. B. and Wilson, B. (1980) 'Primary task and issue-based root definitions in systems studies' *Journal of Applied Systems Analysis*, 7

Cherry, C. (1957) *On Human Communication* Cambridge, Mass.: MIT Press (second edition 1966)

Chestnut, H. (1965) *Systems Engineering Tools* New York: Wiley

Chestnut, H. (1967) *Systems Engineering Methods* New York: Wiley

Chisholm, M. (1975) *Human Geography: Evolution or Revolution* Harmondsworth: Penguin Books

Chorley, R. J. and Kennedy, B. A. (1971) *Physical Geography: A Systems Approach* London: Prentice-Hall International

Churchman, C. W. (1968a) *The Systems Approach* New York: Dell

Churchman, C. W. (1968b) *Challenge to Reason* New York: McGraw-Hill

Churchman, C. W. (1971) *The Design of Inquiring Systems* New York: Basic Books

Churchman, C. W. (1974) 'Philosophical speculations on systems design' *OMEGA*, **2** (4)

Churchman, C. W. (1979) 'Paradise regained: a hope for the future of systems design education' in Bayraktar, B. A. *et al.* (1979) q.v.

Churchman, C. W., Ackoff, R. L. and Arnoff, E. L. (1957) *Introduction to Operations Research* New York: Wiley

Churchman, C. W., Auerbach, L. and Saden, S. (1975) *Thinking for Decisions: Deductive Quantitative Methods* Chicago: Science Research Associates

Churchman, C. W. and Verhulst, M. (Eds.) (1960) *Management Science Models and Techniques*, Vol. 2, Oxford: Pergamon

Clark, P. A. (1972) *Action Research and Organizational Change* London: Harper and Row

Cohen, P. (1968) *Modern Social Theory* London: Heinemann

Collins, L. (Ed.) (1976) *The Use of Models in the Social Sciences* London: Tavistock

Colodny, R. G. (Ed.) (1977) *Logic, Laws and Life* Pittsburg: University Press

Connerton, P. (Ed.) (1976) *Critical Sociology* Harmondsworth: Penguin Books

Cornock, J. S. (1977) 'Understanding supra-institutional problems: systems lessons drawn from an application of the Checkland methodology', 1st International Conference on Applied General Systems Research: Recent developments and trends, SUNY, Binghamton 1977

Cornock, J. S. (1980) 'The use of systems ideas in the domain of the artist' Dissertation, University of Lancaster

Crick, F. (1966) *Of Molecules and Men* Seattle: University of Washington Press

Crombie, A. C. (1953) *Robert Grosseteste and the Origins of Experimental Science 1100–1700* London: Oxford University Press

Crombie, A. C. (1969) *Augustine to Galileo Vol. II: Science in the Later Middle Ages and Early Modern Times* Harmondsworth: Penguin Books (first edition 1959)

Daenzer, W. F. (Ed.) (1976) *Systems Engineering* Cologne: Peter Hanstein

Davies, J. T. (1965) *The Scientific Approach* London: Academic Press

Davies, W. K. D. (Ed.) (1972) *The Conceptual Revolution in Geography* London: University of London Press

Demerath, N. J. and Peterson, R. A. (Eds.) (1967) *System, Change and Conflict* New York: Free Press

Deutsch, K. W. (1963) *The Nerves of Government* New York: The Free Press

Dilthey, W. (1931) *Gesammelte Schriften Vol VIII Weltanschauungslehre* Stuttgart: B. G. Teubner. See also Kluback and Weinbaum, 1957

Doney, W. (Ed.) (1968) *Descartes: A Collection of Critical Essays* London: Macmillan

Dror, Y. (1968) *Public Policymaking Reexamined* San Francisco: Chandler Publishing

Dror, Y. (1971) *Design for Policy Sciences* New York: Elsevier

Drucker, P. (1974) *Management* London: Heinemann

Dunsheath, P. (1961) *A History of Electrical Engineering* London: Faber & Faber

Durbin, P. T. (Ed.) (1978) *Research in Philosophy and Technology* Greenwich, Connecticut: Jai Press

Durkheim, E. (1895) *The Rules of Sociological Method* Translated by S. A. Solovay and J. H. Mueller; edited by G. E. G. Catlin New York: The Free Press, 1964

Eckman, D. P. (Ed.) (1961) *Systems: Research and Design* New York: Wiley

Ehrmann, J. (Ed.) (1970) *Structuralism* New York: Doubleday

Ellul, J. (1965) *The Technological Society* London: Jonathan Cape

Elsasser, W. M. (1966) *Atom and Organism* Princeton: University Press

Emery, F. E. (Ed.) (1969) *Systems Thinking* Harmondsworth: Penguin Books

Emery, F. E. and Trist, E. L. (1960) 'Socio-technical systems' in Churchman and Verhulst (1960) q.v.

Emmet, D. and MacIntyre, A. (Eds.) (1970) *Sociological Theory and Philosophical Analysis* London: Macmillan

Engstrom, E. W. (1957) 'Systems engineering—a growing concept' *Elec. Eng.*, **76**, 113–116

Enke, S. (Ed.) (1967) *Defense Management* Englewood Cliffs: Prentice-Hall

Farrington, B. (1944, 1949) *Greek Science* Harmondsworth: Penguin Books

Farrington, B. (1949, 1961) *Francis Bacon* New York: Schuman; 1961, Collier.

Ferguson, M. (1974) 'Women's magazines: the changing mood' *New Society*, 22.8.74

Feyerabend, P. (1975) *Against Method* London: NLB

Fisher, G. H. (1971) *Cost Considerations in Systems Analysis* New York: Elsevier

Flagle, C. D., Huggins, W. H., and Roy, R. H. (Eds.) (1960) *Operations Research and Systems Engineering* Baltimore: Johns Hopkins Press

Fletcher, R. (1971a) *John Stuart Mill: A Logical Critique of Sociology* London: Michael Joseph

Fletcher, R. (1971b) *The Making of Sociology, Vol. 2: Developments* London: Michael Joseph

Floyd, W. F. and Harris, F. T. C. (Eds.) (1964) *Form and Strategy in Science* Dordrecht: D. Reidel

Forrester, J. W. (1961) *Industrial Dynamics* Cambridge, Mass.: MIT Press

Forrester, J. W. (1969) *Principles of Systems* Cambridge, Mass.: Wright-Allen Press

Foster, M. (1972) 'An introduction to the theory and practice of action research in work organisations' *Human Relations* **25** (6)

Frankel, B. (1974) 'Habermas talking: an interview' *Theory and Society*, **1**, 37

Fremantle, A. (Ed.) (1954) *The Age of Belief: The Medieval Philosophers* (Selections) New York: New American Library (Mentor)

Freund, J. (1979) 'German Sociology in the time of Max Weber' in Bottomore and Nisbet (1979) q.v.

Gardner, H. (1972) *The Quest for Mind: Piaget, Lévi-Strauss and the Structuralist Movement* New York: Knopf

Gellner, E. (1973) *Cause and Meaning in the Social Sciences* London: Routledge and Kegan Paul

Gerard, R. W. (1964) 'Entitation, Animorgs and other systems' in Mesarovic, M. D. q.v.

Gibson, R. E. (1960) 'The recognition of systems engineering' in Flagle, Huggins and Roy (1960) q.v.

Giddens, A. (Ed.) (1974) *Positivism and Sociology* London: Heinemann

Giddens, A. (1976) *New Rules of Sociological Method* London: Hutchinson

Giddens, A. (1978) *Durkheim* London: Fontana/Collins

Giddens, A. (1979) *Central Problems in Social Theory* London: Macmillan

Goldsmith, M. and Mackay, A. (Eds.) (1966) *The Science of Science* (revised edition) Harmondsworth: Penguin Books

Gomez, P., Malik, F., and Oeller, K.-H. (1975) *Systemmethodik: Grundlagen einer Methodik zur Erforschung und Gestaltung komplexer sociotechnischer Systeme* Bern: Haupt

Goode, H. H. and Machol, R. E. (1957) *System Engineering* New York: McGraw-Hill

Gorman, R. A. (1977) *The Dual Vision: Alfred Schutz and the Myth of Phenomenological Social Science* London: Routledge and Kegan Paul

Gosling, W. (1962) *The Design of Engineering Systems* London: Heywood

Gray, W. and Rizzo, N. D. (1973) *Unity Through Diversity: a Festschrift for Ludwig von Bertalanffy* New York: Gordon and Breach

Grene, M. (1974) *The Understanding of Nature* Dordrecht: D. Reidel

Habermas, J. (1970) 'On systematically distorted communication' and 'Towards a theory of communicative competence' *Inquiry*, **13**, 205, 360

Habermas, J. (1971) *Toward a Rational Society* London: Heinemann

Habermas, J. and Luhmann, N. (1971) *Theorie der Gesellschaft oder Sozialtechnologie* Frankfurt: Suhrkamp

Hall, A. D. (1962) *A Methodology for Systems Engineering* Princeton N. J.: Van Nostrand

Hall, A. D. (1969) 'Three dimensional morphology of systems engineering' in Rapp (1974) q.v.

Hall, A. R. (1963) *From Galileo to Newton 1630–1720* London: Collins

Hampshire, S. (Ed.) (1956) *The Age of Reason; the 17th Century Philosophers* (Selections) New York: New American Library (Mentor)

Hartley, R. V. L. (1928) 'Transmission of information' *Bell System Technical Journal*, **7**, 535

Harvey, D. (1969) *Explanation in Geography* London: Arnold

Hesse, M. (1976) 'Models versus paradigms in the natural sciences' in Collins (1976) q.v.

Hitch, C. J. (1955) 'An appreciation of systems analysis' in Optner (1973) q.v.

Hitch, C. J. and McKean, R. N. (1960) *The Economics of Defence in the Nuclear Age* Cambridge, Mass.: Harvard University Press

Hoag, M. W. (1956) 'An introduction to systems analysis' in Optner (1973) q.v.

Hollis, M. (Ed.) (1973) *The Light of Reason: Rationalist Philosophers of the 17th Century* London: Collins/Fontana

Homans, G. C. (1961) *Social Behaviour: Its Elementary Forms* New York: Harcourt Brace

Hoos, I. (1972) *Systems Analysis in Public Policy: A Critique* Berkeley: University of California Press

Hoos, I. (1976) 'Engineers as analysts of social systems: a critical enquiry' *Journal of Systems Engineering*, **4** (2)

Hult, M. and Lennung, S. (1980) 'Towards a definition of action research: a note and a bibliography' *J. of Management Studies*, **17** (2)

Hunt, M. M. (1954) 'Bell Labs 230 Long Range Planners' *Fortune*, May 1954

Husserl, E. (1936) 'The origin of geometry' in Luckmann (1978) q.v.

Hutten, E. H. (1962) *The Origins of Science* London: Allen and Unwin

Jackson, J. A. (Ed.) (1972) *Role: Sociological Studies 4* Cambridge: University Press

Jacob, F. (1974) *The Logic of Living Systems* London: Allen Lane

Jarvie, I. C. (1966) 'The social character of technological problems' in Rapp (1974) q.v.

Jay, M. (1973) *The Dialectical Imagination* London: Heinemann

Jeans, J. (1947) *The Growth of Physical Science* Cambridge: University Press

Jenkins, G. M. (1969) 'The systems approach' *Journal of Systems Engineering*, 1 (1)

Johnson, R. A., Kast, F. E., and Rosenzweig, J. E. (1963) *The Theory and Management of Systems* New York: McGraw-Hill

Jones, J. C. (1967) 'The designing of man-machine systems' in Singleton *et al.* (1967) q.v.

Jones, L. M. (1978) 'The conflicting views of knowledge and control implied by different systems approaches' *Journal of Applied Systems Analysis*, 5 (2)

Jordan, N. (1968) *Themes in Speculative Psychology* London: Tavistock

Karp, W. (1972) 'Isaac Newton' in Mazlish, (1972) q.v.

Kash, M. and Pullman, B. (1962) *Horizons in Biochemistry* New York: Academic Press

Katz, D. and Kahn, R. L. (1966) *The Social Psychology of Organisations* New York: Wiley

Keat, R. and Urry, J. (1976) *Social Theory as Science* London: Routledge and Kegan Paul

Kelleher, G. J. (Ed.) (1970) *The Challenge to Systems Analysis* New York: Wiley

Klir, G. J. (1969) *An Approach to General Systems Theory* New York: Van Nostrand Reinhold

Klir, G. J. (Ed.) (1972) *Trends in General Systems Theory* New York: Wiley

Kluback, W. and Weinbaum, M. (1957) *Dilthey's Philosophy of Existence: Introduction to Weltanschauungslehre* London: Vision Press

Koestler, A. (1945) *The Yogi and the Commissar* London: Jonathan Cape

Koestler, A. (1959) *The Sleepwalkers* London: Hutchinson

Koestler, A. (1964) *The Act of Creation* London: Hutchinson

Koestler, A. (1967) *The Ghost in the Machine* London: Hutchinson

Koestler, A. (1978) *Janus, A Summing Up* London: Hutchinson

Koestler, A. and Smythies, J. R. (Eds.) (1969) *Beyond Reductionism* London: Hutchinson

Kolakowski, L. (1972) *Positivist Philosophy* Harmondsworth: Penguin Books

Korach, M. (1966) 'The science of industry' in Goldsmith and Mackay (1966) q.v.

Kotarbinski, T. (1965) *Praxiology: An Introduction to the Sciences of Efficient Action* Warsaw: PWN and Oxford: Pergamon

Kotarbinski, T. (1966) *Gnosiology: The Scientific approach to the Theory of Knowledge* Oxford: Pergamon

Koyré, A. (1965) *Newtonian Studies* London: Chapman and Hall

Kuhn, A. (1974) *The Logic of Social Systems* San Francisco: Jossey-Bass

Kuhn, T. S. (1962) *The Structure of Scientific Revolutions* Chicago: University Press

Lakatos, I. and Musgrave, A. (Eds.) (1970) *Criticism and the Growth of Knowledge* London: Cambridge University Press

Lane, M. (Ed.) (1970) *Structuralism: a Reader* London: Cape

Lasswell, H. D. and Lerner, D. (Eds.) (1951) *The Policy Sciences: Recent Developments in Scope and Method* Stanford: University Press

Laszlo, E. (1972a) *The Systems View of the World* New York: Gordon and Breach

Laszlo, E. (1972b) *Introduction to Systems Philosophy* New York: Braziller

Lazarsfeld, P. F. (1970) *Main Trends in Sociology* London: Allen and Unwin

306

Lee, A. M. (1970) *Systems Analysis Frameworks* London: Macmillan

Lilienfeld, R. (1978) *The Rise of Systems Theory* New York: Wiley

Litterer, J. A. (1963) *Organizations Vol. 1: Structure and Behaviour; Vol. 2: Systems Control and Adaption* New York: Wiley

Longuet-Higgins, H. C. *et al.* (1972) *The Nature of Mind* (Gifford Lectures 1971/72) Edinburgh: University Press

Loomis, C. P. (1955) Translation of Tönnies's *Gemeinschaft und Gesellschaft* London: Routledge and Kegan Paul

Luckmann, T. (Ed.) (1978) *Phenomenology and Sociology* Harmondsworth: Penguin Books

Lyons, J. (1970) *Chomsky* London: Fontana/Collins

McCarthy, T. (1978) *The Critical Theory of Jürgen Habermas* London: Hutchinson

McCloskey, J. F. and Trefethen, F. N. (Eds.) (1954) *Operations Research for Management* Baltimore: John Hopkins

Machol, R. E. (1965) *Systems Engineering Handbook* New York: McGraw-Hill

MacKay, D. M. (1967) 'Freedom of action in a mechanistic universe' The Twenty-first Eddington Memorial Lecture. Cambridge: University Press

MacKay, D. M. (1970) 'The bankruptcy of determinism' *New Scientist*, 2 July 1970

McKean, R. N. (1958) *Efficiency in Government Through Systems Analysis* New York: Wiley

MacRae, D. G. (1974) *Weber* London: Fontana/Collins

Magee, B. (1971) *Modern British Philosophy* London: Secker and Warburg

Magee, B. (1973) *Popper* London: Fontana/Collins

Magee, B. (1978) *Men of Ideas* London: BBC

Makkreel, R. A. (1975) *Dilthey: Philosopher of the Human Studies* Princeton N. J.: Princeton University Press

Mason, R. O. (1969) 'A dialectical approach to strategic planning' *Management Science*, **15** (8)

Mayr, O. (1970) *The Origins of Feedback Control* Cambridge, Mass.: MIT Press

Mazlish, B. (1972) (Ed.) *Makers of Modern Thought* New York: American Heritage Publishing Company

Medawar, P. B. (1967) *The Art of the Soluble* London: Methuen

Medawar, P. B. (1969) *Induction and Intuition in Scientific Thought* London: Methuen

Medawar, P. B. and Medawar, J. S. (1977) *The Life Science* London: Wildwood House

Merton, R. K. (1957) 'Manifest and latent functions' in Demerath and Peterson (1967) q.v.

Mesarovic, M. D. (1964) *Views on General Systems Theory: Proceedings of the 2nd Systems Symposium at Case Institute* New York: John Wiley

Mesarovic, M. D., Macko, D. and Takahara, Y. (1970) *Theory of Hierarchical, Multilevel Systems* New York: Academic Press

Miles, R. F. (Ed.) (1973) *Systems Concepts* New York: Wiley

Mill, J. S. (1884) Chapter VII of *A System of Logic, Ratiocinative and Inductive* Reprinted in Fletcher (1971a) q.v.

Miller, D. W. and Starr, M. K. (1967) *The Structure of Human Decisions* Englewood Cliffs: Prentice Hall

Miller, E. J. (Ed.) (1976) *Task and Organization* New York: Wiley

Miller, J. G. (1978) *Living Systems* New York: McGraw-Hill

Miller, S. L., and Orgel, L. E. (1974) *The Origins of Life on the Earth* Englewood Cliffs: Prentice-Hall

Mills C. Wright (1959) *The Sociological Imagination* New York: Oxford University Press

Milsum, J. H. (1972) 'The hierarchical basis for general living systems' in Klir (Ed.) (1972) q.v.

Mingers, J. C. (1980) 'Towards an appropriate social theory for applied systems thinking: critical theory and soft systems methodology' *Journal of Applied Systems Analysis*, 7

Moggridge, D. E. (1976) *Keynes* London: Fontana/Collins

Monod, J. (1972) *Chance and Necessity* London: Collins (Published in French 1970)

Montalenti, G. (1974) 'From Aristotle to Democritus via Darwin' in Ayala and Dobzhansky (1974) q.v.

More, L. T. (1934) *Isaac Newton* New York: Charles Scribner's Sons; 1962: Dover Publications

Morris, M. B. (1977) *An Excursion into Creative Sociology* New York: Columbia University Press

Morse, P. M. (1970) 'The history and development of operational research' in Kelleher (1970) q.v.

Nagel, E. (1961) *The Structure of Science* London: Routledge and Kegan Paul

Naughton, J. (1977) 'The Checkland methodology: a readers guide' (2nd edition) Open University Systems Group, Milton Keynes

Naughton, J. (1979a) 'Review of Lilienfeld (1978)' in *Futures*, 11 (2)

Naughton, J. (1979b) 'Functionalism and systems research: a comment' *Journal of Applied Systems Analysis*, 6

Needham, J. (1966) 'Science and society in East and West' in Goldsmith and Mackay (1966) q.v.

de Neufville, R. and Stafford, J. H. (1971) *Systems Analysis for Engineers and Managers* New York: Mc.Graw-Hill

Nyquist, H. (1924) 'Certain factors affecting telegraph speed' *Bell System Technical Journal*, 3, 324

Olby, R. (1976) *The Path to the Double Helix* London: Macmillan

Olmsted, J. M. D. and Olmsted, E. H. (1952) *Claude Bernard and the Experimental Method in Medicine* New York: H. Schuman

O'Neil, J. (1976) *On Critical Theory* London: Heinemann

Optner, S. L. (1965) *Systems Analysis for Business and Industrial Problem-Solving* Englewood Cliffs: Prentice-Hall

Optner, S. L. (Ed.) (1973) *Systems Analysis* Harmondsworth: Penguin Books

Optner, S. L. (1975) *Systems Analysis for Business Management* (third edition) Englewood Cliffs: Prentice-Hall

Orgel, L. E. (1973) *The Origins of Life: Molecules and Natural Selection* London: Chapman and Hall

Ozbekhan, H. (1976) Presentation on 'Complex planning', OR Society Meeting 27 October, London

Ozbekhan, H. (1977) 'The future of Paris: a systems study in strategic urban planning' *Phil. Trans. Roy. Soc. London* A287, 523–544

Pantin, C. P. A. (1968) *The Relations Between the Sciences* London: Cambridge University Press

Parsons, T. (1951) *The Social System* New York: Free Press

Parsons, T. (1977) *Social Systems and the Evolution of Action Theory* New York: Free Press

Pask, G. (1961) *An Approach to Cybernetics* London: Hutchinson

Pattee, H. H. (1968–72) Papers in Waddington (1968–72) q.v.

Pattee, H. H. (1970) 'The problem of biological hierarchy' in Vol. 3 of Waddington (1968–72) q.v.

Pattee, H. H. (Ed.) (1973) *Hierarchy Theory: the Challenge of Complex Systems* New York: Braziller

308

Pettit, P. (1969) *On the Idea of Phenomenology* Dublin: Scepter Books
Phillips, D. C. (1970) 'Organicism in the late nineteenth and early twentieth centuries' *Journal of the History of Ideas*, **31** (3)
Pincus, A. and Minahan, A. (1973) *Social Work and Practice: Model and Method* Itasca, Ill.: Peacock Publishers
Polanyi, M. (1968) 'Life's irreducible structure' *Science*, **160**, 1308
Pollock, S. M. (1972) Review of Hoos (1972) in *Science*, **178** , 739
Popper, K. R. (1945) *The Open Society and its Enemies* (2 volumes) London: Routledge
Popper, K. R. (1957) *The Poverty of Historicism* London: Routledge and Kegan Paul
Popper, K. R. (1959) *The Logic of Scientific Discovery* London: Hutchinson (published in German 1934)
Popper, K. R. (1963) *Conjectures and Refutations: The Growth of Scientific Knowledge* London: Routledge and Kegan Paul (revised edition 1972)
Popper, K. R. (1972) *Objective Knowledge* Oxford: Clarendon Press
Popper, K. R. (1974) 'Autobiography of Karl Popper' in Schilpp (1974) q.v. Published separately as *Unended Quest* (1976) London: Fontana/Collins
Popper, K. R. (1976) 'The logic of the social sciences' in Adorno (1976) q.v.
Pratt, J. R. (1962) 'A "Book Model" of genetic information' in Kash, M. and Pullman, B.
Prévost, P. (1976) ' "Soft" systems methodology, functionalism and the social sciences' *Journal of Applied Systems Analysis* **5** (1)
Pritchard, H. A. (1968) 'Descartes' "Mediations" ' in Doney (1968) q.v.
Quade, E. (1963) 'Military systems analysis' in Optner (1973) q.v.
Quade, E. (1975) *Analysis for Public Decisions* New York: Elsevier
Quade, E. and Boucher, W. I. (Eds.) (1968) *Systems Analysis and Policy Planning: Applications in Defence* New York: Elsevier
Quade, E., Brown, K., Levien, R., Majone, G. and Rakhmankulov, V. (1976) 'Systems analysis: an outline for the state-of-the-art survey publications' IIASA RR-76-16, also *Journal of Applied Systems Analysis*, **5** (2)
Quinton, A. (1974) 'Critical theory: on the Frankfurt school' *Encounter*, October 1974
Randall, J. H. (1957) 'Scientific method in the school of Padua' in Wiener and Noland (1957) q.v.
Rapoport, A. and Horvath, W. J. (1959) 'Thoughts on organization theory' *General Systems*, 4
Rapp, F. (Ed.) (1974) *Contributions to a Philosophy of Technology* Dordrecht: D. Reidel
Ravetz, J. R. (1971) *Scientific Knowledge and its Social Problems* London: Oxford University Press
Rée, J. (1974) *Descartes* London: Allen Lane
Rex, J. (1961) *Key Problems of Sociological Theory* London: Routledge and Kegan Paul
Rex, J. (Ed.) (1974) *Approaches to Sociology* London: Routledge and Kegan Paul
Rice, A. K. (1958) *Productivity and Social Organization: the Ahmedabad Experiment* London: Tavistock
Richie, A. D. (1945) *Civilization, Science and Religion* Harmondsworth: Penguin Books
Robertson, R. (1974) 'Towards the identification of the major axes of sociological analysis' in Rex (1974) q.v.
Roche, M. (1973) *Phenomenology, Language and the Social Sciences* London: Routledge and Kegan Paul

Ronan, C. (1969) *Sir Isaac Newton* London: International Textbook Company Limited

Rosie, A. M. (1966) *Information and Communication Theory* London: Blackie

Roszak, T. (1970) *The Making of a Counter Culture* London: Faber & Faber

Roszak, T. (1973) *Where the Wasteland Ends* London: Faber & Faber

Runciman, W. G. (1963) *Social Science and Political Theory* Cambridge: University Press

Runciman, W. G. (1972) *A Critique of Max Weber's Philosophy of Social Science* Cambridge: University Press

Runciman, W. G. (1978) *Max Weber: Selections in Translation* Cambridge: University Press

Russell, B. (1946) *History of Western Philosophy* London: Allen and Unwin

Ryan, A. (1970) *The Philosophy of the Social Sciences* London: Macmillan

Ryan, A. (Ed.) (1973) *The Philosophy of Social Explanation* London: Oxford University Press

Santillana, G. de (Ed.) (1956) *The Age of Adventure: The Renaissance Philosophers* (Selections) New York: New American Library (Mentor)

Sargent, R. W. H. (1972) 'Forecasts and trends in systems engineering' *The Chemical Engineer*, June 1972

Sayre, K. M. (1976) *Cybernetics and the Philosophy of Mind* London: Routledge and Kegan Paul

Schein, E. H. (1969) *Process Consultation: Its Role in Organization Development* Reading, Mass.: Addison-Wesley

Schilpp, P. A. (Ed.) (1949) *Albert Einstein: Philosopher-Scientist* La Salle, Ill.: Open Court

Schilpp, P. A. (Ed.) (1974) *The Philosophy of Karl Popper* La Salle, Ill.: Open Court

Schoderbek, P. P., Kefalas, A. G. and Schoderbek, C. G. (1975) *Management Systems: Conceptual Considerations* Dallas: Business Publications (Irwin—Dorsey International)

Schrödinger, E. (1944) *What is Life?* London: Cambridge University Press

Schutz, A. (1943) 'The problem of rationality in the social world' in Emmet and MacIntyre (1970) q.v.

Schutz, A. (1962, 1964, 1966) *Collected Papers* Vols. I, II, III The Hague: Nijhoff

Shannon, C. E. and Weaver, W. (1949) *The Mathematical Theory of Communication* Urbana, Ill.: University of Illinois Press

Sherburne, D. W. (1966) *A Key to Whitehead's 'Process and Reality'* New York: Macmillan

Shils, E. A. and Finch, H. A. (Eds.) (1949) Max Weber's *Methodology of the Social Sciences* New York: Free Press

Simon, H. (1962) 'The architecture of complexity', Reprinted in *The Sciences of the Artificial* Cambridge, Mass.: MIT Press (1969)

Singer, C. (1941) *A Short History of Science to the Nineteenth Century* Oxford: University Press

Singh, R. (1979) 'Towards applicable theory in industrial relations using a systems approach' *Journal of Applied Systems Analysis*, **6**

Singleton, W. T., Easterby, R. S. and Whitfield, D. (Eds.) (1967) *The Human Operator in Complex Systems*, London. Taylor and Francis

Sixel, F. W. (1976) 'The Problem of Sense: Habermas-v-Luhmann' in O'Neill (1976) q.v.

Skair, L. (1973) *Organized Knowledge* London: Hart-Davis, MacGibbon

Skolimowski, H. (1966) 'The structure of thinking in technology' in Rapp (1974) q.v.

310

Smith, A. G. R. (1972) *Science and Society in the Sixteenth and Seventeenth Centuries* London: Thames and Hudson

Smith, B. L. R. (1966) *The Rand Corporation: Case Study of a Non-profit Advisory Corporation* Cambridge, Mass.: Harvard University Press

Smith, V. E. (Ed.) (1966) *Philosophical Problems in Biology* New York: St John's University Press

Smuts, J. C. (1926) *Holism and Evolution* London: Macmillan

Smyth, D. S. and Checkland, P. B. (1976) 'Using a systems approach: the structure of root definitions' *Journal of Applied Systems Analysis*, **5** (1)

Sofer, C. (1972) *Organizations in Theory and Practice* London: Heinemann

Sporn, P. (1964) *Foundations of Engineering* Oxford: Pergamon

Stimson, D. H. and Thompson, R. P. (1975) 'The importance of Weltanschauungen in operations research: the case of the school bussing problem' *Management Science*, **21**, No. 10

Susman, G. and Evered, R. D. (1978) 'An assessment of the scientific merits of action research' *Administrative Science Quarterly*, **23**, December

Sutcliffe, F. E. (1968) Introduction to his translation of Descartes' *Discourse on Method* Harmondsworth: Penguin Books

Sztompka, P. (1974) *System and Function: Toward a Theory of Society* New York: Academic Press

Taylor, F. W. (1947) *The Principles of Scientific Management* New York: Harper and Row. (Reprint of texts from early 1900s)

Thompson, A. B. (1967) 'A third generation of synthetic fibre materials', paper to Section B, British Association meeting, Leeds.

Thorpe, W. H. (1974) *Animal Nature and Human Nature* London: Methuen

Thorpe, W. H. (1978) *Purpose in a World of Chance* Oxford,: University Press

Tiryakian, E. A. (1979) 'Emile Durkheim' in Bottomore and Nisbet (1979) q.v.

Toulmin, S. and Goodfield, J. (1962) *The Architecture of Matter* London: Hutchinson

Trappl, R. and Hanika, F. de P. (Eds.) (1975) *Progress in Cybernetics and Systems Research* Volume II Washington: Hemisphere Publications

Trappl, R., Hanika, F. de P. and Pichler, F. R. (Eds.) (1977) *Progress in Cybernetics and Systems Research* Volume V Washington: Hemisphere Publications

Trist, E. L., Higgins, G. W., Murray, L. and Pollack, A. R. (1963) *Organizational Choice: Capabilities of Groups at the Coalface under Changing Technologies* London: Tavistock

Vickers, G. (1965) *The Art of Judgement: A Study of Policy Making* London: Chapman and Hall

Vickers, G. (1968) *Value Systems and Social Process* London: Tavistock Publications

Vickers, G. (1970) *Freedom in a Rocking Boat* London: Allen Lane

Vickers, G. (1973) *Making Institutions Work* London: Associated Business Programmes

Vickers, G. (1974) Letter to the author

Waddington, C. H. (1968–72) *Towards a Theoretical Biology* 4 Vols. Edinburgh: University Press

Waddington, C. H. (1973) *OR in World War 2* London: Elek Science

Waddington, C. H. (1977) *Tools for Thought* London: Jonathan Cape

Warmington, A. (1980) 'Action research: its methods and its implications' *Journal of Applied Systems Analysis*, **7**

Warmington, A., Lupton, T. and Gribbin, C. (1977) *Organizational Behaviour and Performance: An Open Systems Approach to Change* London: Macmillan

Wartofsky, M. W. (1968) *Conceptual Foundations of Scientific Thought* New York: Macmillan

Watson, J. D. (1968) *The Double Helix* London: Weidenfeld & Nicolson

Weaver, W. (1948) 'Science and complexity' *American Scientist*, **36**

Weber, M. (1904) ' "Objectivity" in social science and social policy', in Shils and Finch (1949) q.v.

Whitehead, A. N. (1926) *Science and the Modern World* Cambridge: University Press

Whitehead, A. N. (1929) *Process and Reality* New York: Macmillan

Whyte, L. L., Wilson, A. G. and Wilson, D. (Eds.) (1969) *Hierarchical Structures* New York: Elsevier

Wiener, N. (1948) *Cybernetics* Cambridge, Mass.: MIT Press, and New York: J. Wiley (enlarged edition 1961)

Wiener, N. (1950) *The Human Use of Human Beings* Boston: Houghton Mifflin

Wiener, P. P. and Noland, A. (Eds.) (1957) *Roots of Scientific Thought* New York: Basic Books

Wild, R. (1972) *Management and Production* Harmondsworth: Penguin Books

Williams, T. J. (1961) *Systems Engineering for the Process Industries* New York: McGraw-Hill

Wilson, A. G. (1973) 'From maps to models', in Calder (1973) q.v.

Wilson, B. (1979) 'The design and improvement of management control systems ' *Journal of Applied Systems Analysis*, **6**

Wilson, B. (1980) 'The "Maltese Cross": a tool for information systems analysis and design' *Journal of Applied Systems Analysis*, **7**

Winch, P. (1958) *The Idea of a Social Science* London: Routledge and Kegan Paul

Winch, P. (1974) 'Popper and scientific method in the social sciences' in Schilpp (1974) q.v.

Wisdom, J. O. (1967) 'Rules for making discoveries' in Rapp (1974) q.v.

Woodger, J. H. (1929) *Biological Principles* London: Kegan Paul Trench and Trubner

Wymore, A. W. (1967) *A Mathematical Theory of Systems Engineering* New York: Wiley

Wymore, A. W. (1976) *Systems Engineering Methodology for Interdisciplinary Teams* New York: Wiley

Zilsel, E. (1957) 'The origins of Gilbert's scientific method' in Wiener and Noland (1957) q.v.

Ziman, J. M. (1968) *Public Knowledge: An Essay Concerning the Social Dimension of Science* Cambridge: University Press

Glossary

If soft systems methodology is to be used, discussed, developed and modified, or rejected and replaced, it is important to know precisely what we are dealing with. The terms used in it need to be precisely defined. In particular the words in this glossary, which in their everyday connotations hopefully convey the right general impression, need to be carefully defined so that they may be used seriously, with the rigour usually associated with technical terms in natural science.

The seventy-one terms defined here are important in systems thinking. Those italicized in the definitions are themselves defined. A few extra words outside those directly required in systems thinking have been included; this is in order that the glossary as a whole may be read as an expression of the soft systems paradigm.

Action
An *activity* willed by the doer.

Activity
A neutral term for the carrying out of an act, contrasting with *action* and *behaviour*. The word is used in *'human activity system'* to emphasize that such systems are not descriptions of observed real-world action.

Actor
In *CATWOE* a person who carries out one or more of the activities in the system.

Behaviour
The *activity* of an animal (human or otherwise) which an observer takes to be the result of genetic endowment and social conditioning.

Boundary
In the *formal system model* the area within which the *decision-taking process* of the system has power to make things happen, or prevent them from happening. More generally, a boundary is a distinction made by an observer which marks the difference between an entity he takes to be a *system* and its *environment*.

Bridgeable, or Unbridgeable Gap
In *soft systems methodology* the difference between the expression of the problem situation at stage 2 and the *conceptual models* built at stage 4; it provides the base for the *comparison* at stage 5.

CATWOE
A mnemonic of the six crucial characteristics which should be included in a well-formulated *root definition*.

Climate
A characteristic of a *problem situation*; it is the relation between its elements of *structure* and its elements of *process*.

Communication
The transfer of *information*.

Comparison Stage
In *soft systems methodology*, stage 5, at which the expression of the *problem situation* is compared with the *conceptual models* of *relevant systems*.

Conceptual Model
A systemic account of a *human activity system*, built on the basis of that system's *root definition*, usually in the form of a structured set of verbs in the imperative mood. Such models should contain the minimum necessary *activities* for the system to be the one named in the root definition. Only activities which could be directly carried out should be included—thus, admonishments such as 'succeed' must be avoided.

Conceptual models may be validated or justified only in terms of logic, not by mapping on to the real world, since they do not purport to describe the real world. They may, however, be compared with the *formal system model* in order to check that they are not fundamentally deficient.

Connectivity
In the *formal system model* the property which enables effects to be transmitted through the system. The connectivity may have a physical embodiment (as in an order processing system) or may be a flow of energy (verbal) *information* or influence.

Continuity
In the *formal system model* the property of long-run stability which enables the system to recover stability from some degree of disturbance. The continuity might be guaranteed from outside the system or might derive from its *connectivity* and *control* system.

Control
The process by means of which a whole entity retains its identity and/or performance under changing circumstances. In the *formal system model* the *decision-taking process* ensures that control action is taken in the light of the system's *purpose* or mission and the observed level of the *measure of performance*.

Cultural Feasibility
In *soft systems methodology* (at Stages 5 and 6) one of the criteria which potential changes in the real world must meet if they are to be implemented. The implication is that the culture of a particular *problem situation*, with its unique norms, roles, and values will be able to accept, as meaningful and possible, a certain range of changes. (The other criterion is *systemic desirability*.)

Customer
In *CATWOE* the beneficiary or victim of the system's activity.

Decision-taking Process
In the *formal systems model* the procedures by means of which the system organizes itself, responds to disturbances, and pursues its *purpose*.

Designed System

A man-made entity which an observer elects to treat as a whole having *emergent properties*. Designed systems may be concrete (e.g. 'a tramcar') or abstract (e.g. 'mathematics').

Emergence, Emergent Properties

The principle that whole entities exhibit properties which are meaningful only when attributed to the whole, not to its parts—e.g. the smell of ammonia. Every model of a *human activity system* exhibits properties as a whole entity which derive from its component activities and their structure. but cannot be reduced to them.

Environment

In the *formal system model*, what lies outside the system *boundary*.

Environmental Constraints

In *CATWOE*, impositions which the system takes as given.

Epistemology

A theory concerning means by which we may have and express knowledge of the world.

Formal System Model

A generalized model of any *human activity system* from the point of view: taking purposeful action in pursuit of a purpose. It may be used to test the basic adequacy of *conceptual models*.

Goal

An end which an individual or group of individuals may seek to achieve or which may be attributed to a *system*. Goal is a synonym of *objective*; both differ from *purpose* or mission, in that there is always a 'yes/no' answer to the question: Has the goal been achieved?

Hierarchy

The principle according to which entities meaningfully treated as wholes are built up of smaller entities which are themselves wholes ... and so on. In a hierarchy, *emergent properties* denote the levels.

Human Activity System

A notional *purposive* system which expresses some *purposeful* human activity, activity which could in principle be found in the real world. Such systems are notional in the sense that they are not descriptions of actual real-world activity (which is an exceptionally complex phenomenon) but are intellectual constructs; they are *ideal types* for use in a debate about possible changes which might be introduced into a real-world *problem situation*.

Ideal Type

An intellectual construct to aid thinking (hence a concept of *epistemology*) not a description of something in the real world. The word 'ideal' is not normative, the function of ideal types being to enable comparisons to be made and theories to be developed; but they are usually constructed from empirically observable or historically meaningful components—for example. the individual activities in a *human activity system* are themselves meaningful in the everyday world.

Information
A distinction which reduces uncertainty. In information theory (which is concerned only with the statistics of message transmission) the quantity of information is the number of binary (yes/no) choices which have to be made to achieve a unique selection from the possibilities.

Unfortunately, information theory can say nothing at all about the meaning of the information for the originator or the recipient, although in the everyday world this is the most interesting aspect of information! Outside information theory, the word 'information' is best used to denote a combination of fact plus the meaning an observer attributes to it.

Input
That which is changed by a *transformation process*. Inputs may be concrete (e.g. raw materials) or abstract (e.g. a market need).

Measure of Performance
In the *formal system model*, the indicator whose level signals progress or regress in pursuing the system's *purpose*.

Model
An intellectual construct, descriptive of an entity in which at least one observer has an interest. The observer may wish to relate his model and, if appropriate, its mechanisms, to observables in the world. When this is done it frequently leads—understandably, but not accurately—to descriptions of the world couched in terms of models, as if the world were identical with models of it.

Modelling Language
A class, or set of classes, of elements used to construct *models*. For example, the modelling language suitable for making models of *human activity systems* is all the verbs in the language; an indicator of logical dependency; indicators of flows, concrete or abstract.

Natural System
Part of the world, not man-made, which an observer elects to treat as a whole entity having *emergent properties*. Many 'obvious' natural systems will have easily demonstrated emergent properties (e.g. 'a mouse'); others will be more personal to the observer (e.g. 'all the foxgloves in Lancashire').

Objective
A synonym for *goal*, q.v.

Ontology
A theory of what the world is, or contains. For many people a *positivist stance*, philosophically, yields ontological facts concerning the world.

Output
That which is produced by a *transformation process*. Outputs may be concrete (e.g. manufactured products) or abstract (e.g. the fulfilment of a market need).

Phenomenological Stance
A philosophical position characterized by a readiness to concede primacy to the mental processes of observers rather than to the external world; this contrasts with a *positivist stance*.

Positivist Stance
A philosophical position characterized by a readiness to concede primacy to the given world as known through experimental evidence, contrasting with a *phenomenological stance*.

Problem, Hard
A problem, usually a *real-world problem*, which can be formulated as the search for an efficient means of achieving a defined end.

Problem, Laboratory
A problem which the investigator defines, in terms of form, content, and boundaries. He decides what to take into account and what to leave out; such problems contrast with *real-world problems*.

Problem Owner
The person or persons taken by an investigator to be those likely to gain most from achieved improvement in a *problem situation*. Problem owner is a role in what the investigator defines as the *problem content system*. Very frequently problem owners will actively seek such improvement, but the investigator is free to take as problem owners people who do not recognize the problem situation or are too inarticulate to express their views.

Problem, Real-world
A problem which arises in the everyday world of events and ideas, and may be perceived differently by different people. Such problems are not constructed by the investigator as are *laboratory problems*.

Problem Situation
A nexus of real-world events and ideas which at least one person perceives as problematic: for him other possibilities concerning the situation are worth investigating.

Problem, Soft
A problem, usually a *real-world problem*, which cannot be formulated as a search for an efficient means of achieving a defined end; a problem in which ends, *goals*, *purposes* are themselves problematic.

Problem Solver
A person or persons anxious to bring about improvement in *a problem situation*. Problem solver is a role in what the investigator defines as the *problem-solving system*.

Process
Those elements in a *problem situation* which are characterized by continuous change.

Purpose
An end which can be pursued but never finally achieved (as can an *objective* or *goal*), e.g. 'maintain relationships'. In *hard problems* a purpose sharpens into an objective.
    One of the system characteristics in the *formal system model* is the system's purpose, mission or objective.

Purposeful
Willed; thus *activity* which is purposeful becomes *action*.

Purposive
Describable by an observer as serving a purpose (contrast *purposeful*).

Relevant System
A *human activity system* which an investigator using *soft systems methodology* names as likely to yield insight in later stages of the study. For each relevant system a *root definition* is formulated and a *conceptual model* built.

Resources
In the *formal system model* means, whether concrete or abstract, which are at the disposal of the *decision-taking process* in its seeking to pursue the system's *purpose*.

Rich Picture
The expression of a *problem situation* compiled by an investigator, often by examining elements of *structure*, elements of *process*, and the situation *climate*.

Root Definition
A consise, tightly constructed description of a *human activity system* which states what the system is; what it does is then elaborated in a *conceptual model* which is built on the basis of the definition. Every element in the definition must be reflected in the model derived from it. A well-formulated root definition will make explicit each of the *CATWOE* elements. A completely general root definition embodying *CATWOE* might take the following form:
  A (...O...)—owned system which, under the following environmental constraints which it takes as given: (...E...), transforms this input (...) into this output (...) by means of the following major activities among others: (...
  ......), the transformation being carried out by these actors: (...A...) and directly affecting the following beneficiaries and/or victims (...C...). The world-image which makes this transformation meaningful contains at least the following elements among others: (...W...).

Root Definition, Issue-Based
A *root definition* describing a notional system chosen for its relevance to what the investigator and/or the people in the *problem situation* perceive as matters of contention.

Root Definition, Primary-Task
A *root definition* of a system which carries out some major task manifest in the real world. Such root definitions give would-be neutral accounts of public or 'official' explicit tasks, often ones embodied in an organization or section or department. For example, a primary-task root definition and *conceptual model* of a school examination system would entail such activities as setting examination papers in school, conducting the examination, and marking the resulting scripts.

Structure
Those elements in a *problem situation* which are either permanent or change only slowly and/or occasionally.

Sub-System
Equivalent to *system*, but contained within a larger system.

System
A model of a whole entity; when applied to human activity, the model is

characterized fundamentally in terms of *hierarchical* structure, *emergent properties*, *communication*, and *control*. An observer may choose to relate this model to real-world activity. When applied to natural or man-made entities, the crucial characteristic is the *emergent properties* of the whole.

Systemic Desirability
A criterion for real-world changes debated at stages 5 and 6 of *soft systems methodology*. The implication is that the *systems thinking* of stages 3 and 4 will generate models whose comparison with the expression of the *problem situation* from stage 2 will yield possible changes which this systems analysis recommends as being desirable.

System Owner
In *CATWOE*, the person or persons who could modify or demolish the system.

System, Problem-Content
One of the two systems in terms of which any tackling of a *real-world problem* may be conceptualized by an investigator (the other being the *problem-solving system*). The content system contains the role *problem owner*. The word 'system' is here a *teleological* indication that the investigator will use systems concepts in perceiving the substantive content of the problem.

System, Problem-Solving
One of the two systems in terms of which any tackling of a *real-world problem* may be conceptualized by an investigator (the other being the *problem-content system*). The solving system contains the role *problem solver*. The investigator must define problem-solving and problem-content systems in relation to each other (e.g., if problem-solving resources are one investigator for a month, the problem content had better not be the re-design of the national education system.)

System, Service
A *system* which is conceived as serving another. A *conceptual model* of such a *human activity system* cannot be built until there exists a conceptual model of the system served, since this will dictate the structure and *activity* of the service system.

Systems Methodology, Hard
Systems-based methodology, also known as 'systems engineering', for tackling *real-world problems* in which an objective or end-to-be-achieved can be taken as given. A system is then engineered to achieve the stated objective.

Systems Methodology, Soft
Systems-based methodology for tackling *real-world problems* in which known-to-be-desirable ends cannot be taken as given. Soft systems methodology is based upon a *phenomenological stance*.

Systems Thinking
An *epistemology* which, when applied to human activity is based upon the four basic ideas: *emergence, hierarchy, communication,* and *control* as characteristics of *systems*. When applied to *natural* or *designed systems* the crucial characteristic is the *emergent properties* of the whole.

Teleology
The philosophical doctrine that developments happen as a result of the ends served by them (rather than as a result of prior causes).

Teleonomy
A neutral term (contrast *teleology*) indicating that developments may be described by an observer in terms of the ends served by them. Note: the relation between teleology and teleonomy resembles that between *purposeful* and *purposive*. Thus *conceptual models* of *human activity systems* based on *root definitions* are teleonomic; but they describe activity which, if manifest in the real world, would be teleological.

Transformation Process
In *CATWOE* the core transformation process of a *human activity system*, which can be expressed as the conversion of some *input* into some *output*.

Weltanschauung
In *CATWOE* the (unquestioned) image or model of the world which makes this particular *human activity system* (with its particular *transformation process*) a meaningful one to consider.

Wider System
Equivalent to *system* but containing it.

Name Index

324

Subject Index

(**Bold print** indicates a term defined in the Glossary)